Introduction to Safety Science

A. Kuhlmann

Introduction to Safety Science

With 234 Illustrations

Springer-Verlag
New York Berlin Heidelberg Tokyo

A. Kuhlmann
TÜV Rheinland
Postfach 10 17 50
5000 Köln 1
Federal Republic of Germany

Library of Congress Cataloging in Publication Data
Kuhlmann, A. (Albert)
  Introduction to safety science.
  Bibliography: p.
  Includes index.
  1. Industrial safety. 2. System safety. I. Title.
T55.K85   1985   620.8'6   85-20805

Title of Original German Edition: *Einführung in die
Sicherheitswissenschaft*, © 1981 by Verlag TÜV Rheinland
GmbH, Köln, Federal Republic of Germany.

Typeset by Polyglot Compositors, Singapore.
Printed and bound by R.R. Donnelley and Sons, Harrisonburg,
Virginia.
Printed in the United States of America.

9 8 7 6 5 4 3 2 1

ISBN 0-387-96192-5
Springer-Verlag New York Berlin Heidelberg Tokyo
ISBN 3-540-96192-5
Springer-Verlag Berlin Heidelberg New York Tokyo

Preface

For many years "safety technology" has constituted the essential instrument for the prevention of accidents as a direct result of handling new technology. Its awareness of the interactions prevalent in natural science causes safety technology to act on the basis of actual accidents, and it utilizes to their fullest extent any means provided by the engineering sciences. Man proceeds in a general direction towards preservation and improvement, thus working towards the optimization of the technical design. However, a new set of basic problems presented itself the moment new large-scale technologies were introduced into the areas of processing, energy, and traffic, thereby creating a considerable amount of additional danger potential. This also signified the end of an era when safety technology could be practiced chiefly on the basis of accident statistics.

For ethical reasons it became necessary that a credible prognosis as to the type and effect of accidents took the place, or at least supplemented, the hitherto practiced purely reactive methods. The realization that the available means of safety technology were no longer sufficient in a highly technologized environment spurred the demand for entirely new concepts which would eventually lead to a higher degree of safety. A decisive step had to be taken away from a purely technical approach and towards and all-encompassing look at accident systems, because man had become aware of the fact that accidents will always be a part of the interaction between man, technology, and environment. The individual factors in their totality act in concert, and complex situations will produce human errors which are ultimately responsible for any and all damage or loss. Such an effort towards an all-encompassing analysis of these interactions as an important factor in the design and utilization of technology will undoubtedly create a demand for specialized scientific procedures and methods to form a "science of safety" which could serve as an intellectual basis for safety-technology-related action.

After the concept of "science of safety" (or safety science) has been introduced, people would ask me about its actual meaning and import. This book is an attempt to introduce the science of safety in a comprehensive form and in its fundamental characteristics. Its intrinsic interdisciplinary character made it necessary to enlist the help of a group of experts from various and vastly different fields. Working towards the goal I had set for myself together with my collaborators proved to be a great pleasure and a valuable experience for me, and I wish to express my heartfelt gratitude to these ladies and gentlemen. With this publication I would like to address students as well as experts in the fields of planning, design, and expertise alike. However, I also hope to be able to provide new ideas to the researcher dedicated to this particular field. My greatest wish, however, is that the book might find a positive echo among my fellow professionals.

A. Kuhlmann

Contents

Chapter 1

Functions and Goals of Safety Science

The science of safety is concerned with the safety from possible dangers connected with the utilization of technology, however not with safety or security in a military or social sense nor with the safety from sickness or disease which is unrelated to and not caused by handling technology. Such a limitation is mandated from the point of view of methodology as well as content, although "technical" safety must be regarded as an integral part of human life. The above-mentioned interdependence becomes evident the moment one has to exactly define the degree of safety which must be guaranteed when handling technology. This degree of safety is not exclusively determined by the current state of the art in technology but also by the desires and aspirations of the individual and of society. Therefore, ethical, humane, and social aspects are of considerable importance.

The ultimate goal of the science of safety is to hold to an absolute minimum any damaging effects from handling modern technology, or, at least, to keep them within tolerable limits. In the above context, damaging effects may mean accidents caused by technology as well as other destruction or loss, e.g., through environmental pollution caused by a technical installation.

As far as life, health, property, or ideal values (e.g., the beauty of nature) are endangered by undesirable side effects of technology, avoiding such damage as far as possible is a task dictated by ethical considerations. In many cases, however, it will be realistic economical considerations which prompt us to aspire to the same goal. It is certainly not within the province of the science of safety to engage in moralizing about the relative merits of duty and inclination; however, it is necessary to point out that in many cases personal interest and ethical considerations may often coincide.

The peculiar function of the science of safety in achieving this aim is the acquisition and summarization of knowledge regarding the conditions and design of safety in handling technical systems, as well as regarding the various possibilities of protection from their inherent dangers and to incorporate these findings and acquired knowledge into safety engineering. The science of safety, in short, is the research and doctrine of safety. So far it is to be assumed, but must yet be proven in the following, that the science of safety actually and inherently constitutes a separate field of scientific endeavor. There is, especially, the necessity of finding special interrelationships between accidents, their causes, their consequences, and the degree of probability of their occurrence which, without the science of safety, would remain unrecognized.

The science of safety must recognize its responsibility of being able to answer the questions raised by the public concerning the safety of technology and the demand voiced by society for protection against the hazards of technology on the basis of its competency. It must thus create a climate of trust in technology and strengthen it whenever possible. The following two practical examples are meant to demonstrate this necessity:

(a) German law determines that sufficient safety measures have been instituted for "technical installations requiring supervision" whenever the installation meets the technical rules and recognized regulations. However, these rules and regulations, according to current practice, are promulgated only after learning wisdom by experience. Nowhere is it stated the amount of accidents which must happen and how much damage of what degree of severity must occur before new safety regulations will be put into effect. Therein lies a special responsibility on the part of the science of safety to see to it that, if we have to learn the hard way, we have not to pay dearly for our wisdom.

(b) The extraordinary dangers which an accident in a certain large-scale technological environment may trigger have prompted those responsible to make the granting of a permit or of a license for such an installation dependent on the essentially stricter condition of requiring that the "measure of precaution necessary in accordance with the present state of the art in technology" has actually been taken. It is a special task of safety science to determine "the present state of the art in technology" under the aspect of safety and to submit it to critical examination.

Safety science should be looked upon as an interfacultative and interdisciplinary branch of science which, among other things, also encompasses accident research in its conventional sense. The foremost aim of the latter has been and still is to avoid harmful side effects of technology by helping to recognize the causes and possible consequences of accidents. In this way accident research contributes towards improvement of the technical object itself as well as to the manner of its utilization.

However, despite the achievements and successes, which are significant in many fields, a certain feeling of uneasiness is now pervading the public, which demands that additional and new avenues be explored. This, in turn, has led to an effort to make available any new scientific findings, knowledge, and methods of research devoted to the prevention of accidents in the world of technology. The path which finally led to the science of safety constitutes an inevitable and necessary process which, now and in the future, must be carefully shaped and controlled.

Progress in mathematics has provided effective instruments for this. Of special importance are probability theory, mathematical statistics, and cybernetics. An essential basis and starting point for safety science is constituted by the system-oriented approach. It describes the interaction of man and machine, and of both embedded in the environment as a "man–machine–environment–system" (MMES). The occurrence of accidents is exclusively regarded as a sequence of exceptional states within the MMES. An analysis of the MMES must uncover their causes and must provide the possibility of an evaluative comparison between a given admissible degree of endangerment on the one hand and actual danger on the other. The degree of endangerment is a function of the totality of system characteristics within a MMES. These are represented by a number of variables, among others:

—target factors of the system, i.e., suitability for the purpose, availability, ease of maintenance, service life, etc.;
—the boundary conditions under which a system can be operated, namely given conditions of nature, human settlement, the economy, politics, etc.

Also, as far as safety itself is concerned, the following variables are most essential:

—the quality of the technical components, e.g., material behavior and component quality;
—the demands made on man himself, e.g., measured in ergonometric or physiological parameters.

An efficient system analysis, after all, is only possible whenever one succeeds in incorporating human behavior as well—if possible, expressed in numerical data. Some very successful starts in that direction have already been made.

Two procedures are available to system analysis:

(a) the empirical system analysis which determines the degree of endangerment and the properties (variables) of a MMES by statistical (empiric) means and by the analysis of accidents which have already occurred;
(b) the theoretical system analysis which deducts the degree of endangerment of a MMES from the interaction of its various components theoretically. This includes a theoretical analysis of possible accidents.

The voluminous analytical procedures for the determination of causes of accidents and damage afford the possibility of instituting purposeful measures, which are staged according to priorities, and of using them for the promotion of increased safety in technical installations. This also offers the possibility of dissociating oneself from purely indicative protective measures which are often inefficient and thus more costly. This may concern the design in technology as well as the skill and quality of the personnel operating the installation. Also concerned are the storage of necessary spare parts and other logistics-related consequences.

The possibility of theoretically implementing alternative considerations relating to the safety of technical systems may result in the increased capability of supplanting technologies which have already been introduced with other and safer concepts. Hitherto this process has mainly been set into motion whenever actual damage had been sustained and not during the preliminary planning and conceptualization phases where it would have made much more sense.

For quantitative safety considerations—despite the voluminous instrumentation available through mathematics—any premature expectation of success in this field has to be dampened because

—the necessary input data are not yet available in many cases, or
—in the case of complex MMESs, the computational expenditure for a mass problem of this kind often demands an unrealistic degree of simplification—sometimes even in such cases where the exact structure of interdependence is fully known.

At present the value of a quantitative safety determination of a specific MMES is often limited by the fact that its incorporation into a socio-ecological total system cannot be evaluated in a convincing manner. Evaluation methods relating to safety science will only be sufficient whenever man, his social system, and the ecosphere enveloping both are adequately protected by setting limits which must be respected under any circumstances. Whenever technical plants or systems, either individually or in their totality, pose a long-range threat to either the ecosphere or man and his social system, further research must be directed towards a technology assessment. Since each MMES will in some way influence ecosphere and social system, criteria of relevancy must be set up to determine the point when, in each particular case, such far-reaching research activity becomes necessary.

Chapter 2

Special Definitions

Safety science is still at the beginning of its development. This is already made clear by the fact that many basic concepts it must use have not been fully determined; neither have they been universally accepted. Within the framework of standardization, first attempts to move in this direction are discernible. What we should do now is to give them our wholehearted support so that the introduction of special definitions may be implemented with all possible speed. This is why this book will refer to previous work in this field whenever possible.

The conclusive determination of its own system of definitions constitutes a not unimportant task on the part of the science of safety. In working towards this end, one should at all times keep in mind that the scientific use of language should be based on living language and not the other way around, namely that language should first be subjected to scientific standardization before becoming part of everyday speech. It would never do to lend differing connotations to words which, also in elevated language, are used with the same connotation. Pascal's warning should be taken to heart: Never try to explain that which is clear from the outset.

2.1 Damage

In the legal sense, "damage" denotes any impairment of a person with relation to his or her inalienable rights. Damage within the meaning of safety science constitutes the impairment of any person as far as his or her inalienable rights are concerned, whereby such impairment has been caused by chemical/physical effects stemming from the utilization of technology. "Person," in this context, may either denote a real person or a body politic or even entities lacking immediate legal status as, e.g., the neighborhood, the public, etc.

It serves our purpose best if we regard "damage" as a basic dimension of the science of safety, the magnitude of which can be measured in monetary value, human lives, number of injured persons, or any other suitable unit. The concepts of "magnitude," "dimension," and "unit" are used in this connection in the same way they are used in physics ([2-1] and [2-2]). There, e.g., length and time are dimensions expressed in the units meter and kilometer or second and hour, respectively. The strictly defined concept of dimension demands that the units which are used to determine the extent of damage run into one another by multiplying them with figures. This is in accordance with the efforts of developing a monetary value or equivalent for any type of damage sustained. Examples of this can be found in civil law, e.g., the payment of an insurance settlement by an insurance company, and within the legal domain affecting public bodies, e.g., in the shape of the impairment of a person's ability to earn a living because of injuries sustained, or in the domain of economy in the shape of efforts made to determine a monetary value for a person killed in an automobile accident.

From an ethical point of view, however, there is a certain amount of hesitation when contemplating the task of determining a monetary compensation for death by accident, bodily injury, or the loss of idealistic values. Thus defining a dimension called "damage" should be extended in such a way as to permit incommensurable quantities like bodily injury and damage to property to be included in the dimension of "damage."

2.2 Danger, Endangerment, and Safety

The meaning of these words, despite a necessary degree of flexibility in its application, is clear and sharply defined and does not require any further explanation or definition. In case one wishes to express certain nuances in the various usages of the concepts of "danger" and "endangerment," this can be achieved by the following paraphrases:

—"Danger" may emanate from the utilization of a technical system and means the—often temporarily limited—possibility of damage to individuals or property.
—"Endangerment" may arise for individuals and property alike whenever a technical system is being utilized and human beings or property are located within the system's effective range.

The concepts of danger and endangerment incorporate the possibility— but not the certainty—of damage and thus induce the element of probability. In accordance with the foregoing, the definition of "safety" encompasses the degree of certitude that such possible damage will never occur.

2.3 Accident-like Occurrences Involving Damage

In accordance with the understanding that the science of safety, besides concerning itself with accident-like occurrences, should also concern itself with chronic occurrences of damage, the initial concepts of "damage" and "danger" have been selected in such a way that they will encompass both. Also, the word "occurrence" is not meant to convey the meaning of an "isolated" happening of relatively short duration; it simply expresses a certain event for which the expected frequency should be noted.

If one intentionally refers to accident-like occurrences, the definitions "accident," "case of imperilment," "incident," or similar definitions should be used. Due to the necessary flexibility of such words in language, the various meanings of these words cannot be sharply delineated. They will be used in the following sense (based on actual usage in insurance law):

— An accident is an unexpected, temporarily limited occurrence entailing damage to life and limb or property.
— A case of imperilment is an unexpected, temporarily limited occurrence entailing the danger of damage to life and limb or property. "Case of imperilment" is an umbrella definition for "accident."
— An incident is an unexpected, temporarily limited occurrence within a technical system in which it cannot from the outset be excluded that a case of imperilment is actually occurring.
— A disturbance is an undesirable, unexpected, and temporarily limited event (interrupted service) within a technical system. "Disturbance" is an umbrella definition of "incident."

"Case of damage," in insurance terminology, denotes the actual occurrence of the damage which previously had been considered as only a possibility. Thus this concept, still, comprises the characteristic of a lack of foreseeability but no longer that of limitation in time, and thus it no longer comes within the framework of the above definitions. For example, contracting a hearing defect as a consequence of exposure to high noise levels constitutes a "case of damage"; however, it does not fall under the concept of "accident," because it was caused by a prolonged period of exposure to excessive noise levels and thus has no connection with a given accident, case of imperilment, incident, etc.

2.4 Source of Danger, Danger Field, Danger Potential, and Potential of Endangerment

Differentiating between "case of endangerment" and "incident" again stresses the necessity of distinguishing between the source of danger at the location of a technical system and the spreading of said danger to such locations where it affects things which may sustain damage.

If the source of endangerment can be located and defined, then the locality-dependent endangerment will define in its immediate environment a danger field which—after having determined a valid unit of measurement—may then be delineated by lines of an equal degree of endangerment. Several danger field will be allocated to one single source of danger whenever there are various types of loss or damage with a varying degree of spread or propagation of said danger.

The potential of endangering by a technical installation constitutes the totality of loss of human life and damage to property in the environment of said plant or system, which may be foreseen under certain clearly defined circumstances and conditions. These defined conditions relate to normal operation of the technical installation, to impaired operational conditions influenced by hypothetical incident conditions, as well as to the conditions prevalent after utilization of said technical plant.

The danger potential of a technical system equals the upper limit of its combined potentials of endangerment.

2.5 Risk

Individual and Global Risk. From a linguistic point of view, the concept of "risk" has practically the same connotation as "venture" or "hazard." In the economic environment, however, it includes the possibility of loss as well as the expectation of gain. He who participates in a game of chance is "risking" his stake in the hope of winning a larger sum of money. In general, however, there is a consensus of opinion that the word "risk" has something to do with the possibility of loss, damage, injuries, death, etc. Thus, in insurance language, "risk" means a measure of the amount of money which must be set aside from the first moment on in order to be able to provide for recompense—should any loss be sustained—in the way of repair, replacement, or indemnity. Consequently it will be safe to regard "risk"—in its more general usage within the terminology of safety science—as a "certain level of danger" relating to a group of possibly affected people and to a period of time in the course of which such a loss or damage constitutes a distinct possibility.

The level of danger emanating from a source of danger or a group of such sources and which concerns a group of people comprising a number N of individuals within a period of time Δt, can be measured for any given type of damage by means of the expectation value

$$E(S) = \int S \, dF(S). \tag{2-1}$$

Here, S stands for the amount of damage or magnitude of loss, and $F(S)$ is its distribution function (cumulative probability function).

From the expectation value $E(S)$ for a certain type of damage, the general form of individual risk can be defined as follows:

$$R = \frac{E(S)}{N\,\Delta t} \quad \left[\frac{\text{unit of loss or damage sustained}}{\text{individual} \times \text{unit of time}}\right], \qquad (2\text{-}2)$$

e.g.

$$\left[\frac{\text{monetary unit}}{\text{individual} \times \text{unit of time}}\right].$$

This relates to the group of individuals and is considered as expected loss expressed in the number of dead, injured, in monetary value, etc., per person and unit time.

If the type of damage is defined in such a manner that it only contains events each involving one and the same amount of loss or damage S_K, then the total amount of damage is the product from S_K and the number n of singular events resulting in this damage. From Equation (2-1) we obtain the following result:

$$\int S\,dF(S) = \sum_n S_K n P_{Sn}, \qquad (2\text{-}3)$$

where P_{Sn} stands for the degree of probability with which—within the group of persons observed—the exact number n of such events will occur during a period of time. Consequently, for individual risk this translates into

$$R = S_K \frac{\sum\limits_n n P_{Sn}}{N \times \Delta t} = S_K H_S. \qquad (2\text{-}4)$$

Here, H_S is the mean frequency of the event of loss or damage per unit of time. This corresponds to the accepted definition of "individual risk":

risk = amount of damage × mean frequency of occurrence
of such loss or damage.

If, during the given unit of time, only one event or occurrence of loss or damage is possible per individual, or if the frequency of such an occurrence is so low that the simultaneous happening of several events of loss or damage can be discounted, the mean frequency of loss or damage equals the probability of incidence P_S of the occurrence for one person during the given unit of time. For "individual risk" it follows (expressed in numbers):

$$R = S_K P_S, \qquad (2\text{-}5)$$

which means

risk = amount of damage × probability of occurrence.

Here it should be pointed out that the risk according to Equation (2-2) represents a mean value with regard to a group of individuals considered. Thus, when making statements concerning individual risk, it should always be specified to which group of persons the mean value refers.

Furthermore, it will be necessary to consider whether the observation period Δt contained in Equation (2-2) refers either to the exposure time within the area of danger or to any part of the lifespan of the individual. Whenever R^* represents the individual risk during the time a person dwells within the area of danger, and if the relative proportion of this exposure time of the person's lifespan Δt enclosing it is indicated by δ, then the share ΔR of the risk considered of the total risk of the individual under real life conditions amounts to

$$\Delta R = R^* \delta. \qquad (2\text{-}6)$$

If reference to an individual person is not desired, the result, instead of Equation (2-2), will be the global risk of a group of individuals:

$$R_{\text{global}} = \frac{E(S)}{\Delta t} \left[\frac{\text{unit of damage}}{\text{unit of time}} \right], \qquad (2\text{-}7)$$

e.g.

$$\left[\frac{\text{monetary unit}}{\text{unit of time}} \right].$$

Here R_{global} constitutes the degree of expectation of danger for the group of persons considered per unit time. If type of damage and source of danger are equally delineated, then the following relation exists between the individual risk—for purposes of distinction, hereafter called R_{ind}—and global risk of a group of persons comprising N individuals:

$$R_{\text{global}} = R_{\text{ind}} \cdot N. \qquad (2\text{-}8)$$

In other words,

global risk = individual risk × number of persons within the area of observation.

It is obvious that, as practiced in the determination of the individual risk for real persons, the same risk can also be applied to legal entities or any other unit.

In the determination of the individual risks, we must first consider that—according to Section 2.4—damages of a different nature are measured in noncommensurable units. It is common practice to make a distinction between the different types of loss or damage, namely death, injury, and loss of property.

Risk Fields. The concept of risk as introduced above allows for the possibility of quantitatively describing the danger fields which can be ascribed to a locally determinable source of danger [2-3]. From that source of danger, effects will be felt in its environment, e.g., because of pressure waves, hazardous substances, etc., which will lead to site-dependent effective conditions $WB(r)$. Taking as a basis a certain type or class of event resulting in the same amount of damage S_K, Equation (2-4) permits us to deduct an individual risk dependent on the site r, i.e., an "impressed risk field"

$$R(r) = S_K H_S[WB(r)] = S_K H_S(r). \tag{2-9}$$

Here H_S constitutes the mean frequency of the occurrence of damage per person and unit time, provided the person in question is subject to the effective conditions $WB(r)$. With regard to this person, we must take the average over a population group which will dwell within the danger area or hazard area. The exposure times to be expected must form part of the consideration. The "impressed risk field" defined by $R(r)$ is dependent upon the conditions at the source of danger and upon general condition of the location, e.g., climate and weather, but not upon the actual population distribution there. From the density of population $B(r)$ we can derive the density of the population risk

$$r_B(r) = R(r)B(r) \tag{2-10}$$

and, upon integration over the considered area, a population risk allocated to the source of danger within a certain environment:

$$R_B = \int r_B(r)\, df$$
$$= S_K \int H_S(r)B(r)\, df. \tag{2-11}$$

In conclusion it may be summarily stated that the "risk" concept with its probability-theory-oriented formulation can be used in a variety of ways and requires unequivocal framework conditions.

Chapter 3

Cybernetics and the Science of Safety

3.1 The Basics of Cybernetics

According to Wiener, cybernetics is the science of "control and communication in the animal and in the machine" [3-1]. While the concept of cybernetics was used by Ampere during the past century, cybernetics in its present form can essentially be traced back to Norbert Wiener. He and several others devoted themselves to control processes in technology as well as in the human organism. It was found that in both fields the problems, as far as their structure and possibilities of mathematical description are concerned, were similar. Wiener had already pointed out that the development of suitable methods might well contribute to a better understanding of complex social and economic processes. This is also true for ecological systems. Wiener's approach [3-1] has had a fruitful effect on many disciplines of science. Figure 3.1 illustrates several crucial applications.

In accordance with a broad application of cybernetics, there were several attempts during its development phase to define it and delimit it against related fields like systems theory or informatics.

In order to utilize the cybernetic approach for the science of safety, it would help if one would regard cybernetics as the "general science of cybernetic systems." Its tasks and goals within the meaning of this treatise can be described as follows:

—Cybernetics deals with systems consisting of elements and the relations of elements among each other and with the environment.
—What should be examined and analyzed is the time-related dynamic behavior of complex systems. Of great interest is any information as to how target-oriented behavior is achieved in existing systems by means of control processes or how such behavior can be achieved in systems which man is designing and in which he intervenes.
—Cybernetics has been designed to bring out analogous phenomena in systems of a varying nature and to formulate them in a similar manner.

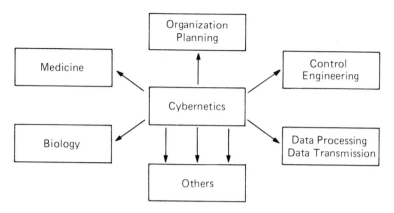

Figure 3.1 Application fields of cybernetics.

This makes it possible to cross the boundaries between various disciplines when describing the behavior of systems, and thus lays the foundation of the interdisciplinary character of cybernetics.

The cybernetic description of systems of different types is characterized by concepts and definitions like feedback, delay time, stochastic processes, and stability. These aspects will, for the time being, be demonstrated on the control loop as an example of a simple cybernetic system, but one that contains all typical properties. Thereby the importance of the probability calculus and communication in cybernetics can be clearly explained. This is then followed by a general representation of cybernetic systems.

The Control Loop as a Cybernetic System. Figure 3.2 shows in schematic manner a liquid-level controller as an example of a simple technical control loop [3-4]. The system to be controlled, the so-called controlled system, is a liquid container and its contents.

The liquid level X' in the container is to be kept at the specified value, X. The varying amount of liquid drained, Q_a, acts as a disturbance variable. Equalization is achieved by a measuring device determining a deviation of the actual level X' from the specified level X. This deviation, via a correcting device acting as a regulator, is converted into a change of the control quantity Y. This modification, in turn, effects a modification of Q_e which will correct the liquid level. If A is the section through the container, the equation for the change in time of X' is

$$\frac{dX'}{dt} = \frac{Q_e(Y)}{A} - \frac{Q_a}{A}.$$

(3-1)

The liquid level is not constant, but demonstrates a nonforeseeable

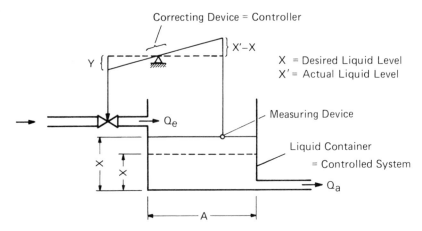

Figure 3.2 Controlling the level of a liquid: A = cross section of the container; Q_e = amount of liquid added per unit of time; Q_a = amount of liquid drained per unit of time.

time-dependent behavior of the form

$$X'(t) = X + \Delta X'(t). \tag{3-2}$$

Feeding back the information on the actual liquid level to the correcting device acting as a regulator and the adaptation of the added liquid amount Q_e will guarantee that the deviation of the liquid level from the specified value will remain negligible, as demanded by the function of the container.

Up to this point, the variations in the drained amount of liquid Q_a have been regarded as the decisive disturbing quantity; but even if Q_a is assumed to remain constant, a controlling problem will still remain, since the amount of liquid added to the container can only be regulated with a limited amount of accuracy. Without any controlling function, the liquid level will change in time with a high degree of probability, so that a corrective action will become mandatory. An analogous consideration holds for the measuring device which has a measuring error used to ascertain the exact liquid level.

The above example of a simple technical controller leads us to a general consideration of the control circuit. If a system is to fulfill its function, a certain type of behavior must be expected of it: namely the desired behavior denoted by the input X. The system can be influenced by the correcting variable Y. It is to be assumed that, acting on a system, whose exact behavior in time is unknown, will trigger unforeseeable changes in the system's behavior. The actual behavior X does not equal the desired behavior, but deviates from the latter by an amount $X(t)$ which, in the course of time, will undergo certain changes:

$$X \rightarrow Y \rightarrow X' = X + \Delta X'(t). \tag{3-3}$$

If it is desired that the actual behavior deviate from the desired behavior only by an amount permissible for the function of the system in question, a correction of the deviation by means of a reverse measure will be necessary. This is based on the premise that, first, the actual behavior X is being monitored by a corresponding monitoring device and is constantly checked inside the controller against the desired behavior, and, second, each deviation detected will trigger a modification of the correcting variable Y. By this feedback a control circuit is established.

If the system to be controlled is called according to common usage, a control system, and if the fact is taken into consideration that, in principle, the monitoring device as well as the controller are subject to disturbances, the result from the above will be the general illustration of a control system according to Figure 3.3.

In connection with Figure 3.3, it must also be taken into consideration that the monitoring action is erroneous, so that the controller is ΔX only supplied with information on the actual behavior which, however, incorporates an error X''.

Even if the desired behavior is firmly predetermined, the actual behavior generated via the control circuit is, as mentioned before, a time-dependent quantity $X = X(t)$. Due to the random disturbance factors, this function cannot be accurately predicted; any number of actual courses are possible. With that, the control circuit generates a random process. A typical form of the variation equation for the actual behavior is available, analogous to the level controller, as

$$X = f_y(Y) + f_z(Z_{RS}), \tag{3-4}$$

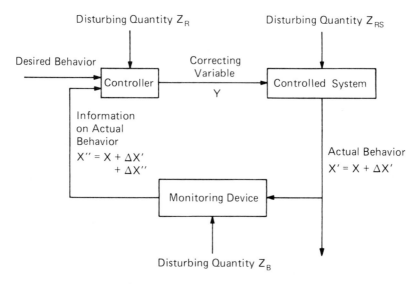

Figure 3.3 Schematic representation of a control loop.

where X stands for the time derivative of X, Z_{RS} stands for the disturbing quantity of the control system, and f_y and f_z are certain functions which depend on the actual situation. However, via the feedback within the control circuit, the correcting variable, in turn, is dependent on the desired value plus the generally erroneous information $X'' = X' + \Delta X''$ on the actual behavior. For the random process initiated by the control circuit, we find the following equation:

$$X \rightarrow \dot{X}' = f(X, X' + \Delta X'') + f_z(Z_{RS}). \qquad (3\text{-}5)$$

However, the above formulation does not yet take into consideration that delays in time must likewise be taken into account in the monitoring of the specified behavior in the conversion of the information into variations of the correcting variable within the controller and in the reaction of the control system to these variations. If the delay times T_{RS}, T_B, and T_R are allocated to the control system, the monitoring device, and the control device, respectively, the following result will supplant the above equations:

$$Y(t) = Y[X, X'(t - T_R - T_B) + \Delta X''], \qquad (3\text{-}6a)$$

$$\begin{aligned} \dot{X}' &= f_y[Y(t - T_{RS})] + f_z(Z_{RS}) \\ &= [X, X'(t - T_{RS} - T_B - T_R) + \Delta X''] + f_z(Z_{RS}). \qquad (3\text{-}6b) \end{aligned}$$

Instead of such a simple variation equation for the derivative of X over time, a system of equations with higher derivatives may take its place, depending on the type of control system. In the present case, however, we are not interested in the special form of the variation equation, but only in the fact that the actual value enters in a time-delayed manner and that the equation contains random elements. It is the task of a control circuit to regulate the generated random process in such a way that any variations of X remain within predetermined limits close to the desired value X, and that the temporal mean value $\bar{X}'$ remain as close as possible to the specified value X. The mean value $\bar{X}'$, in this case, constitutes a state of stability. The control circuit has the tendency of correcting any deviations from the above-mentioned state of stability.

Now one might ask the question: When will a control circuit become unstable?, i.e. When will it be incapable of fulfilling its controlling function any longer so that the system to be controlled becomes nonfunctioning? This will be the case whenever the disturbance factors acting on the actual behavior are of such a magnitude as to overcome the counteracting regulating forces. Even if that is not the case, time delays in feedback may lead to instabilities. If, e.g., disturbance factors display a certain periodicity, they may hype the system into a state of instability even if their energy is very low. This is the case if the delay inherent to the control loop is in the vicinity of half an oscillation time of the disturbance factor. In such a case, controlling forces activated at the wrong time may even increase the disturbing influences (resonance).

A controller may in some cases compensate for the time delay inherent in the control circuit by working prognostically. For this purpose—based on the knowledge of the controlled system and the analysis of the actual behavior—the controller determines a time-dependent expected value $X'(t)$ for the future. As the prognosis period expands, the accuracy will naturally decrease. The efficiency of such a controller, which is called an intelligent controller, depends on its capability of storing and analyzing data. An intelligent controller may be a technical device equipped with a microprocessor or a process computer or it may be an individual human being or organization of men respectively. The essential statements of the above representations will not change whenever the desired behavior is not predetermined as constant but as time dependent. If the behavior of the controlled system is described not by a simple variable X but by several variables $X_1, X_2, \ldots, X_n$, X in the representations must be regarded as a vector having the components $X_1, X_2, \ldots, X_n$. The same applies for the correcting variable. Likewise, the enumerated principles will remain the same if the controlling mechanism is not designed in such a way that the system to be controlled approaches a predetermined behavior as closely as possible, but that certain limit values are not exceeded. Before the above is supplemented by various examples for different control circuits, a few basic remarks are in order.

It has been explained earlier that, in a control circuit, the behavior of the system to be regulated is not as a rule, deterministic, but that the time behavior constitutes a random process. This is an important reason for the fundamental significance of probability theory and statistics within the framework of cybernetics. Likewise, controlling and feedback are inseparably linked with the transmission of information and thus with information and communication theory. In many cases the latter is regarded as an integral part of cybernetics.

Control circuits of various types demonstrate similar behavior whenever they can be characterized by equation systems of the same kind for their corresponding quantities. In such a case their real elements will be of no importance. This, e.g., affords the possibility of replacing a mechanical control circuit by an electrical circuit and studying the behavior of the former by means of the latter. Examples of this may be found in the literature [3-5]. Even for Wiener, one of the essential approaches in his collaboration with medical men was to develop mechanical, electrical, or even computational models for the simulation of control processes in living organisms.

Now we give a few examples of different types of control circuits. They are meant to underline the general character of the above-mentioned basic elements of rationalization in cybernetic thinking.

A. Autopilot in an Airplane (Figure 3.4). Modern airplanes are equipped with autopilots. The pilot just sets the course and the cruising altitude. The

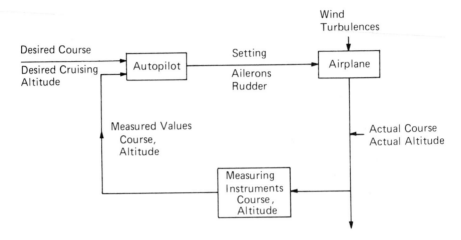

Figure 3.4 Schematic drawing of controlling an airplane by means of an autopilot.

controlled system in this particular case is the airplane itself; the controller is the autopilot; the measuring instrument for both the course and cruising altitude is the monitoring device. The disturbance factors are wind and turbulence. While the desired course is a predetermined quantity, in a properly functioning autopilot the actual course varies around the course in such a way that the time mean value $\bar{X}'$ of the actual course practically equals the desired course. This has been demonstrated in Figure 3.5. The above also relates to the cruising altitude in a corresponding manner.

The autopilot itself requires controlling in order to avoid negative effects of possible defects. The pilot, e.g., may watch whether the course or the altitude displays a tendency to systematically deviate from the desired value, as demonstrated in Figure 3.6. If this process is scrutinized more closely, we see the formation of a moving average $\hat{X}'(t)$ over time intervals of suitable lengths. This can be realized by measurements with the help of an integrator, either by connecting the integrator to a measuring instrument or by utilizing the human brain. Such a moving average can be expressed in general terms as follows:

$$\hat{X}'(t) = \frac{1}{\tau} \int_0^\tau X'(t - t')\, dt'. \tag{3-7}$$

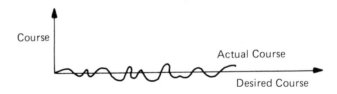

Figure 3.5 Actual course with properly functioning autopilot.

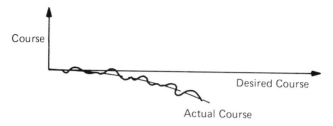

Figure 3.6 Actual course with defective autopilot.

At this point a test must be made to determine whether the above-mentioned moving average deviates excessively from the theoretical course. If this turns out to be the case, the pilot switches to manual control. The autopilot is replaced by the pilot (Figure 3.7). Within the framework of cybernetics, this means that a second control circuit with the controller acting as a controlled system is superior to the control circuit with the autopilot. Many actual systems consist of a multitude of interacting superior and subordinate feedback control circuits in order to be able to make corrections to individual parts when they malfunction.

B. Motor Vehicle with Driver (Figure 3.8). The desired behavior in this case is the controlling of a motor vehicle according to the rules of traffic without any accidents and endangerment of other traffic. The driver, in his capacity as a controller, achieves this by actuating the gas and brake pedals and the steering wheel, respectively. Disturbing influences are derived from the prevailing situation in traffic, e.g., unexpected obstacles, from the state of

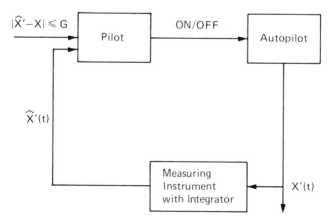

Figure 3.7 Schematic drawing of the supervision of the autopilot by the pilot acting as a control circuit.

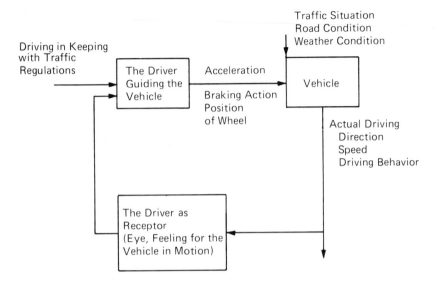

Figure 3.8 Vehicle–driver system as a control circuit.

the road, and from weather conditions. The driver not only acts as a controller but also as a monitor by using his eyes and utilizing his feeling for the moving vehicle. The information perceived through the eye and the sensory nerves is transmitted to the brain and put together as a composite picture of the prevailing traffic situation. On the basis of this picture and stored driving experience, commands are transmitted from the brain via the nervous system to hand and foot for the actuation of gas and brake pedals and the wheel.

If an unexpected obstacle comes up, the driver reacts with a certain amount of delay which we call "(delayed) reaction time." This delay is the consequence of the fact that the transmission of information to the brain and the processing of this information in the brain require a certain amount of time. A further delay is caused by the reaction time of the brake when the brake pedal has been actuated. These delays, it is true, are usually of magnitudes well under a second. However, they may be of decisive influence in determining whether the control system, under the given circumstances, fails to function and will lead to an accident. Also, other types of control circuit malfunction may contribute to an accident. Here are a few examples:

—Failure to recognize an obstacle because of impaired vision. This corresponds to a defect of excessive magnitude in the monitoring device.
—Failure to recognize the obstacle because of distraction. This corresponds to disturbing influences acting on the monitoring device.
—Incorrect reaction because of the influence of alcohol. This corresponds to disturbing influences on the controller.

General Representation of Cybernetic Systems. The control circuit which has been used in this connection to illustrate several basic concepts in cybernetics constitutes an especially simplified version of a cybernetic system. Referring to this special case, a general cybernetic system can be defined as a structure consisting of system elements (SE) with input quantities E and output quantities A (Figure 3.9).

Due to unknown influence factors which, in general, may be called disturbances, the relationship between input E and output A is not unique. To any realization of input E a possible multitude of realizations of output A is allocated, whereby the latter cannot be described deterministically, but only statistically. The structure of a cybernetic system is determined by the relations between various elements inasmuch as the output of one element is at the same time the input of one or several other elements. These input and output quantities may be energy flow, transport of mass, action, and information. For the simpler type of control circuit in a formulation like the one above, the result will be the representation according to Figure 3.10. A more general example of a cybernetic system is shown in Figure 3.11. The rectangle symbolizes the boundary of the system against the environment which acts on the system by means of known input and also by means of disturbances, whereby the course of these is

Figure 3.9 System element of a cybernetic system.

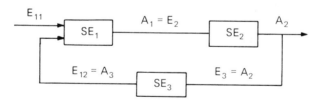

Figure 3.10 Schematic representation of the control circuit as a cybernetic system.

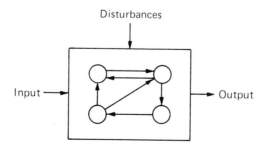

Figure 3.11 Schematic representation of a general cybernetic system consisting of four system elements.

unknown. The outputs of the system into the environment are variable in time and, because of the inner structure of the system and their dependence on time, constitute random processes.

A feedback situation is given whenever at least one closed chain of inputs and outputs of elements exists which, emanating from one element, ultimately leads back to that same element.

A cybernetic system is a dynamic system within which random processes are evolving. Target-oriented behavior of a cybernetic system entails reducing the multitude of random processes; that is, a behavior in accordance with the predetermined target while adapting to the environmental conditions.

3.2 Cybernetic Approach of the Science of Safety

A first approach of cybernetics to the safety complex of a technical system is the frequent utilization of control circuits in order to keep the behavior of a machine on a predetermined path. The selection of the technical methods meant to achieve safety is the task of safety engineering and not the science of safety. The latter has a far greater scope.

The interlinking of cybernetics and the science of safety results from the fact that technical installations are operated and monitored by human beings who are involved within the effective range of that same installation. They are using the machine but are likewise exposed to its dangers. Human behavior as well as the behavior of the machine, on the other hand, are dependent on the conditions of their environment. The environment, in turn, is often influenced by them in various ways, e.g., by the production of waste, sewage, noise, and alien substances into the air. Man, on the other hand, is able to influence these environmental factors. Each and every technical installation is thus embedded in a man–machine–environment–system characterized by mutual interaction (Figure 3.12).

The safety of technical installations always constitutes the safety of complex technical systems; the task of creating safety is an interdisciplinary one. It consists of the task of formulating safety requirements and to shape

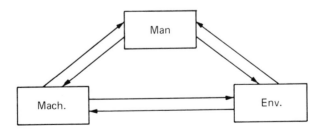

Figure 3.12 Basic cybernetic schematic design of the man–machine–environment–system (MMES).

man–machine–environment–systems in such a way that they will behave in a target-oriented manner. Thereby the objects of the science of safety often turn out to be realizations of the objects of cybernetics, whereby one of its special targets is the prevention of endangerment through technology. The science of safety, in proving its existence theoretically and in elaborating its methods, may make use of cybernetics.

Behind the schematic drawing shown in Figure 3.12 there is an unlimited number of overlapping subsystems and feedback processes. Then, for each individual case, meaningful representation depends upon the problem posed and on the scientific approach. One can obtain a certain survey of interest for the basic approach of the science of safety by looking at various spheres of action. One can then distinguish between a local or intracompany level of effect, a regional, and a supraregional or global sphere of effects. This is shown in Figure 3.13 and explained thereafter.

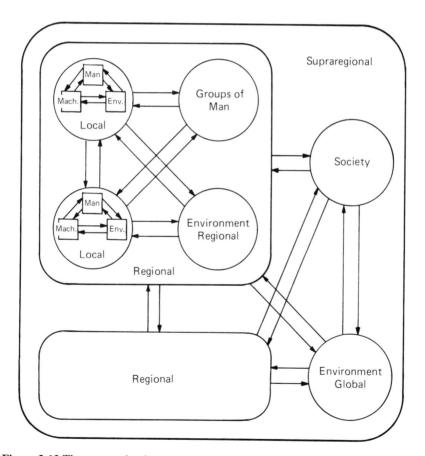

Figure 3.13 The connection between local, regional, and global effect levels within the man–machine–environment–system (MMES).

The local level is characterized by the direct contact of man with technical installations in his household, in traffic, and in industry. Thus, in the industrial area, the local level is the intracompany level. Due to accidents, incidents, or permanent stress by the local environment, persons or property may sustain damage. Accidents and incidents may have technical origins. This, seen from the point of view of cybernetics, means that a technical installation, despite correct handling and operation, may not behave as expected because of other disturbance factors like manufacturing defects, harmful substances in the atmosphere, or climatic conditions. Another cause may be a lack of coordination in the interaction between man and machine. Defects like these should be minimized by operator-oriented design of the installation in accordance with ergonomic factors. This would mean adapting the requirements to man. On the other hand, man—through proper selection, training, and motivation—may be induced to behave properly and protect himself purposefully. Such protection, e.g., may also include protection against stress through air or noise pollution. It must be taken into consideration, however, that man's behavior is never uniform or unchanging but displays variations from one human being to the other and also varies over time. It is influenced by the state of man's environment. Any improvement in the environment thus often also means an adaptation of the demands made on man.

In general, the local and intraplant effect level is embedded in a regional environment, characterized by the social and societal infrastructure, the regional status of environmental stress, and the climatic conditions. This regional environment influences the behavior of the individual man–machine–environmental–systems located within its range. The regional level of danger results to a large degree from the sum total of the influence factors of the individual man–machine–environment–systems in this region. Among these are the health hazards present at the workplace affecting the general state of health, as well as the risks which are present due to the possibility of incidents in technical installations in a larger area. Finally, air pollution, noise emissions, and waste and sewage which are introduced into the environment are included. While on the local level the individual man–machine–environment–system is the object of the science of safety, on a regional level it is the structure of the technical installations already existing or in the planning stage and their effects on the individual subsystems.

The supraregional level coincides with the consideration of any promoted or applied technology within a country and the present state of the art in technology, of ergonomy, safety at the place of work, and environmental protection. It retroactively influences the realizable structures of man–machine–environment–systems on a regional and local level. The behavior of the regional and local systems determines the extent of the total damage to society based on endangerment and impairment of the environment. Due to the international interlinking of economy and technology, and because of the fact that the spreading of substances endangering the environment through the air, water, and foodstuffs does not stop at national borders, one might also speak of a global level: man, because of his innate intelligence, is able to interfere in a MMES and shape its structure, thereby reducing the dangers emanating from technology on all three levels.

Local means of controlling this process include licensing procedures by government agencies as well as installation-related requirements and safety measures prescribed by the law. The regional level corresponds to city and regional planning, and the supraregional level to the promotion or prevention of technologies. As far as endangerment through technical installations is concerned, the local level corresponds to the level of risk at the individual installation and is described by an area of risk; to the regional level, the overlapping of risks represented by the risk survey; and to the supraregional level, the statistical allocation of risk levels to certain technologies.

If safety is to be rendered practicable, the all-encompassing system of interactions in which each technical device and each local man–machine–environment–system is embedded must be dissolved. This should be effected via interfaces which have been agreed upon previously. Within the economical-technical area, the existing system of standards can be utilized to that end. For environmental protection, maximum emission levels, installation-specific emission limits, etc., are available as controlling factors. In the area of safety proper corresponding design requirements, etc, are formulated. Laws, rules, and regulations devolving from the stress environment of social forces and technical development as well as monitoring systems will be indispensable.

Safety as a Cybernetic Problem. The designing and engineering of safety into technical systems is as such already a part of cybernetic processes, the basic abstraction of which can be illustrated by means of a control circuit (Figure 3.14). Technical installations must meet certain safety standards so that during the erection and operation of a plant, they will be able to limit side effects in the form of accidents, damage, and environmental pollution. Safety requirements must already be converted into concrete measures during the planning and construction phases of technical installations. With the help of the outer control circuit shown in Figure 3.14, safety regulations can be monitored and improvements achieved by observing the side effects in existing installations. However, the above approach to safety engineering is no longer sufficient in the present environment of modern technology. The inevitable time delay until practicable results can be obtained by monitoring would no longer be acceptable. Some of the reasons for this are the following:

—Technology is developing so fast that experience gleaned from accident and damage statistics is often meager and insufficient for a credible projection into the future.

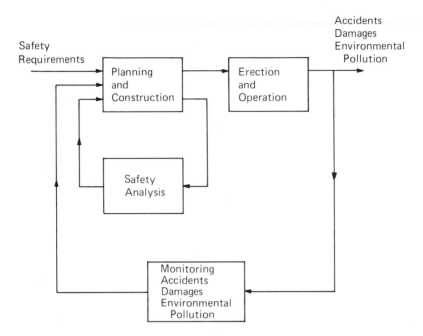

Figure 3.14 Safety engineering as a cybernetic process.

—The danger potential and the complexity of technical installations are increasing. "Waiting" (or rather, "hoping") for experience would mean that, in the interim, a considerable amount of damage would have to be accepted. Such a solution would be devoid of any justification.

—In some cases, side effects of technical installations may become visible only after a considerable time, namely in the form of irreversible damage to the health of man and his environment. Late consequences caused by radioactive material or toxic chemicals are only one example.

—Severe accidents in installations of considerable danger potential are rare occurrences for which a sufficient number of observations is not possible.

For the above reasons it is imperative that prognoses already be established about the levels of endangerment—which must be expected during the erection and operation of a technical installation—during the planning phase and that it be determined whether these prognoses will remain within predetermined boundaries or not. If necessary, existing plans should be revised and the design improved. Similar examinations and tests can be conducted on existing installations. Their foremost aim is a prognosis of the effects of modifications even before the latter are put into effect. For establishing such prognoses, methods of safety analysis have been developed which will be explained below.

Understanding the approach to safety as a cybernetic problem brings out another very important aspect. Precise control demands an exact description and delimiting of the desired and the actual behavior of the system. This corresponds to passing from a qualitative description of the endangerment by a technical installation to its quantitative description. The qualitative description, no doubt, will lead to constructive and organizational safety measures and has already made great contributions to constant improvement in safety engineering. However, as far as the realization of technical installations is concerned, it leaves too wide a margin. This margin, in specific cases, is of such a magnitude as to permit the subjugation of the safety requirements to the primary considerations of usefulness, and thus unequivocal safety-related criteria cannot be applied. The desired behavior under the aspects of safety is thus inaccurately defined. This, in turn, will result in a fuzzy delimited comparison between the desired and the actual state of affairs. The latter is also the reason why some people may in good conscience describe a certain actual state as "unsafe" while others will describe the same state as "safe." Therefore, in specific cases, the necessity of setting precise safety requirements in the form of standards for endangerment measures will arise. For this it must be taken into consideration that each positive man–machine–environment–system is subject to influencing factors of which the designer or organizer knows absolutely nothing, nor is he aware of their time behavior; also, even the reaction of the monitored system is not unique. Possible uncertainties may, e.g., arise from human behavior, the weather, qualitative changes in the environment, etc. The actual behavior of an MMES is therefore—as was mentioned before—a random process. Endangerment measures used in prognoses must thus be of a statistical nature and must have the character of mathematical expectations. Risk evaluation as a product of the mathematical expectation of frequency of damage and the expected volume of damage is of the aforementioned type. In order to close the control circuit, a more exact definition of endangerment measures must be accompanied by adequate monitoring facilities. Such monitoring facilities and means of surveillance are statistics, e.g., accident statistics, or measuring technologies for the determination of water, air, and noise pollution.

In principle, the quantitatively exact definition of a safety requirement by means of endangerment measures is preferable to a qualitative requirement. However, the quantitative definition is truly preferable only if it yields an accurate and true value for the required amount of safety, and if no essential aspects are being neglected. Otherwise this would render any control faulty. The quantification of the level of endangerment and the prognosis of endangerment are closely interwoven, in itself the qualitative description of the level of danger also permits important prognostic statements.

In conclusion one may say that the prognosis of endangerment can never replace the monitoring of the actual level of danger. Both must contribute to the controlling of the technical development. From the standpoint of cybernetics, statistics on accidents, incidents, and damage are monitoring methods in the sense of cybernetics which will only lead to concrete results after a certain amount of delay and can only be helpful whenever they are suitably employed and interpreted.

Safety Actions from a Cybernetic Point of View. From the point of view of cybernetics, the disturbing influences of a system are considered potential sources of endangerment. Their time cannot be controlled nor individually predicted. They pose a danger to the system's stability. A system, in turn, has two possibilities of absorbing the effects of such disturbances: compensation by feedback and buffering or shielding, respectively. In real systems, combinations of the two are often applied.

Feedback is realized by technical devices as well as by arrangements of the type where an operator of a unit reacts to deviations of the measuring data from the desired values. As far as the types and modes of operation of the disturbances are concerned, the system to be controlled is regarded as an "unknown quantity," i.e., in the language of cybernetics, a "black box."

Buffering or shielding, when suitably applied, will absorb the effects of disturbing influences. However, this requires knowledge of the type and mode of operation of these disturbances. A typical example of buffering are shock absorbers cushioning the effects of irregularities in the road surface motor vehicle.

3.3 Systems Engineering as a Cybernetic Instrument for Assessing Technical Plants

Tasks and Objectives of Systems Engineering. Systems engineering includes the "systematic application of technical tools and scientifically justified methods in the planning of modifiable, target-oriented, and complex systems" [6-6]. According to Ropohl [3-8], systems engineering which is oriented toward practical applications has its roots in Bertalanffy's approach—which, later on, became known as "general system theory" [3-21]—and in cybernetics. Both systems approaches have led to today's concept of cybernetics as described in Section 3.1. Thus we may regard systems engineering as the conversion of cybernetic systems approach into practical activity in designing complex systems.

Systems engineering regards the objects to be designed as well as the development and realization processes themselves as a cybernetic system. Systems engineering is confronted by two essential problems:

—How to organize a purposeful and target-oriented cooperation of decision-making bodies, work groups, and individuals, especially from different disciplines of science, for the development and realization of objects?

—How to obtain sufficient material as a basis for decision-making concerning the evaluation of planned objects and the selection of possibilities of realization?

The first question is directed towards a suitable system management, starting from problem analysis via planning of objectives, the selection of the suitable realization alternatives to the licensing, erection, and operation of technical systems. System management tools include networks, bar graphs, etc. In principle, the chain of steps should be subdivided into phases in such a way that at some given points a check of the decisions will be possible. It is assumed that failures may occur in all phases. Therefore suitable control mechanisms must be built into the procedure so that failure can be recognized without delay. The entire work on one single object is understood as a multifeedback cybernetic process. Figure 3.15 gives a basic diagrammatic presentation of the procedural steps from the first time a problem is posed to the selection of a suitable solution alternative. The instrument providing data for the decision-making processes for planned objects and for the selection of realization alternatives is system analysis. On the one hand, it must be capable of permitting an

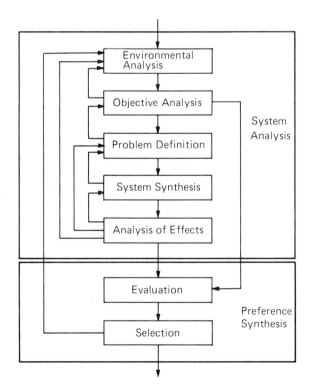

Figure 3.15 Procedural structure of a system of actions.

evaluation of the performance of the planned object with a view to the desired usefulness and its intended function and, on the other hand, it must provide a credible statement of the behavior of the object with regard to additional requirements, especially those relating to safety and environmental protection. For the latter aim it is of the utmost importance that this information be obtained as early as possible during the planning phase. This explains the special importance attached to prognostic methods in systems engineering. These are either based on the projection of empiric results or on a theoretical analysis of the structure of the object to be evaluated being known. An important instrument of theoretical system analysis is simulation using a formal mathematical or logical model. Generally known methods of simulation include methods of operations research, like linear and dynamic programming. These methods, however, also encompass theoretical reliability tests which are experiencing increasingly greater acceptance in technology.

Within the framework of the science of safety, the methods of system analysis which are especially important in systems engineering are those suited for the evaluation of all kinds of safety aspects, especially the frequency of incidents and their impact. These methods will be described later in their capacity of working tools. At this point we will discuss their methodological basis and the limits of their application.

Problem-Related System Description. Any system analysis, whether empiric or theoretical, must be preceded by a determination of the objectives of that analysis. From this, then, are derived the definition of the system as well as the system description which relates to the objective of analysis.

In accordance with the basic abstraction of the cybernetic system, the following must in essence be achieved (Figure 3.16):

—Delimitation of the system to be examined against the environment.
—Determination of which input values from the environment and output values into the environment are important.
—Determination of which other disturbances from the environment must be taken into consideration.
—Clarification of how the system is described: by which elements having which properties.
—Determination of which type of interactions between these elements must be considered.

Each concrete description of a system is the result of a subjective way of looking at it. This fact should also be taken into account. Thus no single and generally valid description of a system will be possible from which an object-related system description could be derived according to simple rules. Obtaining a description suitable for system analysis itself is an

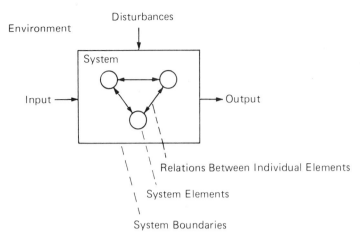

Figure 3.16 Diagrammatic layout concerning problem-related system description.

analytical process *per se* and actually constitutes the first step in system analysis. For this, partly competitive requirements must be established:

—The system description must be as simple as possible in order to obtain a clear and unobstructed survey and to limit the necessary expenditure.
—The description must be complete to such a degree that it contains all the important components, variables, and their relations which are needed to characterize the problem to be solved and without which the object of analysis cannot be achieved.
—For large-scale systems, the description must be of a nature which, via the definition of suitable interfaces, permits a breakdown into subsystems that may be subjected to separate analyses.

The problem-related system description may be represented by written text, by a drawing, or in the form of mathematical equations. In each case, however, the system description is based on a model-like idea. The model serves for the simplification of the real system to be analyzed. A model which fails to convey the actual problem will cause the information gained through system analysis to be incomplete or incorrectly interpreted. During the phase of acquiring an object-related system description, the course is already set towards a "correct" system analysis.

Empirical System Analysis. In empirical system analysis, information on the behavior of systems is obtained by observing and measuring as well as interpreting.

Empirical system analysis is a tool

—for the estimation of safety-related parameters of objects or components which, in turn, may be utilized in system analysis for the prognosis of the behavior of man–machine–environment–systems, and—for the direct empirical determination of a relation between the frequency and severity of accidents and losses, on one hand, and the modifiable properties of the examined systems, on the other hand.

For carrying out empirical system analyses there are essentially the following possibilities:

—the retrospective evaluation of the behavior of actually existing systems;
—experiments conducted in existing systems under laboratory conditions;
—experiments on physical models of the systems to be analyzed (experimental simulation).

The first group includes evaluations and analyses of statistical recordings on realized systems, including epidemiological examinations. These systems may be individual technical installations or even technical systems including their environment, e.g., street traffic in a certain region. Safety-related recordings are accident and loss statistics, recordings on incidents in systems and parts of an installation, in some cases even statistics on diseases, causes of death, and environmental impact. These must be linked with other descriptive characteristics where an interrelationship with directly safety-related data is suspected. The retrospective evaluation of data on actually existing systems incorporates the advantage of mirroring their behavior under actual conditions. However, it also is saddled with the disadvantage that the impact of all interference factors, and modification of individual system parameters is impossible or at least extremely limited. The derivation of accurate statements on the relations between system behavior and system parameters under these circumstances is often difficult in the extreme.

If experiments on existing systems are carried out under laboratory conditions, the objects of the examination are individual or several technical devices representing the totality of comparable devices. Such experiments include tests on test stands, crash tests of vehicles, and experiments conducted on test persons. The advantage of tests conducted under experimental conditions is that the relevant input variables can be systematically modified and disturbances eliminated. The difficulty with this method is, however, in many cases, the setting of close-to-real boundary conditions in order to be able to apply the lab results to actual conditions.

Lab experiments can also be conducted on physical models of the systems to be analyzed. The previously mentioned crash tests become part of that group whenever they are conducted using dummies designed to simulate the behavior of the human body. This group also includes experiments using test persons on simulators, as practiced in aeronautics. These simulators are copies of actual cockpits; the various flight situations are created with the help of video technology. Likewise, tests inside a wind tunnel, using smaller scale versions of actual technical devices, belong to this particular group. Experimental simulation has the advantage that the parameters of the systems to be tested can be modified along with the input values or other environmental influence factors. Unfortunately, this advantage has to be paid for by a greater amount of uncertainty about its applicability to actual systems, since the objects tested are themselves only models.

Within the framework of system analysis as an instrument of systems engineering, empirical results are of value only as far as they permit prognoses on the expected behavior of systems for the future. The data on which a system analysis is based constitutes a sort of spot check from the basic totality of the data of all comparable actual or imaginable systems and is thus prone to random variations. Thus mathematical statistics play an eminent role in the evaluation of these data. It permits statements on confidence intervals for derived characteristics and of significance limits for dependencies. Both represent a measure of the degree of uncertainty of the interpretations derived from these data, and knowledge of them is indispensable if wrong conclusions are to be avoided.

Mathematical statistics constitutes a broad instrumentarium for the evaluation of measuring and monitoring series. In retrospective evaluations of statistical recordings on systems which have been realized, especially of interest are those methods of mathematical statistics which also permit multivariate analysis of the loss and accident frequencies. They include such dependencies of modifiable and influenceable system properties or other disturbances as, e.g., weather conditions in road traffic. Mathematical-statistical methods of the above type are variance analysis, regression analysis, and factor analysis [3-9], [3-10], [3-11], [3-12], [3-13], [3-20].

With the help of variance analysis, it is determined how far the variance of observed variable X can be traced back to suspected influence factors. These influence factors may be qualitative or quantitative variables. Variance analysis is based on the assumption that, in addition to data of the observed variable X, data on other suspected influence factors are also present in a measuring series, whereby these influence factors can be classified in such a way that each observed value of X can be associated to a class i. In the case of a simple variance analysis with one additional influence quantity, the following equation will result for X:

$$X = \mu + \zeta_i + \epsilon, \tag{3-8}$$

where μ is the mean of X, ζ is the distance of the class mean value of X in the i-class of mean value μ, and ϵ is a random component which cannot be explained by means of the additional influence quantity. What is determined is ζ, and the residual standard deviation $\sigma_\sigma \cdot \sigma_\epsilon$ is the standard deviation of ϵ and the residual of X which cannot be traced back to a change in the additional influence quantity. Regression analysis serves to explicitly present the observed quantity X as a function of additionally observed quantitative influence factors $Z_1, \ldots, Z_n$ in the following equation:

$$X = a_0 + a_1 Z_1 + a_2 Z_2 + \cdots + a_n Z_n + \epsilon \qquad (3\text{-}9)$$

or in the general form

$$X = a_0 + a_1 f_1(Z_1) + a_2 f_2(Z_2) + \cdots + f_n(Z_n). \qquad (3\text{-}10)$$

What must now be determined are the regression coefficients a_0, $a_1, \ldots, a_n$ and the residual variations σ_ϵ. Factor analysis is helpful in discovering, in a group of parallel observation series, common influence factors which, for the moment, are not explicitly determined and measured. Analogous to regression analysis, this representation yields the following equation:

$$X_k = a_{1k} Z_1 + \cdots + a_{mk} Z_m + a_k U_k. \qquad (3\text{-}11)$$

Here, however, only the X_k are available as an observed quantity and the meaning of the common "factors" $Z_1, Z_2, \ldots, Z_m$ will have to be derived from the analysis.

In general it must be taken into account that the statistical evaluation of observed and measuring data will, for the moment, only yield "statistical" dependencies, i.e., that observed quantities are correlated. These correlations may be traced back to common third causes; however, this need not be so everywhere. In order to decide whether a statistical dependence is also a causal one, an expert interpretation is indispensable.

Theoretical System Analysis. The object of theoretical system analysis is to formulate statements on the behavior of a system based on its structure and its relationship with the environment using the laws of natural science, mathematics, and logic. In theoretical system analysis one must rely on empirically obtained information relating to system elements or subsystems. Theoretical system analysis is of decisive importance in that it will permit a statement on the expected behavior of a system even at a point in time when the system—at least at the time when the structure of the system in its entirety is being decided—is not accessible for an experimental examination and sufficient experience gained through retrospective observations in comparable systems is not feasible. In modern large-scale installations, however, such is often the case. Within the framework of safety considerations, credible statements on dangers emanating from technical devices are mandatory. Theoretical system analysis, which is utilized for this purpose, may refer to specific technologies. In that case—known as technology assessment—its object is to recognize the impact on society and environment at an early stage of a technological development. On a social level, it is designed to afford decision-making bodies the possibility of deciding on the propagation of specific technologies and their steering in a specified direction.

However, theoretical system analysis may also refer to individual technical installations. This, under safety aspects, is called safety analysis. In other words, system analysis can be technology oriented (technology assessment) or installation oriented (safety analysis).

Theoretical system analysis may be of a quantitative or a qualitative nature. In each case, quantitative analysis must be preceded by a qualitative analysis in order to discover relevant problems and recognize the areas where qualitative analysis must be complemented by quantitative analysis. Even then, a certain restriction will be necessary because of the scarcity of the possibilities at one's disposal, e.g., mathematical models, data, and computers. The instrument of quantitative theoretical system analysis is mathematical simulation. Due to its importance for modern safety analysis, this will be treated separately in the following section.

3.4 Mathematical Simulation

Mathematical simulation is the most important instrument for conducting quantitative theoretical system analyses. Quantification and mathematical simulation as its tool correspond to scientific/technical thinking and to the precision in describing facts which, in general, is expected from it. Its fundamental significance for the designability of technical safety has already been derived from the cybernetic approach. The ability to simulate even complex systems is constantly improved because of the steadily increasing speed and size of digital computer systems. In this manner, mathematical simulation has increasingly gained importance as a tool for modern safety analyses. It will be important not only to recognize the possibilities but also perceive the prerequisites and limitations of this tool in order to evaluate the results obtained and to avoid mistakes during simulation.

In a mathematical simulation, the behavior of a system is not examined on the system itself or in a physical model but with the help of a mathematical model. For this the behavior of the system is traced back to known logical, mathematical, scientific, and technical laws as well as to the known behavior of partial systems. The prerequisite in each case is that the system behavior of interest and the structure of the system including input and output quantities can be quantitatively described by means of mathematical equations. The value of mathematical models and simulations carried out with their help resides in the fact that the input data and the system parameters can be changed easily and that the influence of such modifications on the behavior can be studied. Mathematical simulation, in essence, can be of value for three types of projects:

—It can be utilized for supporting the interpretation of empirical examinations of the behavior of a specific system. Influence factors and states of the system important in variations and fluctuations should be determined in this manner.

—It can be utilized as a complement to experimental examinations in order to reduce the number of necessary experiments.

—When utilized during the planning phase, it may help in the acquisition of additional information needed for a decision to be made on functional and possible design variations.

When using mathematical simulation as an instrument of safety analysis, the last aspect is of special interest.

In accordance with the basic model of the cybernetic system, the variables which are symbolically represented in Figure 3.17 enter to a mathematical model. These magnitudes may be continuously modifiable or discrete. They also include logical variables of values 0 and 1.

Legend:

X: Input variables.
Z: Disturbances.
α: System parameters which are independent of input X.
W: System parameters which are variable as intermediate quantities and which describe dependency of the system state on input variables and, possibly, on the time. These, through their relations, and together with the above-mentioned system parameters, represent the inner structure of the system.
Y: Output quantities.

Whenever the described variables have several components, they are to be treated as vectors.

The mathematical model consists of a system of equations into which the above-mentioned quantities are entering. Here a distinction is made between static and dynamic models. Dynamic models observe the time-dependent system behavior; static models will only monitor individual states of the system at constant input values. If the output vector Y consists of n components of the value Y_v and the vector W of m components W, the static model, in a simple case, can be expressed by an implicit system of equations of the following type:

$$f_v(X, Z, \alpha, W, Y) = 0, \qquad v = 1, 2, \ldots, n;$$
$$g_\mu(X, Z, \alpha, W, Y) = 0, \qquad \mu = 1, 2, \ldots, m. \qquad (3\text{-}12)$$

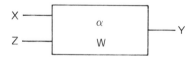

Figure 3.17 Types of variables in mathematical models.

The system of equations can be complemented by additional boundary conditions. In general, not only the variables themselves, but also ordinary and partial differential quotients and the number of equations will increase correspondingly. The system of equations will retain its formal representation if the derivatives themselves are regarded as variables. The static solution of the system of equations entails the determination of the state of equilibrium with fixed input values. If the reaction period of a dynamic system is small in comparison with the period of time in which the input values are changing, the dynamic time flow can also be obtained in the shape of a sequence of states of equilibrium (equilibrium conditions) with the help of static models. The time flow is then a quasistatic one.

For dynamic models, the time behavior of X, W, and Y is entering the system of equations, as a rule; especially the derivatives of the intermediate quantities W and the output variable Y are absorbed in time. The boundary conditions are complemented by initial conditions of the state at a given time t_0.

The solution of the equations which is determined by the mathematical model may be effected by explicit solvability for the output quantities or by means of numerical methods. In complex, especially dynamic, models, the explicit type of solvability is an exception.

The cybernetic system approach demonstrates that the behavior of complex systems cannot be unequivocally determined but is subject to statistical variations. These can either be traced back to outside influence factors known as disturbances impacting the behavior of the system components or to the fact that input quantities are subject to statistical fluctuations. Correspondingly, the output quantities are also subject to statistical variations. Therefore, mathematical system models must be checked as to the way and to what extent they take into consideration these statistical variations.

In deterministic models, the statistically distributed input or system quantities are represented by individual values. Their selection may be effected according to various criteria, depending on whether an unfavorable or a characteristically average system behavior is to be simulated. Correspondingly, the resulting values of the output quantities also assume the character of limit or mean values. The results must be evaluated accordingly.

In probabilistic models, the statistical variations are at least partially included in the simulation by the input and system quantities. Thus the result will not consist of individual values of output quantities, but of statements or data on the statistical distribution of output quantities. Therefore, in principle, two possibilities for determining the statistical distribution of the output quantities are available which, as the case may be, may also be applied in a mixed fashion. On one hand, the distribution parameters of the output quantities are directly derived from the distribution parameters of the input and the system quantities according to the rules of the probability calculus. Therefore it is assumed that the statistical distributions of the input or system quantities, and also the output quantities, can be described by only a few distribution parameters. This corresponds to our diagrammatic representation in Figure 3.18.

The second possibility is the type of simulation using the so-called Monte Carlo method [3-11]. Here individual histories of the described systems are created in the model by selecting for each "game," by means of a quantity, one set of values due to the distribution of the input and system quantities. Each individual "game" will then result in a concrete set of values of the output quantities. The statistical distribution pattern of the output quantities can then be derived from a large number of such "games." This corresponds to Figure 3.19. With this method, a description of the statistical distribution by distribution parameters is not necessary, but will be helpful in carrying it out with a minimum expenditure.

The Monte Carlo method therefore simulates by means of a system model an individual sampling value. For evaluation of the results, the known procedures of mathematical statistics can be used. A very important instrument in Monte-Carlo-type simulation is the randomizer. It generates random values within the numerical interval $(0, 1)$. The allocation of the random number to a specific value of the random variables is effected via a given distribution function in accordance with Figure 3.20.

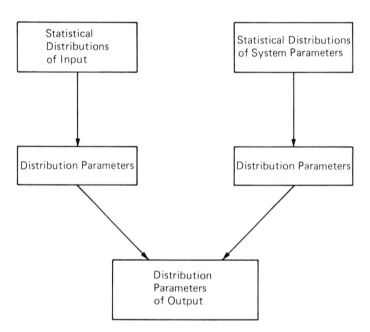

Figure 3.18 Representation of the explicit computation of the distribution parameters of the output in probabilistic models.

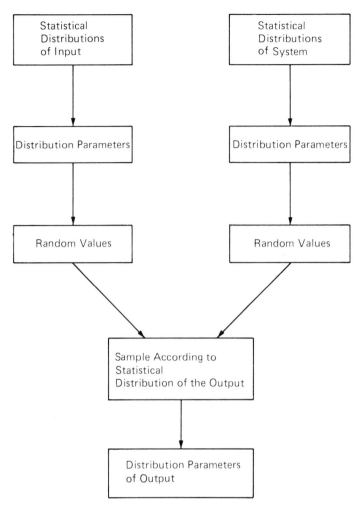

Figure 3.19 Representation of the statistical distribution of output by means of the Monte-Carlo simulation.

In summary, it may be stated that for mathematical simulation—depending on the type of utilization—the entire spectrum of modern mathematics will be needed. The most important disciplines are ordinary and partial differential equations, linear algebra and matrix algebra, Boole's algebra, and numerical mathematics as well as probability theory and statistics.

The mathematical simulation which is used for the prognosis of safety behavior of technical installations during the planning phase—if necessary, by including interactions within the MMES—is based on a mathematical model for which a solution must be found and which, in turn,

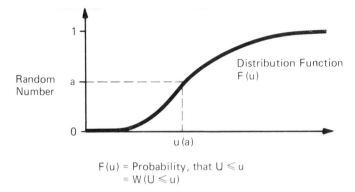

$$F(u) = \text{Probability, that } U \leqslant u$$
$$= W(U \leqslant u)$$

Figure 3.20 Determination of random values for a predetermined distribution by means of the random number generator.

must be converted into a (computation program). The quality of the computed results depends on the quality of the model, the quality of the available data, the formal correctness of the program itself, and its proper use.

In developing and using mathematical simulation a few very important rules should be observed:

— Each mathematical model constitutes a simplification of the system to be evaluated. A certain degree of simplification already resides in the system delimiting against its environment. These delimitations replace interfaces to the environment, partly in the form of boundary conditions. Other simplifications are the aggregation of system elements and the elimination of elements and relations between these elements which are not considered relevant. Limitations by available computational programs often force further delimitations of partial systems. Nevertheless, a certain degree of problem-related completeness must be guaranteed.

— The mathematical model should be as simple as possible so that it will remain surveyable as far as its structure and the given boundary conditions are concerned. A model should never be more detailed than the available data will permit. Otherwise, a false impression of accuracy is created.

— If the data introduced into the model are doubtful, the possible influence of these doubtful values on the results must be carefully examined. If necessary, data should be estimated towards the "safe" side.

— The results of a mathematical simulation must be checked for plausibility. If the results do not correspond with experience, any deviations must be checked. The compounded examination carried out by an expert constitutes an effective control mechanism against flaws in the model, in the program, and against faulty input. Close cooperation

between installation and simulation specialists is absolutely indispensable.
—Whenever possible, the computational model, on which the simulation is based, and the program should be formed on the basis of observations of other available experiments or operational flows, and a constant check should be kept up by periodically comparing measurements and computations.

At this point, attention should be directed towards the special set of problems connected with models used for the determination of reliability values and failure probabilities in complex systems. Within these models, failures of various elements are, in general, considered independently of one another. This may lead to grave misjudgments towards the "unsafe" side. Whenever there is a failure of different elements, caused by one common factor, we call this a "common mode case." However, situations like this may also arise because the behavior of the system elements is not independent from disturbance factors excluded from the model. A similar situation may arise whenever conditions in the environment of a technical installation lead to a shift in the failure rates obtained through lab experimentation or operational recordings.

Chapter 4

Safety Analysis

4.1 Tasks and Aims

As stated in the preceding chapter, safety analysis of a technical system is systems analysis governed by safety engineering aspects. Its essential purpose is to ascertain hazards which could result from a technical system and to deal with questions of possible reduction of such hazards. In this context the analysis will view the technical system as the sum of all elements in a state of reciprocal action and the conditions resulting from the combined action of these elements. When dealing with the technical system, the aim is to show the structures, i.e., to identify the control and regulating mechanisms, to understand states of equilibrium, to ascertain problems, and to furnish a clear picture of the consequences of failures of regulating variables.

This system-related method facilitates analysis of a technical installation on the basis of planning documents and actual experiences with component and subsystem behavior as well as familiar laws of mathematics and physics. In the case of new installations, the analysis should be initiated at an early stage and should proceed segment by segment, parallel to the planning process. This method facilitates use of a retroactive process between the system designer and the safety expert and can result in a system design geared to safety aspects and effective use of means for providing safety improvement. Safety analysis represents a tool reaching beyond component-related considerations and utilization of accident and loss statistics, permitting prognosis of system behavior from the vantage point of safety aspects. Safety analysis starts with the premise that in a technical system, problems which result in loss to persons and property can arise. The primary task of safety analysis is to respond to the following questions:

—Which incidents should be viewed as significant?
—Which incidents could be avoided by a suitable change in system structure?

—Which incidents cannot be excluded completely and what is the expected frequency?

—What will be the type and extent of effects of these incidents on persons and property in the immediate and the less immediate vicinity and what safety measures must be planned?

Safety analysis data input consists of facts and figures on the production process used, on the materials handled, the technical design of the installation, the operations organization, and the environment. The sum total defines the object under analysis. For the analysis proper, systematic representation of the installation with due consideration of safety aspects as well as selection of work methods and evaluation criteria are significant.

The primary aim of safety analysis is the optimal overall solution based on the principle of plant concept by operationally safe design. The interaction of all steps is shown in Figure 4.1. The optimal solution will have been attained when the degree of safety is sufficient to cover externally specified requirements (for example, risk boundary values) and when the economic costs of safety requirements remain as low as possible. In the event that the plant design is such that adequate safety is not possible at economically justifiable costs, changes in the design concept must be considered.

The scope of safety analysis ranges from a rough qualitative appraisal of sources of hazards, via detailed qualitative analysis, to quantitative analysis with figures concerning the expected frequency and consequences of incidents. The scope must be governed by the definition of the task. Here the following must be taken into account:

—The initial position of the analysis must be problem oriented and technically meaningful. For example, a plant should not be tested from the viewpoint of the fire hazard emanating from it if such a risk is not significant in practice. Determinations of specifications of the initial analysis position call for high requirements, since by means of these specifications the key points for all subsequent steps are established, and neglecting important aspects could have serious consequences.

—The analysis steps to be initiated according to the definition of the initial position must be put in logical sequence. Results obtained must not be contradictory and must lead to a self-contained statement with concrete values.

—There must be a reasonable ratio of analysis costs to statements which can be obtained. Possibly a cost effectiveness survey could be made parallel to the analysis so that the depth of penetration into the problem can be determined from this point of view as well.

In practice the procedure will start with a rough qualitative analysis to ascertain sources of hazards as a basis for more detailed investigation. Then, according to need, the scope and depth of analysis can be gradually

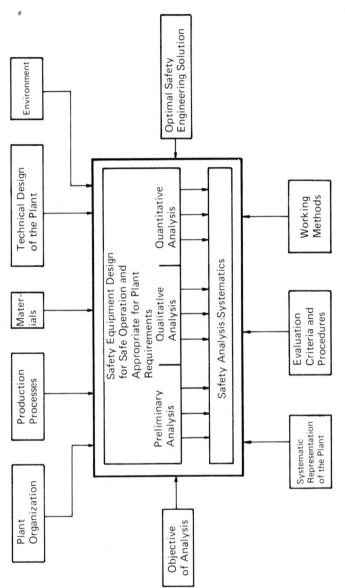

Figure 4.1 Diagram of safety analysis.

expanded. Quantitative analysis will, as a rule, relate only to that part of the plant which is particularly important with respect to safety engineering.

Figures 4.2 (a) and (b) show the sequence of the procedure, indicating the points where decisions governing step-by-step action are made. Depending on plant definition, in simple cases a component-oriented analysis may suffice, one which is complete once the determination is made whether requirements specified for components and their design correspond. In many instances the analysis used will be the one which takes into account the materials employed, process parameters, and measuring and control technology. This analysis is indicated when new technical processes are used, when particularly complex interactions are considered, when a higher risk potential exists in the plant, or when economic reasons favor such a course. In a subsequent step the analysis can be in-depth, to the point of the determination of the logic system structure. This will be necessary when particularly detailed results are required or when a transition to a quantitative analysis must be created. Quantitative analysis presumes an adequate data base.

Safety analysis is completed when weak points have been identified and when proposals for their neutralization (elimination by suitable measures) have been made. The sections which follow explain the equipment for execution of safety analysis and provide examples for practical procedure.

4.2 Safety Analysis Work Methods

Safety analysis uses logic structure representative of possible incidents. Such work methods as fault tree, incident analysis, and decision table are suitable for this purpose. Computation rules for determination of expected frequency of incidents must be formulated accordingly. In a broad sense all mathematical simulation methods which are suited for determination of stress states in technical installations and their parts become aids in safety analysis. These will be described in partial detail later. Here characteristic work methods which are of direct significance with respect to system-related and prognostic consideration of safety analysis will be discussed first.

4.2.1 Mathematical Principles

Concept of Probability and Rules of Computation. The definition of probability is determined as application-oriented. Logic difficulties resulting as compared to exact mathematical definition on quantity-theoretical basis, can be ignored in the present context. If A is the designation of an event as the result of an "experiment" with clearly described objects under

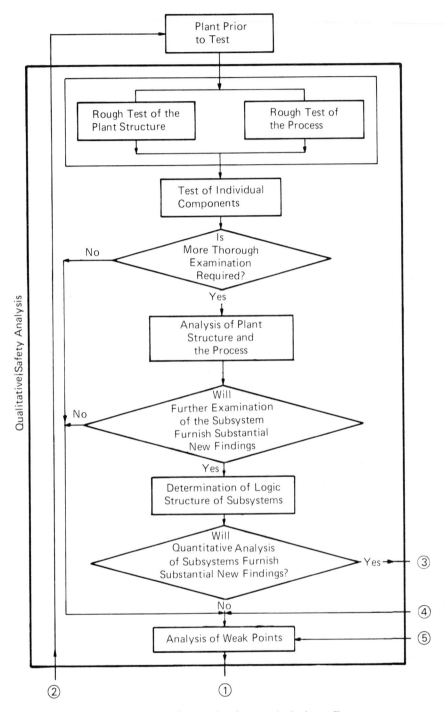

Figure 4.2(a) Steps of safety analysis (part I).

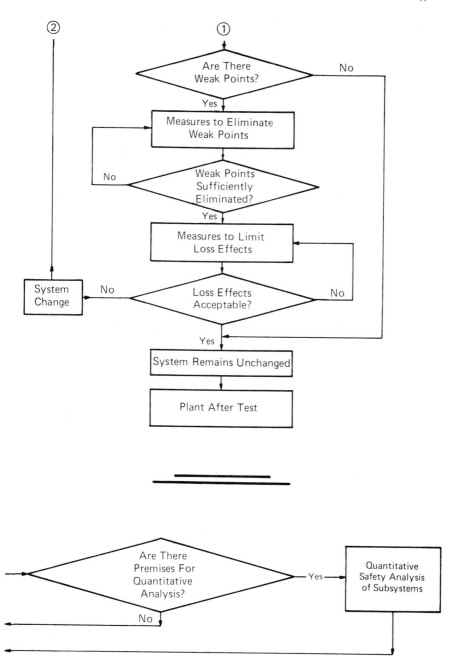

Figure 4.2(b) Steps of safety analysis (part II).

specified boundary conditions, the probability for the occurrence of the result A is

$$W(A) = \lim_{N \to \infty} \frac{N(A)}{N}. \qquad (4\text{-}1)$$

Here N is the total number of experiments and $N(A)$ is the number of experiments with the result A as incident. The incident probability is the boundary value of relative incident frequency when the number of experiments moves toward ∞. Incidents in this sense are, for example, the function or malfunction of a safety installation or its components, a specific failure in a technical system as a result of a trigger event, the appearance of manufacturing errors of a specific nature in the case of specific products, or the observance or nonobservance of boundary values as well as various conditions in general of the object under observation. Probabilities can assume only values between 0 and 1, so that

$$0 \le W(A) \le 1 \qquad (4\text{-}2)$$

is generally valid.

If B is a further incident, the conditional probability $W(A|B)$ for the occurrence of A on the secondary condition that B occurs as well can be derived as

$$W(A|B) = \lim_{N_B \to \infty} \frac{N_B(A)}{N_B}. \qquad (4\text{-}3)$$

Here N_B is the total number of experiments with incident B occurring and $N_B(A)$ is the number of experiments where, in addition to B, the incident A occurs as well. In the event that incidents A and B exclude each other, then $W(A|B) = 0$ is valid. If incidents A and B are independent of each other, i.e., if the occurrence of incident A does not depend on whether or not B occurs, then $A(A|B) = W(A)$ is valid.

If, under given conditions, several incidents $A_1, \ldots, A_n$ are possible, incident linkages can be derived, which in turn are again incidents. The incident which consists of at least one of the incidents $A_1, \ldots, A_n$ occurring is designated as "OR"-linkage, represented by the symbol $\vee$:

$$A_\vee = A_1 \vee A_2 \vee \cdots \vee A_n. \qquad (4\text{-}4)$$

The incident which consists of incidents $A_1, \ldots, A_n$ occurring simultaneously under the conditions observed is designated as an "AND" linkage, represented by the symbol $\wedge$:

$$A_\wedge = A_1 \wedge A_2 \wedge \cdots \wedge A_n. \qquad (4\text{-}5)$$

Figure 4.3 shows these relations for two initial incidents A and B from an incident quantity M. The incident linkages correspond to computation rules for the assigned probabilities.

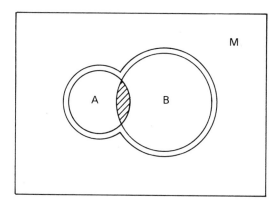

Figure 4.3 Diagram of the "OR" and "AND" linkage. The surface surrounded by the double line $A \lor B$ "OR"; area with slanted lines $A \land B$ "AND".

For each of the initial incidents A and B,

$$W(A \land B) = W(A|B)W(B), \qquad (4\text{-}6)$$
$$W(A \lor B) = W(A) + W(B) - W(A \land B) \qquad (4\text{-}7)$$

is generally valid. For independent incidents, $W(A|B) = W(A)$ and thus

$$W(A \land B) = W(A) \cdot W(B). \qquad (4\text{-}8)$$

For several independent incidents, the following is valid:

$$W(A_1 \land \text{---} \land A_n) = W(A_1) \cdot W(A_2) \text{---} W(A_n)$$
$$= \prod_i W(A_i). \qquad (4\text{-}9)$$

For incidents which are mutually exclusive

$$W(A|B) = 0$$

and thus

$$W(A \lor B) = W(A) + W(B); \qquad (4\text{-}10)$$

and in the cases of several incidents which are mutually exclusive, the result is

$$W(A_1 \lor \text{---} \lor A_n) = W(A_1) + W(A_2) + \cdots + W(A_n)$$
$$= \sum_i W(A_i). \qquad (4\text{-}11)$$

For every incident A, a complementary incident can be cited. It consists of the nonoccurrence of incident A. It is also designated as negation of A and is written $\bar{A}$. In general,

$$W(A) + W(\bar{A}) = 1 \qquad (4\text{-}12)$$

is valid.

When the negation is applied to "AND", or "OR" linkages, the so-called Morgan rules are the result:

$$\overline{A_1 \vee A_2 \vee \ldots \vee A_n} = \overline{A}_1 \wedge \overline{A}_2 \wedge \ldots \wedge \overline{A}_n; \qquad (4\text{-}13)$$

$$\overline{A_1 \wedge A_2 \wedge \ldots \wedge A_n} = \overline{A}_1 \vee \overline{A}_2 \vee \ldots \vee \overline{A}_n. \qquad (4\text{-}14)$$

These rules permit return of the "OR" linkage to the "AND" linkage and vice versa. From this a further important rule can be derived for independent incidents $A_1, \ldots, A_n$. With $A_1, \ldots, A_n$, the complementary incidents $\overline{A}_1, \ldots, \overline{A}_n$ are also independent, and by using (4-12), (4-13), and (4-9), the result is

$$W(A_1 \vee A_2 \vee \ldots \vee A_n) = 1 - [1 - W(A_1)] \cdot [1 - W(A_2)] \ldots [1 - W(A_n)]$$

$$= 1 - \prod_i [1 - W(A_i)]. \qquad (4\text{-}15)$$

These are the basic rules for probability computation in the context of safety analysis.

System Functions. Examination of technical systems requires representation of logic structures by means of mathematical functions. The following is a description of its basic principles.

Consider, for example, a technical system S with components K_1, $K_2, \ldots, K_n$. The behavior required for the given functions is expected of the components and the complete system. Deviation of the component K_i from the desired behavior defines the undesirable incident A_i; the occurrence of the desired behavior is the incident $\overline{A}_i$.

Correspondingly A_s and $\overline{A}_s$ are the designations for incident alternatives of the complete system. The behavior of the complete system depends on the behavior of the components $K_1, \ldots, K_n$; this is shown in diagram form by Figure 4.4. Analysis of the inner structure of such incident systems and the derivation of the occurrence probabilities for system incidents A_s or $\overline{A}_s$

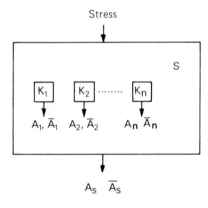

Figure 4.4 Diagram of an incident system.

from the occurrence probability of the component events is a frequent task of safety analysis. System functions are used for the purpose of description.

To arrive at the definition of a system function, to every component K_i assign a variable Z_i with the value 1 when the incident A_i occurs and the value 0 when the incident $\bar{A}_i$ occurs. The incident A_i is then identical to the incident $Z_i = 1$; the incident $\bar{A}_i$ is identical to the incident $Z_i = 0$. This type of variable is called a Boolean variable. Correspondingly, for the system S, a Boolean variable Z_s can be determined by assigning the value $Z_s = 1$ to the occurrence of the undesired event A_s, and the value $Z_s = 0$ to the complementary event A_s. In summary, the following complementation is obtained:

$$\begin{aligned} A_i &\to Z_i = 1, \\ \bar{A}_i &\to Z_i = 0, \\ A_s &\to Z_s = 1, \\ \bar{A}_s &\to Z_s = 0. \end{aligned} \tag{4-16}$$

Correspondingly, the following is valid for occurrence probabilities q and p:

$$\begin{aligned} q_i &= W(A_i) = W(Z_i = 1), \\ p_i &= W(\bar{A}_i) = W(Z_i = 0) = 1 - q_i, \\ q_s &= W(A_s) = W(Z_s = 1), \\ p_s &= W(\bar{A}_s) = W(Z_s = 0) = 1 - q_s. \end{aligned} \tag{4-17}$$

If Z_a, Z_b, and Z are random state variables, then

$$Z' = Z_a \wedge Z_{b'},$$

$$Z' = Z_a \vee Z_{b'},$$

$$Z' = \bar{Z}$$

designate linkage rules between the state variables which can also be designated as "AND", "OR", and negations. Value assignment for Z' in the case of given values of Z_a, Z_b, and Z appears in the second column of Table 4.1. Linkages can be shown in a simple manner mathematically by means of fundamental operations as shown in the third column of Table 4.1. By means of these Boolean linkage rules, the state variables become the elements of a Boolean notation. The Boolean linkage of state variables provides a Boolean function which, in turn, again represents a state variable which can have only the values 0 and 1. Complex Boolean functions and corresponding mathematical representations can be derived from the linkages shown in Figure 4.1 by repeated application and combination. The computation rules for determination of Boolean functions of state variables correspond to computation rules for determination of probabilities of system states. Those are also entered in the third column of Table 4.1.

Table 4.1 Boolean linkages of state variables and corresponding rules of computation

Designation	Values of state variables		Rules of computations
"AND"	Z_a	0 0 1 1	State variable
$Z' = Z_a \wedge Z_b$	Z_b	0 1 0 1	$Z' = Z_a \cdot Z_b$
	Z'	0 0 1 1	Probabilities $W(Z') = W(Z_A)W(Z_b|Z_a)$
"OR"	Z_a	0 0 1 1	State variable
$Z' = Z_a \vee Z_b$	Z_b	0 1 0 1	$Z' = Z_a + Z_b - Z_aZ_b$
	Z'	0 1 1 1	Probabilities $W(Z') = W(Z_a) + W(Z_b)$ $- W(Z_a)W(Z_b|Z_a)$
Negation	Z	0 1	State variable
$Z' = \bar{Z}$	$\bar{Z}$	1 0	$Z' = 1 - Z$
			Probabilities $W(Z') = 1 - (W)Z$

The vector $(Z_1, Z_2, \ldots, Z_n)$ of the state variable of components $K_1, \ldots, K_n$ is designated as the state vector of the system. The system function $\varphi_S(Z_1, Z_2, \ldots, Z_n)$ of the system is defined by

$$Z_s = \varphi_S(Z_1, Z_2, \ldots, Z_n). \tag{4-18}$$

The system function assigns the value $Z_s = 1$ or $Z_s = 0$ to every value set of the state vector, according to whether the "state" has been assigned the system incident A_s or $\bar{A}_s$. The system function can be represented as a Boolean function as well as mathematically and depends on Z_i. The mathematical form represents the logic structure of the system in formula. When the states of components are independent of each other, and if a suitable mathematical representation of φ_S is selected with Z_i replaced by q_i, the probability q_s that in the overall system the incident A_s, respectively the state $Z_s = 1$, occurs is

$$q_s = \varphi_S(q_1, \ldots, q_n). \tag{4-19}$$

The Boolean linkages of the state variables Z_i are equivalent to corresponding incident linkages. For example, $A_s = A_1 \wedge A_2 \wedge \ldots \wedge A_n$ means that the undesirable system incident will occur only when all components behave in an undesirable way. The statement $Z_s = Z_1 \wedge Z_2 \wedge \ldots \wedge Z_2$ indicates that the state variable Z_s of the system is exactly 1 when the state variable Z_i of the components are 1. These statements are equivalent.

Thus the following connection results for the system description:

> Linkage of incidents A_i, $\bar{A}_i$

> Equivalent Boolean function of state variables Z_i

> Equivalent mathematical system function $\varphi_S(Z_i, \ldots, Z_n)$

> Incident probability $q_s = \varphi_S(q_1, q_2, \ldots, q_n)$

For several basic forms of component linkage, the Boolean system function, the mathematical system function, and the probability q_s are shown in Table 4.2. The assigned graphic presentation as it is used in the case of fault trees is completed (see Section 4.2.2).

In the case of the "AND" component linkage, the system failure occurs precisely when all components fail; in the case of the "OR" linkage, the system failure occurs when at least one of the components fails. The other linkages are combinations of both.

In the formulation of the Boolean linkage functions, the name of the component is used to designate the state variables, for example, $K_1, \ldots, K_n$ instead of $Z_1, \ldots, Z_n$. A component K_i can also represent a partial system made up of subcomponents and its behavior can be described by certain linkage rules. In reliability theory, the state variables are frequently defined in a complementary manner by assigning $Z = 0$ to the failure and the value $Z = 1$ to the desired behavior. In standardized procedures used in safety analysis, the form used here has been successful (see Section 4.2.2) [4-10], [4-11].

Distribution Functions and Statistical Characteristic Quantities. The occurrence probabilities of specific incidents involving components or subsystems of a complete system depend on properties with statistically distributed values and which can be described by distribution functions. In addition to the distribution function, a few types of characteristic data which are of particular interest in safety analysis will be discussed below.

The stochastic variable X and the parameter x are given. If the parameter x is permitted to run continuously from $-\infty$ to $+\infty$ and the probability for values $x \leq x$ is determined, the distribution function

$$F(x) = W(X \leq x) \qquad (4\text{-}20)$$

is obtained.

If X can assume discrete values only, the distribution function $F(x)$ is a step function as shown in Figure 4.5(a). Figure 4.5(b) shows a continuous

Table 4.2 Boolean system functions, mathematical system functions, and occurrence probabilities for the system behavior for several basic forms of component linkage. The graphic symbols correspond to [4-10]

Designations, graphic symbols	Boolean function	Mathematical system function, probability q_S
"AND" & K_2 K_n	$Z_s = Z_1 \wedge Z_2 \wedge \ldots \wedge Z_n$	$\varphi_S = \prod_i Z_i$ $q_S = \prod_i q_i$
"OR" ≥ 1 K_1 K_n	$Z_s = Z_1 \vee Z_2 \vee \ldots \vee Z_n$	$\varphi_S = 1 - \prod_i (1 - Z_i)$ $q_S = 1 - \prod_i (1 - q_i)$ $\approx \sum_i q_i$, in case $q_i \ll 1$
& ≥ 1 ≥ 1 K_{11} $K_1 n_1$ K_{m1} K_{mn_m}	$Z_s = (Z_{11} \vee Z_{12} \vee \ldots \vee Z_{1n_1})$ $\wedge \ldots$ . . . $\wedge (Z_{m1} \vee Z_{m2} \vee \ldots \vee Z_{mn_m})$	$\varphi_S = \prod_{j=1}^{m} \left[1 - \prod_{i=1}^{n_j} (1 - Z_{ji}) \right]$ $q_S = \prod_{j=1}^{m} \left[1 - \prod_{i=1}^{n_j} (1 - q_{ji}) \right]$
≥ 1 & & K_{11} $K_1 n_1$ K_{m1} K_{mn_m}	$Z_s = (Z_{11} \wedge Z_{\vee 2} \wedge \ldots \wedge Z_{1n_1})$ $\vee \ldots$ . . . $\vee (Z_{m1} \wedge Z_{m2} \wedge \ldots \wedge Z_{mn_m})$	$\varphi_S = 1 - \prod_{j=1}^{m} \left(1 - \prod_{i=1}^{n_j} Z_{ji} \right)$ $q_S = 1 - \prod_{j=1}^{m} \left(1 - \prod_{i=1}^{n_j} q_{ji} \right)$

distribution function. If continuous distribution functions can be continuously differentiated, the probability density function [Figure 4.5(c)] results as

$$f(x) = \frac{dF(x)}{dx}. \tag{4-21}$$

Instead of (4-20), it can also be written as

$$F(x) = W(X \leq x) = \int_{-\infty}^{x} f(x')dx'. \tag{4-22}$$

The probability that the stochastic variable is within given boundaries x_1

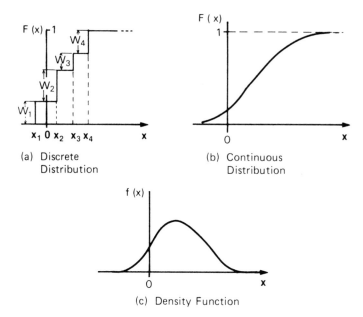

(a) Discrete
    Distribution

(b) Continuous
    Distribution

(c) Density Function

Figure 4.5 Diagram representation of distribution and density functions.

and x_2 is then characterized by

$$W(x_1 \leq X \leq x_2) = F(x_2) - F(x_1)$$

$$= \int_{x^1}^{x^2} f(x)dx. \qquad (4\text{-}23)$$

The expected or mean value of the stochastic variables X is

$$\mu = \int_{-\infty}^{+\infty} x dF(x). \qquad (4\text{-}24)$$

For discreet stochastic variables which assume the values x_i with the probability W_i, Equation (4-24) becomes

$$\mu = \sum_i x_i \cdot W_i. \qquad (4\text{-}25)$$

For continuously differentiated distribution functions, the result is

$$\mu = \int_{-\infty}^{+\infty} x f(x) dx. \qquad (4\text{-}26)$$

In the case of such functions, the variation σ, which is obtained from the variance σ^2, is of interest.

$$\sigma^2 = \int_{-\infty}^{+\infty} (x - \mu)^2 f(x) \, dx. \qquad (4\text{-}27)$$

A series of probability distributions of importance in practice is already completely described by the parameters μ and σ^2, and their course is determined.

In the case of time-dependent processes, the question concerning the life of the components of a system arises frequently. The distribution function $F(t)$ is then assigned to the stochastic variable "lifetime T".

$$F(t) = W(\text{lifetime } T \leq t), \tag{4-28}$$

where $F(t) = 0$ is set for $t < 0$. The survival probability is given by

$$1 - F(t) = W(\text{lifetime } T > t). \tag{4-29}$$

The *failure rate* λ assigned to $F(t)$ is defined as a function of

$$\lambda(t) = \frac{1}{1 - F(t)} \cdot \frac{dF(t)}{dt} = \frac{1}{1 - F(t)} f(t), \tag{4-30}$$

$$Ft = 1 - \exp\left(-\int_0^t \lambda(t')\,dt'\right). \tag{4-31}$$

Application of this function always requires verification of whether the results obtained do not contradict basic laws of physics. It would be nonsense if, for example, a distribution function were selected to describe failures due to wear, with the function leading to a declining failure rate dependent on time.

Following these general statements on distribution functions, it is time to deal with functions which have special significance in the evaluation of data and for the determination of statistical characteristics from data. These functions will be normal distribution, logarithmic Gauss distribution, and Weibull distribution. The form of the distribution and density function as well as the failure rate are shown in Table 4.3.

The course of the normal distribution corresponds to that of the distribution and density function according to Figures 4.5(b) and 4.5(c). The failure rate is a function increasing continuously with the variable x. The application is possible in many instances, for example, in cases of failure due to wear.

Another important function in the data evaluation is the logarithmic Gauss distribution. It is characterized by the $x > 0$ area of application and is therefore well suited for data whose appearance above zero is set by natural givens (dimension, times, extent). According to the magnitude of parameters v and x, the failure rate increases or decreases. In the application of this function, the physical significance of the incident must be checked carefully.

Due to its versatility, the Weibull distribution is very well suited for many data evaluation applications. However, the exact determination of the three parameters which are its determinants sometimes creates considerable difficulties, in particular in the case of insufficiently large data base [4-8]. According to the magnitude and combination of parameters, the failure rate can increase or decline. Therefore a critical check is required prior to application. An exceptional case in the Weibull distribution is the exponential distribution with $\alpha = 0$ and $\gamma = 1$. Here, with $1/\beta = \lambda$, the distribution function is

$$F(x) = 1 - \exp(-\lambda x), \qquad (4\text{-}32)$$

the probability density function is

$$f(x) = \lambda \, \exp(-\lambda x), \qquad (4\text{-}33)$$

and the failure rate is

$$\lambda(x) = \lambda = \text{const.} \qquad (4\text{-}34)$$

The exponential distribution is very widely used, due to its simple construction with $\lambda = $ const and the resulting ease of handling. In principle this formulation is wrong with respect to physics in the case of components subject to wear. This will also be true when a time dependence cannot be readily derived from available data. In the short term the error can of course be neglected. This is shown clearly by the so-called "bathtub life curve," Figure 4.6. There the course of the failure rate $\lambda(t)$ is shown as time dependent. Range I covers the period in which early failures were observed, for example, due to poor manufacturing quality or unforeseen operational conditions. Next is range II, where overlapping of two behavior phenomena is apparent: decreasing early failure and increasing failure due to wear.

The result of the overlapping is that, for a certain period, the failure rate $\lambda(t)$ can be viewed as practically constant. Range III deals with the period of failures due to wear. The failure rate increases with time.

The probability that an operationally sound component lasting from 0 to time t_i also lasts through the period t_i to $t_i + \Delta t$ is

$$W(\Delta t) = \frac{W(\Delta t + t_i)}{W(t_i)}. \qquad (4\text{-}35)$$

In the case of an exponential distribution

$$W(t_i) = 1 - [1 - \exp(-\lambda t_i)]$$

and

$$W(\Delta t + t_i) = 1 - [1 - \exp(-\lambda(\Delta t + t_i))].$$

Therefore the result becomes

$$W(\Delta t) = \exp(-\lambda \, \Delta t - \lambda t_i + \lambda t_i),$$
$$W(\Delta t) = \exp(-\lambda \, \Delta t). \qquad (4\text{-}36)$$

This shows that the probability of lasting over a certain period Δt apparently

Table 4.3 Normal distribution, logarithmic Gauss distribution, and Weibull distribution

Distribution designation	Distribution parameter	Distribution and density function, failure rate	
normal distribution	mean value μ from x	distribution function	$\dfrac{1}{\sigma\sqrt{2\pi}}\displaystyle\int_{-\infty}^{x}\exp[-(x'-\mu)^2/(2\sigma^2)]dx'$
area of application	variation σ from x	density function	$\dfrac{1}{\sigma\sqrt{2\pi}}\exp[-(x-\mu)^2/(2\sigma^2)]$
$-\infty < x < \infty$		failure rate	$\dfrac{\exp[-(x-\mu)^2/(2\sigma^2)]}{\displaystyle\int_{x}^{\infty}\exp[-(x'-\mu)^2/(2\sigma^2)]dx'}$
logarithmic Gauss distribution	mean value of v from $\ln x$	distribution function	$\dfrac{1}{\tau\sqrt{2\pi}}\displaystyle\int_{o}^{x}\exp[-(\ln x'-v)^2/(2\tau^2)]\dfrac{dx'}{x'}$
area of application	variation τ from $\ln x$	density function	$\dfrac{1}{\tau\sqrt{2\pi}}\exp[-(\ln x-v)^2/(2\tau^2)]\dfrac{1}{x}$

$0 < x < \infty$	failure rate	$\dfrac{1}{x} \cdot \dfrac{\exp[-(\ln x - \nu)^2/(2\tau^2)]}{\int_x^{\infty} \exp[-(\ln x' - \nu)^2/(2\tau^2)]\,\dfrac{dx'}{x'}}$
	mean value of x $\mu = \exp(\nu + \tau^2/2)$ variation of x $\sigma = \exp(\nu + \tau^2/2)\,\sqrt{e^{\tau^2} - 1}$	
Weibull distribution	position parameter α	distribution function	$1 - \exp[-[(x - a)/\beta]^{\gamma}]$
area of application	scale parameter $\beta > 0$	density function	$\dfrac{\gamma}{\beta}[(x - a)/\beta]^{\gamma - 1}\exp[-(x - a)/\beta]^{\gamma}]$
$\alpha \leq x$	form parameter $\gamma > 0$		
	mean value of x $\mu = \beta\Gamma(1 + 1/\gamma) + \alpha$	failure rate	$\dfrac{\gamma}{\beta}[(x - a)/\beta]^{\gamma - 1}$
	variation of x $\sigma = \beta\sqrt{\Gamma(1 + 2/\gamma) - \Gamma^2(1 + \tfrac{1}{\gamma})}$ ($\Gamma(t)$ gamma function)		

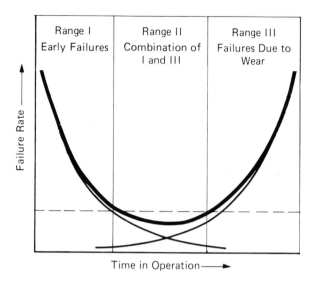

Figure 4.6 Bathtub curve–time-dependent failure rate.

becomes independent of age as long as λ = const is valid for the failure rate. The stress applied up to the time element under observation has no effect.

4.2.2 Analysis and Prognosis Procedures

System behavior analysis and prognosis can be executed by means of various procedures. The procedures generally described in the literature can be traced back to three standard types, namely failure effect analysis [4-9], fault tree analysis [4-10], and incident progression analysis [4-11].

The three procedures will be discussed, as well as the concept of the decision table technique, which is also a good tool but has rarely been discussed in the literature in connection with this application. To begin, the customary analysis techniques [4-9], [4-10], [4-11] will be discussed in alphabetical order. This will serve to delineate and distinguish the procedures.

— *Failure effect analysis* is a process for investigating the types of failures of all components of a system and their effects on the system. No failure combinations will be discussed.
— *Fault tree analysis* serves to ascertain logic linkage of component and subsystem failures leading to an undesired incident. The undesired incident must be a given. Then all causes which could lead to the incident are investigated.
— *Incident progression analysis* serves to ascertain undesired events with a common specific incident trigger.

Failure Effect Analysis. Failure effect analysis is essentially a qualitative prognosis method. Failure effect analysis is not used to investigate failure combinations. It serves the purpose of systematic and complete analysis of a system. This analysis will help in deciding what other more far-reaching investigations (for example, the fault tree analysis) are required. To the extent to which the listed failure effects can be documented by failure frequency or failure rates, this analysis will be the basis for a quantitative prognosis procedure. As a rule, failure effect analysis is the basis for fault tree analysis or the decision table technique. Evaluation of component failure effects can follow various criteria. From the viewpoint of safety engineering, safety measures or design changes can be derived from this evaluation. Failure effect analysis is a suitable analysis technique for system proposal checks since it permits execution without mathematical effort. An example will be used to describe the method of failure effect analysis.

EXAMPLE. In the case of a magnetic cableway, an investigation is carried out to ascertain whether the measures provided for breakdowns are adequate guarantees for safe operation. Table 4.4 lists components and types of failure and failure consequences, the resulting vehicle reactions, and the planned protections.

Fault Tree Analysis. In the case of fault tree analysis, an undesired event is a given for the system to be evaluated; for example, power failure or interruption of the pumping medium in a system. Subsequently all logic linkages, failure combinations of components, or subsystem failures which could lead to the preset undesired incident are compiled. This compilation provides the so-called fault tree. The fault tree symbols are standardized [4-10]. Several established standard symbols are shown in Figure 4.7. Fault tree analysis is suited for the evaluation of simple as well as complex systems. The primary aim is systematic identification of all possible failure combinations leading to the undesired event. The premise is precise knowledge of the system to be evaluated. For this reason, prior to the fault tree analysis, the analyst must make a qualitative analysis of the system in such a way that, in addition to an understanding of the process, detailed knowledge is acquired concerning the operation of components and subsystems. Reference should be made here to several basic aspects of setting up fault trees. In the determination of failure combinations of components, the premise is that the failures are unrelated incidents. In operation, however, components are likely to be subject to failures due to a common cause, so-called common mode failures (Chapter 5). These have to be considered in the setting up of the fault tree. The second aspect concerns human effect factors on the system under investigation. Insofar

Table 4.4 Example of failure effect analysis

No.	Component	Type of defect	Consequence	Vehicle reaction	Protection
1	electr. drive	failure (e.g. black-out, transformer switch, converter, etc.)	no current in stator	vehicle flattens out	driver applies brakes by means of controlled mechanical brake —additional control by GPU (automatic) and control station (control station operator) with emergency brake if necessary
2	electr. drive	d.c.-a.c. converter error in set point	max. thrust	max. vehicle acceleration	same as 1
3	electr. drive	out of phase due to corresp. failures	(see, also, Section 4.1.2)	(4.1.2 ?)	
4	telemetry control station/ transformer substation	failure power supply	drive disconnected from all control channels	pulse block d.c.-a.c. converter due to failure of order "telemetry i.O.," vehicle flattens out	same as 1.
5	telemetry control station/ transformer substation	defective command thrust = max in the activated output channel	max. thrust	veh. accelerates to V_{max} and is braked via profile computer and brakes applied below V_{max} accelerates again, and so forth	two-point control response around V_{max} not critical

6	line connection control station transformer substation	interruption	emergency out d.c.-a.c. converter (zero signal current principle)	same as 1	same as 1
7	telemetry vehicle/ control station	failure power supply	no info. transmission vehicle/control station except for VHF radio	pulse block d.c.-a.c. converter due to failure of "telemetry i.O." vehicle flattens out	braking by operator by means of controlled mechanical brake additional control control by GPU (automatic)
8	telemetry vehicle/ control station	defective command thrust = max. in the activated output channel (HF operation)	same as 5	same as 5	same as 5
9	route profile computer	failure power supply	computer area failure $V_{perm}/V_{max}/V_{limit}$ lights in vehicle failure V_{actual} V_{max}-control	manual thrust setting bypassing computer	quasiundisturbed, normal drive to station
10	route profile computer	software off-routine	defective V_{perm}/V_{max} V_{limit}-indicator in vehicle and control station and/or failure control to V_{actual} V_{max} and/or undefined thrust command	undisturbed undefined thrust preset	quasiundisturbed normal drive to station as in 7

No.	Designation and Symbol	Comments
1	Standard Input 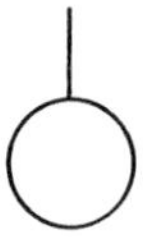	The symbol represents a functional element, when primary failure is possible. Characteristic quantities of the primary failure and those for the failure time of the functional element are assigned to the symbol.
2	Non-linkage 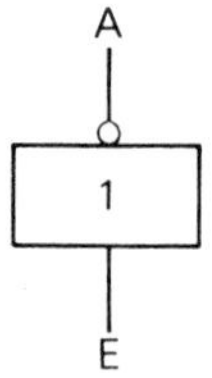	The NON-linkage represents negation. If the input E of the linkage is "O", the issue A is "1" and vice versa.
3	OR-linkage	The function table below applies to two inputs of this linkgae. The linkage can have any number of inputs.
4	AND-linkage	For two inputs of this linkage the function table below applies. The linkage can have any number of inputs.

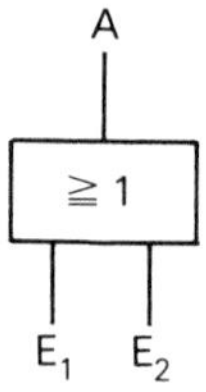

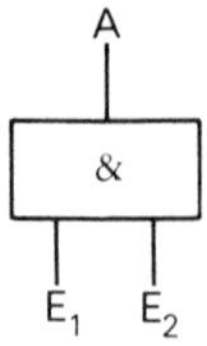

Non-linkage Function Table

E	A
1	0
0	1

OR-linkage Function Table

E_1	E_2	A
1	1	1
1	0	1
0	1	1
0	0	0

AND-linkage Function Table

E_1	E_2	A
1	1	1
1	0	0
0	1	0
0	0	0

Figure 4.7 Standard symbols for graphic representation of fault trees [4-10].

No.	Designation and Symbol	Comments
5	Remark 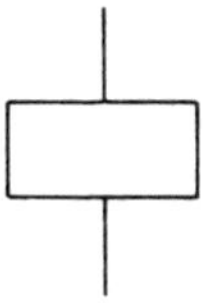	Descriptions of input, output, or linkages are entered in rectangles.
6	Transfer Input  Transfer Output 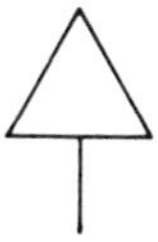	The transfer symbol serves to terminate the fault tree, respectively the fault tree will be continued at another point.

Figure 4.7 (*Continued*)

as technical measures do not prevent it (blocking), possible faulty operation of the component by man must be included in the fault tree. Another problem is due to the fact that the fault tree is a static system, and this must be considered in time-dependent processes. In such instances, several fault trees will be set up for the dynamic process. Frequently, however, it will suffice to enter time dependence of individual components in the fault tree.

EXAMPLE: The following is a demonstration of the fault tree analysis method using the example of an emergency cooling system of a water-boiler nuclear reactor. For the sake of clarity, only the mechanical components will be considered.

The system (Figure 4.8) basically consists of a fore-pump which provides the high-pressure and low-pressure section with water from a condensation chamber. In case of a minor leak in the reactor cooling circulation, water loss must be compensated by means of the high-pressure pump (HP pump) feeding water from the condensation chamber to keep the reactor continuously cooled. In case of failure of the high-pressure feed, reactor pressure can be reduced by means of the subsystem "automatic pressure relief" (not shown in Figure 4.8, but included in Table 4.5 as X_{11}) to such

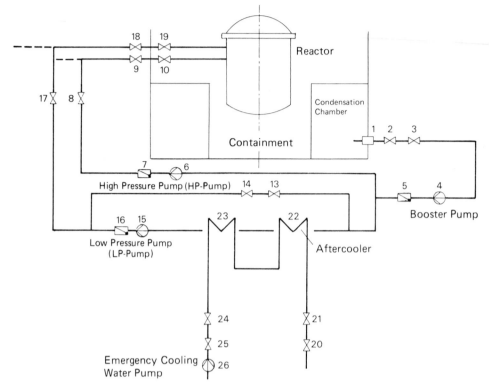

Figure 4.8 Diagram of an emergency cooling system.

an extent that the low-pressure pump can effect the feed. The stored heat and the disintegration heat still present in the switched-off reactor will be evacuated via the after-cooler (Figure 4.8, positions 22 and 23). The undesired event is the failure of the residual heat evacuation of the reactor due to failure of all emergency cooling systems. The example assumes that only one emergency cooling system is available. Thus the undesired event of this system's failure is equivalent to "failure of the after-heat evacuation."

For fault tree determination, all independent events which lead to system failure are subject to the search. For this purpose all relevant components are viewed as having failed one after another or in groups, and the effect on system function is investigated. Table 4.5 lists all components X with indication according to Figure 4.8 and shows the type of failure.

On the basis of the knowledge of which failure combinations lead to the undesired event "failure of after-heat evacuation," the fault tree can be set up from Table 4.5 according to the Boolean rules explained in Section 4.2.1.

Table 4.5 Emergency cooling system: component list (Figure 4.8)

X	Designation	Failure type
1	penetration shell	breaks
2	slide	closed
3	slide	closed
4	fore-pump	off
5	return valve	closed
6	high-pressure pump	off
7	return valve	closed
8	high-pressure motor slide	closed
9	penetration armature	closed
10	penetration armature	closed
11	automatic pressure relief	fails
12	pipeline	breaks
13	low-pressure motor slide	open
14	low-pressure motor slide	open
15	low-pressure pump	off
16	return valve	closed
17	motor valve	closed
18	penetration armature	closed
19	penetration armature	closed
20	motor valve	closed
21	motor valve	closed
22	after-cooler I	defect
23	after-cooler II	defect
24	motor valve	closed
25	motor valve	closed
26	auxiliary cooling–water pump	off
27	pipeline	breaks

In Figures 4.9(a) and 4.9(b), the fault tree of the system is shown. When the fault tree is written as a Boolean function, the result is as follows:

$$
\begin{aligned}
V_{13} &= X_{13} \wedge X_{14}, \\
V_{12} &= X_{22} \vee X_{23}, \\
V_{11} &= X_{20} \vee X_{21} \vee X_{24} \vee X_{25} \vee X_{26}, \\
V_{10} &= X_{15} \vee X_{16} \vee X_{17} \vee X_{18} \vee X_{19}, \\
V_9 &= V_{13} \vee V_{12} \vee V_{11} \vee V_{10} \vee X_{27}, \\
V_8 &= X_1 \vee X_2 \vee X_3 \\
V_7 &= X_6 \vee X_7 \vee X_8 \vee X_9 \vee X_{10}, \qquad (4\text{-}37) \\
V_6 &= X_4 \vee X_5, \\
V_5 &= V_8 \vee V_6, \\
V_4 &= X_{11} \wedge V_7, \\
V_3 &= X_{12} \vee V_5, \\
V_2 &= V_9 \wedge V_7, \\
V_1 &= V_4 \vee V_3 \vee V_2.
\end{aligned}
$$

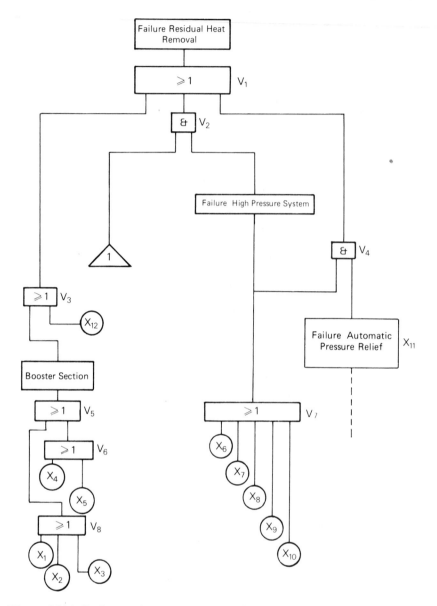

Figure 4.9(a) Fault tree for emergency cooling system according to Figure 4.8 (part I).

Here X_i and V_i are to be understood as state variables. The function values are 1-failed, 0-operational. X_i are to be viewed as the state variables for components of the same name. V_i are the state variables derived from component groups according to the above linkages, and V_1 are the state variables of the system involved. $V_1 = 1$ represents the undesired event "failure of after-heat evacuation."

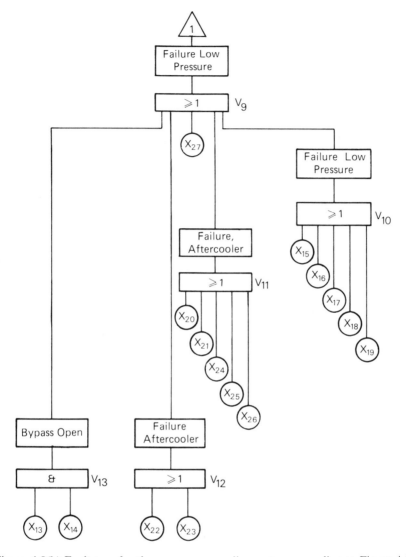

Figure 4.9(b) Fault tree for the emergency cooling system according to Figure 4.8 (part II).

The system fault tree in Figure 4.9(a) shows that the components X_1 to X_5 and X_{12} are critical since they are within a chain of "OR" linkages. The failure of one of these components is sufficient cause for the failure of the after-heat evacuation. Furthermore, when it is considered that comparably low failure probabilities are to be expected with respect to components X_1 and X_{12}, the components X_2, X_3, X_4, and X_5 emerge as the weak points of the system.

When the analysis is completed by setting up the fault tree, the prognosis procedure is qualitative. Up to this step the fault tree analysis provides qualitative indications concerning the weak points in the system structure. In the event that it is possible to assign a failure rate, failure time, and repair time to component X_i, then the fault tree analysis can be continued as a quantitative prognosis procedure. These data are used to compute the probability of the occurrence of the undesired event.

Incident Sequence Analysis. Incident sequence analysis, frequently referred to as incident analysis, is a procedure resembling fault tree analysis. While in the case of fault tree analysis, the undesired incident is a given, in the case of incident sequence analysis, all undesired events as well as combinations of events with a common incident trigger are the subjects of the investigation. Incident sequence analysis is suited to evaluation of any kind of installation. The condition for its execution is the same as in the case of fault tree analysis, namely exact knowledge of the system to be evaluated. For this reason, prior to an incident sequence analysis, a qualitative system analysis is required, dealing with system functions, environmental conditions, auxiliary equipment, component behavior, organization, and other system behavior. In the case of an incident sequence analysis, it is assumed that after an initial incident (e.g., component failure), there are consequential incidents due to time-dependent processes. The logic linkages of these events are represented by means of standardized symbols (Figure 4.10) and show the so-called incident progression diagram. At first it contains only qualitative information. The conditions are specified concerning an incident cause which via a given incident sequence can lead to given effects [4-11].

For the example of the reactor emergency cooling system (Figure 4.8) for which the fault tree had already been set up, the incident sequence

No.	Symbol	Statement
1	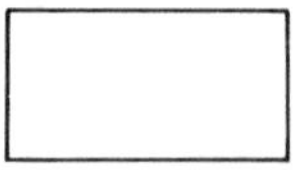	Initial incident, trigger incident. Intermediate incident. Incident failure effect. End of incident.
2		Effect line with comment.

Figure 4.10 Symbols for the graphic presentation of incident sequence diagrams according to [4-11].

No.	Symbol	Statement
3		Effect delay.
4	**4.1** 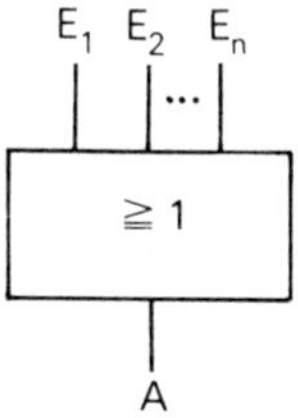	"OR" (inclusive OR) Incident disjunction $E_1, E_2 \ldots E_n$. $A = E_1 \vee E_2 \vee \ldots E_n$. A is present when one or more of the incidents E_1 to E_n are present.
	4.2 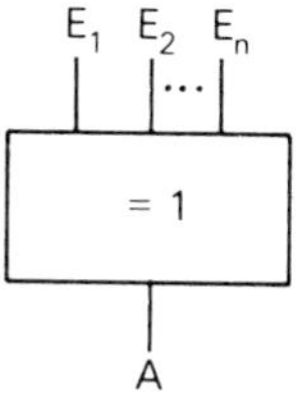	"OR" (exclusive OR) Disjunction of the mutually exclusive incidents $E_1, E_2 \ldots E_n$. A is present when one and only one of the E_1 to E_n is present.
5	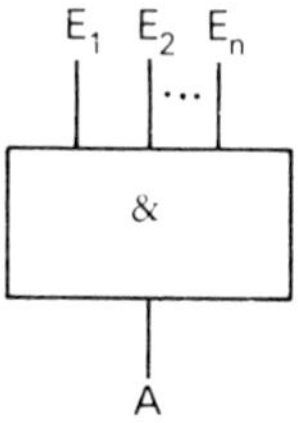	"AND" Conjunction of incidents $E_1, E_2 \ldots E_n$. $A = E_1 \wedge E_2 \wedge \ldots E_n$. The incidents $E_1, E_2 \ldots E_n$ are generally statistically dependent on each other. A is present when E_1 and E_2 and. $.E_n$ (all incidents) are present.
6	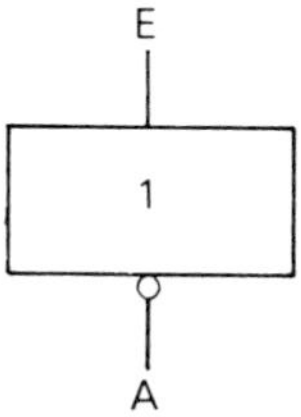	"NON" Negation of the incident E. $A = \overline{E}$. A is present when E is not present ($\overline{E}$) and vice versa.

Figure 4.10 (*Continued*)

No.	Symbol	Statement
7 7.1	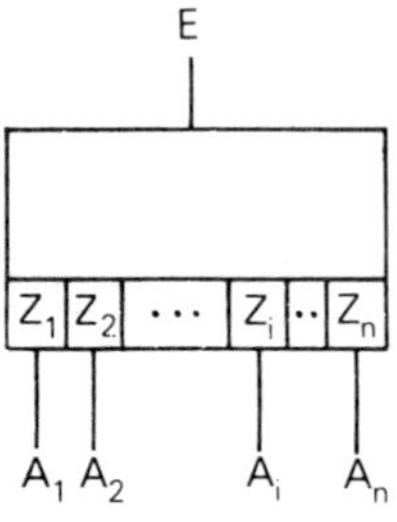	Multiple branching Incident E leads to function requirement of a unit with several possible states. Branching of incident E by conjunction with disjunctive states Z_i. $A_i = E \wedge Z_i$.
7.2	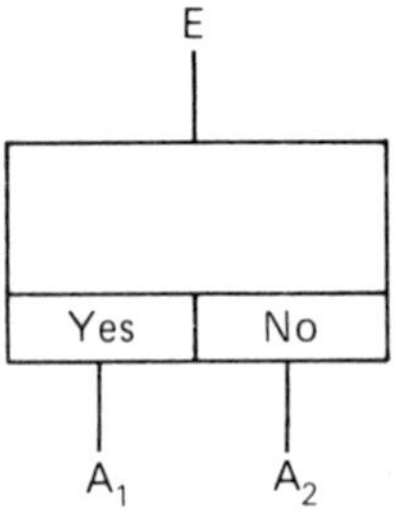	Simple branching Incident E leads to a function requirement of a unit with 2 possible disjunctive states. Branching of E by conjunction with the state Z_1 (yes) and state Z_2 (no) of the unit. $A_1 = E \wedge Z_1;\ A_2 = E \wedge Z_2$ The branching of the incident E can also occur by fulfillment, respectively non-fulfillment of a physical criterion described in the field.
8		Transfer, respectively continuation symbol. The incident progression diagram is discontinued by means of this symbol, respectively continued at another point.

Figure 4.10 (*Continued*)

diagram is given in Figures 4.11 (a) and 4.11 (b). It is based on subsystems; component failures are not considered. The example of an incident sequence diagram provides a good overview of the incident "minor leak in the cooling system" of a water-boiler nuclear reactor.

The following incidents led to the failure of the residual heat removal system as observed in the fault tree: instrument failure, automatic residual heat removal heat exchanger failure, failure of reactor protection and operators to start the automatic system, failure of high-pressure emergency cooling and of automatic pressure relief, failure of high-pressure and low-pressure emergency cooling, failure of auxiliary cooling-water system.

The incident sequence diagram can be evaluated quantitatively in a manner similar to that of the fault tree. For this, failure probabilities must

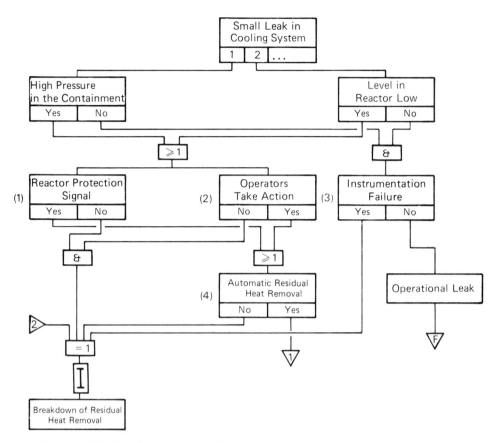

Figure 4.11(a) Incident sequence diagram of a case of a minor leak in a cooling system (part I).

be computed for subsystem incidents and the failure probability of the entire system must be determined according to Boolean algebra. If W is a state variable, where $W = 1$ again designates failure, Boolean rules serve to provide the system function for after-heat evacuation as

$$W_9 = (W_1 \wedge W_2) \vee W_3 \vee W_4 \vee (W_5 \wedge W_6) \vee (W_5 \wedge W_7) \vee W_8. \quad (4\text{-}38)$$

If W is interpreted as failure probability, and if for every partial failure a failure probability of, e.g., 10^{-2} is assumed, then, with consideration for the fact that only mutually exclusive incidents are linked, the following failure probability is obtained for after-heat evacuation:

$$W_9 = 10^{-4} + 10^{-2} + 10^{-2} + 10^{-4} + 10^{-2}, \quad (4\text{-}39)$$
$$W_9 \approx 3 \times 10^{-2}.$$

Decision Table Technique. Fault tree analysis and incident sequence analysis are methods leading to the representation of the logic structure of

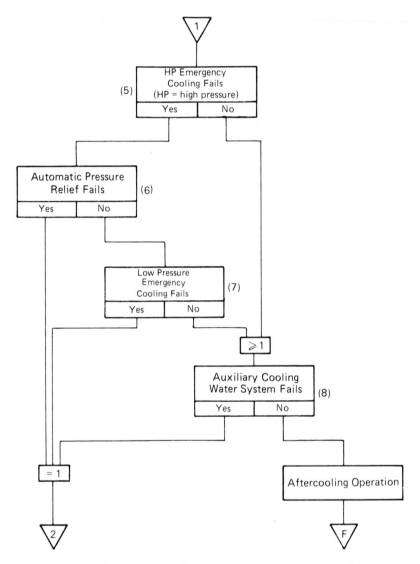

Figure 4.11(b) Incident sequence diagram of a case of a minor leak in a cooling system (part II).

a system. The decision table technique deals with a working method particularly suited for finding an unknown logic structure by starting from an undesired incident [4-5], [4-6], [4-7]. It connects directly with the logic structure formulation as a mathematical function of state variables as discussed in Section 4.2.1. In its customary form, a decision table includes four areas, as shown by Table 4.6 (a).

In the decision table the components of the system are entered in the upper left area, one below the other. In the lower left area the undesired

Table 4.6 Example of a decision table for five components

Initial form

	Component states	"AND"
1. Component	1 1 1 1 1 1 1 1 1 1 1 1 1 1 1 1 0 0 0 0 0 0 0 0 0 0 0 0 0 0 0 0	
2. Component	1 1 1 1 1 1 1 1 0 0 0 0 0 0 0 0 1 1 1 1 1 1 1 1 0 0 0 0 0 0 0 0	
3. Component	1 1 1 1 0 0 0 0 1 1 1 1 0 0 0 0 1 1 1 1 0 0 0 0 1 1 1 1 0 0 0 0	
4. Component	1 1 0 0 1 1 0 0 1 1 0 0 1 1 0 0 1 1 0 0 1 1 0 0 1 1 0 0 1 1 0 0	
5. Component	1 0 1 0 1 0 1 0 1 0 1 0 1 0 1 0 1 0 1 0 1 0 1 0 1 0 1 0 1 0 1 0	
System failure	X X X X X X X X	

(a)   ⟶ "OR"

Compression   Compression
1. Step   2. Step

(b)			(c)	
1. Component	1 1 1 1 1 1 1 1		1. Component	1 1 1 1
2. Component	1 1 1 1 0 0 0 0		2. Component	1 1 0 0
3. Component	1 1 0 0 1 1 0 0		3. Component	1 0 1 0
4. Component	1 1 1 0 0 0 0 0		4. Component	1 - 0 0
5. Component	1 0 0 0 1 0 1 0		5. Component	- 0 - -

incident of the system failure is given, for example, "failure inert gas supply." The upper right area includes, column by column, the individual varying system states possible in view of the number of components. Since, according to the definition, the state variables may have only two values, "0" and "1," here, in the initial form in the case of n components, only 2^n columns are involved. An example of the system is shown in Table 4.6(a) for five components, whose varying state combinations are $2^n = 2^5 = 32$ and are in 32 columns.

As agreed, the symbol "1" denotes "failed," the symbol "0" denotes "intact." If it is found that components contribute to all possible system states which lead to system failure, it will be advisable to place them at the beginning of the table. In Table 4.6(a) this applies to the first component. Therefore, in further obeservations, only those states will be considered for which a "1" is entered with the first component; the number of combinations to be dealt with further is cut in half.

The next step is to check which combinations by column bring about the undesired incident. The corresponding columns are marked by X in the lower right area, as shown in the initial form in Table 4.6(a). In this manner the compressed form, according to Table 4.6(b), containing all columns which bring about the undesired incident are obtained. When columns appear which differ only with respect to the state of a component, further simplification is possible. This means that the state of this component is insignificant in both columns. In Table 4.6 this is shown by way of an example in the third and fourth columns. This fact is taken into account by designating the state of component four by a dash in the subsequent compression in Table 4.6(c). With the compressed table form obtained by a further step, Table 4.6(c) provides a comprehensive picture of the logic structure in view of the undesired incident. The second column shows, for example, that for the occurrence of the undesired incident, the first and second components must have failed, that the third and fifth components could be intact, and that the state of the fourth component is of no importance.

In the decision table the component states of one column are connected by "AND," and the columns are connected by "OR." Starting from the basic table form according to Table 4.6, the result is a system function representation in the so-called "disjunctive" normal form with

$$Z_s = \sum_{JE} \prod_i Z_1^{Z_{ij}} (1 - Z_i)^{1 - Z_{ij}}, \qquad (4\text{-}40)$$

where Z_{ij} is the state variable Z_i in the j column. The sum is to be extended over columns j_E which lead to the undesired incident. If the component states are not dependent on each other, then by inserting "failure probabilities" q_i for the components instead of Z_i, the result is the occurrence probability q_s for the undesired system incident with

$$q_s = \sum_{j_E} \prod_i q_i^{Z_{ij}} (1 - q_i)^{1 - Z_{ij}}. \qquad (4\text{-}41)$$

If $q_i \ll 1$ so that $1 - q_i \approx 1$, then

$$q_s \approx \sum_{i_E} \prod_i q_i^{Z_{ij}} = \sum_{j_E} \prod_i Z_{ij} = 1 \, q_i. \qquad (4\text{-}42)$$

It is generally valid that the right-hand side of Equation (4-42) is always greater than the right-hand side of Equation (4-41). The use of Equation (4-41) therefore entails estimating on the safe side in every case.

In the same way the system function can be obtained from the compressed Table 4.6(c), while the product in every column is to be taken only over lines for which no dash is entered in the table.

To the "AND" or "OR" linkages of component states considered in Section 4.2.1 and represented in Table 4.2, specific structures are assigned in the decision tables. "AND" means that the undesired system incident will occur only when all components under observation have failed. If, in the example, five components are again used as point of departure, then the compressed decision table according to Table 4.7(a) will correspond to the state equation represented as the Boolean system function

$$Z_s = Z_1 \wedge Z_2 \wedge \ldots \wedge Z_n. \qquad (4\text{-}43)$$

"OR" means that the undesired system incident occurs when at least one of the components fails. To the corresponding system function

$$Z_s = Z_1 \vee Z_2 \vee \ldots \vee Z_n \qquad (4\text{-}44)$$

a compressed decision table according to Table 4.7(b) will apply. If failure

Table 4.7 Compressed decision tables

1. Component 1
2. Component 1
3. Component 1
4. Component 1
5. Component 1

(a) "AND" linkage of 5 components

1. Component	1	0	0	0	0
2. Component	–	1	0	0	0
3. Component	–	–	1	0	0
4. Component	–	–	–	1	0
5. Component	–	–	–	–	1

(b) "OR" linkage of 5 components

1. Component	1	–	–	–	–
2. Component	–	1	–	–	–
3. Component	–	–	1	–	–
4. Component	–	–	–	1	–
5. Component	–	–	–	–	1

(c) "OR" linkage of 5 components in the case of low failure probabilities

probabilities q_i of the components are very low, Table 4.7(b) can be replaced by Table 4.7(c). Combinations of "AND" and "OR" linkages will correspond to a combination of decision tables made up of parts which in turn are structured according to Table 4.7.

Beyond the mathematical statement, the decision table provides the first practical indications concerning system behavior and possible weak points. Thus, the example according to Table 4.6 shows that the first component has special significance for system behavior if its failure probability is not sufficiently low. In the fourth column of Table 4.6(c) only the first component appears whose failure by itself could lead to the undesired incident. On the other hand, the case will probably be that system failure brought about by incident progression according to the first column becomes highly improbable. For this, four components would have to fail at the same time.

Application of the decision table technique to determine the logic structure of a system will be explained further in Chapter 5 by means of practical examples.

Combination of Behavior in the Execution of Safety Analysis. Complete detection of hazards emanating from a technical system is, in principle, outside technical system methodological possibilities. Attaining a high degree of comprehensiveness in weak-point detection requires the application of various analysis methods. The sequence of the methods selected should, subject to identified hazard sources, facilitate gradual in-depth system observation.

The start of safety analysis must be systematic comprehension of the system concerned, so that by means of the analyst's experience and in conjunction with knowledge set forth, for example, in technical rules, the particular hazard sources can be ascertained. The resources to be used can be a preliminary hazard analysis tailored to the system particulars, as described in Section 4.3, or the failure effect analysis described above. Both methods derive their orientation from hazardous conditions which can originate from component failure. A method which, in the determination of hazardous conditions also considers overall system behavior, is incident sequence analysis.

When undesired incidents which can bring about failures are investigated in this manner, then, by means of fault tree analysis or the decision table technique, system behavior in the case of such incidents can be examined. If the required data are available, weak-point analysis can be continued quantitatively.

4.3 Qualitative Safety Analysis

Qualitative safety analysis is designed to use system considerations of a technical plant to acquire comprehensive understanding of all the factors of importance for system safety and of those which could work to its detriment. This requires identifying all components relevant to safety and checking them to see what hazardous system conditions and resulting failures are possible. In this context the determination of frequency or occurrence possibilities of system states is waived. The statement potential of a qualitative safety analysis is limited by available evaluation criteria. The bases for analysis result evaluations are experiences under comparable conditions and conclusions derived, as well as requirements for similar systems, subsystems, or components. In the following analysis, processes will be discussed which have been found satisfactory in technical safety evaluations of major technical plants. Safety appraisals of technical plants should first provide statements of whether and under what conditions safe operation is possible in principle. Beyond this, at a later stage, the safety of the plant during start-up must be determined concretely. The progression of a qualitative safety analysis is shown by the example of a technical process plant in Figure 4.12.

Following the completion of the first draft design of the planned installation, the necessary requirements derived from experience must be established. Applicable laws and technical rules must be observed. In the event that requirements based on experience cannot adequately guarantee system safety, special attention must be devoted to failure effect on the environment. Qualitative analysis will be continued step by step until unequivocal assessment is possible. In the event that qualitative analysis does not furnish a clear statement concerning system safety, and if the expected danger for the environment due to the behavior of the system under consideration requires it, quantitative analysis should follow qualitative analysis. As a rule, quantitative analysis concerns only particularly critical subsystems indicated in the scope of qualitative analysis.

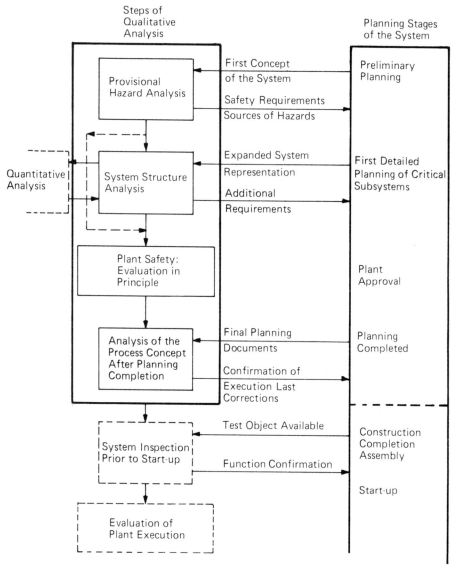

Figure 4.12 Sequence of qualitative safety analysis of a technical process plant from plant planning to initial operation.

4.3.1 Qualitative Safety Analysis Procedure

A series of methods [4-17], [4-18], [4-19] were developed for qualitative safety analysis and found to be successful. The choice of the analysis procedure, the special methods included, the application sequence, and the required extent of work depend on the project. The following must be considered:

—the hazard extent expected from the technical installation;
—the size of the installation;
—the information available;
—the stage of concretization;
—the time and cost requirement.

In principle, the determination of safety requirements for the planned installation is done by means of the preliminary hazard analysis (Figure 4.12).

For components or parts which over the course of this work are found to show a particular degree of risk, a *system structure analysis* (Figure 4.12) will be executed after detailed planning, if required, and basic safety of the parts will be evaluated as related to system failure. Here it may be necessary to extend the analysis by quantitative method (Section 4.4). Both analysis steps facilitate the decision of whether and under what conditions safe operation is possible.

Due to safety requirements ascertained with respect to the plant, essential system changes might result. In the event that, with due consideration for necessary changes, the planning of the plant is completed, *analysis after planning completion* takes place. The completed planning documents are checked to see whether the technical design observes specifications and can attain the safety aim.

Here it should be pointed out that inspection programs for production, start-up, and current operation are derived from safety analysis. In the following the individual operational steps will be explained and applied to an example. For this a leaching solution settling tank is selected as subsystem of a petrochemical plant. This system will be described first.

Butene from naphtha cracking is to be returned into the process, following intermediate storage in the tank system, in the form of liquid fuel. In the case of direct use of butene, there is corrosion due to washing liquid residue from the separation process. Therefore, to avoid damage, a leaching solution settling tank is included in the cycle. This serves to separate washing liquid and butene and is added ahead of a heating system. In operation according to plan, the pumps deliver the mixture from the tanks to the settling tank. During the retention period, controlled by the tank volume and level, separation takes place. The leaching solution collects in the lower part of the tank and is fed from there discontinuously to the removal system. The liquid butene above is fed to the heating system of the naphtha cracking installation.

The block diagram (Figure 4.13) shows the separation of the system under investigation from the complete system. By way of inlets the subsystem "separation butene/washing liquid" has the line from the crude-butene tank and the line from the low-pressure steam network for heating the settling tank, and by way of outlets, the line to the purified butene heating system, the lines to the removal system, namely the line for

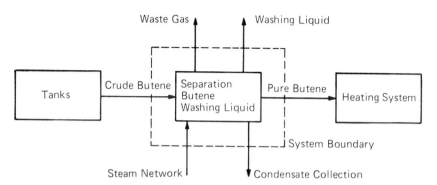

Figure 4.13 Representation of the system "separation butene/washing liquid" in block diagram form.

discontinuously removed washing liquid, the safety valve pressure release line, as well as the line for condensate removal from the tank steam heating.

Figure 4.14 shows the pipeline and instrument flow chart (PI flow chart) of the system under examination in simplified form. This figure shows the essential components.

The most important operational data of the system under investigation are given in Table 4.8.

Preliminary Hazard Analysis. "Preliminary hazard analysis" will be executed in two steps:

(a) Basic check of the entire installation for observance of applicable technical regulations.
(b) Determination of incidents which might result from deviation from operation as specified, and determination of parts which require further analysis in view of such incidents.

The work papers for the analysis are the pipeline and instrument flow charts (hereafter referred to as PI flow charts), process descriptions, and operating manuals. In checking the technical system to see whether applicable technical regulations have been observed, a division into subsystems may be practical. The subdivision will be dictated by technical regulations. The procedure will be as follows:

—Establishment of characteristics of the technical installation, or the subsystems, with respect to assignment to controls.
—Determination of applicable laws, regulations, and technical regulations.
—Review of documents to check the observance of governing safety principles.

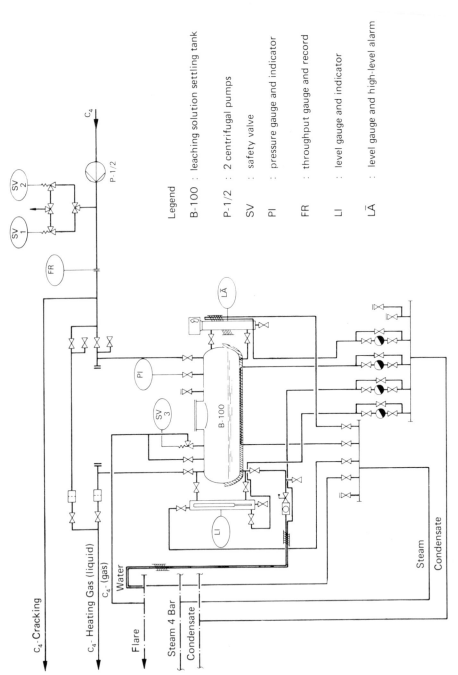

Figure 4.14 Pipeline and instrument flow chart of the butene leaching solution settling tank.

Table 4.8 Operation data of the system "separation butene/washing liquid"

Leaching solution settling tank B-100

	Tank	Heater
design excess pressure	23 bar	10 bar
design temperatures	60°C	160°C
content	1720 l	259 l
material	H II	St 35.8

Pressure generator

two centrifugal pumps P-1/2
switched parallel
precompression $\qquad P_{e,v} = 5$ bar
max. feed pressure $\qquad \Delta p = 25.8/27.28$ bar
max. feed $\qquad \dot{m} = 26.8/62$ Mg/h
(at $P_e = 23$ bar)

Steam heat

max. heating temp. $\qquad t = 164°C$
(at steam net cut-out
pressure $P_e = 6$ bar)

Medium

butene with minor parts washing liquid
(mainly leaching solutions)
boiling range of C_4-butene isomers at 1 bar $\qquad -7.1°$ to $+3.7°C$
steam pressure $\qquad\qquad$ at 0°C $\qquad P_D = 1.22$ bar
$\qquad\qquad\qquad\qquad$ at 50°C $\qquad P_D = 6.1$ bar
$\qquad\qquad\qquad\qquad\qquad$ 1.6 $\div$ 10 vol.%
ignition range—air
health hazard $\qquad\qquad$ Vapors have slight narcotic effect. Suffocation and damage
$\qquad\qquad\qquad\qquad\qquad$ to nervous system possible. Contact with liquid causes cold
$\qquad\qquad\qquad\qquad\qquad$ injury. MAK-value (max. concentration work place) not
$\qquad\qquad\qquad\qquad\qquad$ determined.

—Establishment of a list showing the shortcomings relating to technical safety requirements.

This analysis section will be followed by a review of possible incidents involving the system and which could become hazardous, for example, due to fire, explosion, or release of toxic materials. In this context energy sources and media should be noted in particular. The system will be subdivided into subsystems which permit definite evaluation of possible hazards. The subsystems to be evaluated must permit ready separation from each other with respect to the set-up and mode of operation, while all reciprocal actions must remain identifiable.

The analysis will be continued beyond the scope of the preliminary hazard analysis only for such a subsystem in which one of the previously defined incidents could occur. In the case of these subsystems, operating conditions under which an incident is conceivable are ascertained and the undesired events are determined.

Preliminary Hazard Analysis of the Leaching Solution Settling Tank. The leaching solution settling tank is a pressure tank in which butene is in intermediary storage as pressure gas in liquid state. For safety design, legal provisions and technical regulations must be known and observed [4-27], [4-28], [4-29], [4-30].

Review of the planning documents (PI-flow chart, technical data, operating manuals) shows that technical safety equipment of the leaching solution settling tank is adequate for the intended operation with respect to formal requirements. Separate evidence will be necessary to confirm that safety valves have been correctly dimensioned and that the possibility of inadmissible electrostatic charge of butene is avoided. In the case of butene release, there is considerable fire or explosion risk in view of the low ignition limits in the air of 1.6–10 vol.%. Health hazards due to butene effects occur only in the case of higher concentration. Incidents such as fire and explosion can also be caused by interaction between system components, which means that tank observation alone is not enough. Therefore the system must undergo further analysis. "Discharge of butene" is viewed as an undesired event which can lead to an incident.

System Structure Analysis. After the identification of subsystems to be examined and the definition of undesired events within the context of preliminary hazard analysis, events which lead to incidents are investigated. These event sequences can be represented as logic structure in a block diagram, a flow diagram, a fault tree, or a decision table. In the presentation which follows (Table 4.9.), a decision table was used. It contains, column by column, the combinations of system states which lead to the undesired event. The presentation permits qualitative identification of weak points in the system. In general, for example, the probability of a system state will decline with the growing number of failed components. The logic structure presentation could form the basis for further quantitative analyses.

System Structure of the Leaching Solution Settling Tank. The preliminary hazard analysis of the leaching solution settling tank showed the undesired event "discharge of butene in gas form." Figure 4.15 shows the PI flow chart with identification of components by numbers. In contrast to the diagram in Figure 4.14, the result of the preliminary hazard analysis has been entered: the components 31 (temperature sensor, temperature alarm) and 32 (control valve steam line). The naming of the components as well as the statement of the assigned operation and failure state is included in

Table 4.9 State in operation and failure: components according to Figure 4.15

No. (see Figure 4.14)	Designation	Operation state ($\cong 0$)	Failure state ($\cong 1$)
1	centrifugal pumps P-1/2	deliver	do not deliver
2	safety valve SV1	responds	does not respond
3	safety valve SV2	responds	does not respond
4	alternating lock	operational	defective
5	volume measure FR	measurement of delivery volume	no/erroneous measurement
6	main slide	open	closed
7	main slide	open	closed
8	check valve	closed	open
9	safety valve SV3	operational	defective
10	check valve	closed	open
11	pressure indicator	operational	defective
12	main slide	open	closed
14	liquid-level gauge	operational	defective
15	Alarm LA 9012	operational	defective
16	check valve	open	closed
17	check valve	open	closed
18	main slide without dummy flange	closed	open, leak
19	liquid-level indicator LI	operational	defective
20	slide	open	closed
21	slide	open	closed
22	slide	closed	open, leak
23	quick-action stop	operational	defective
24	viewing glass	tight	leak
25	main slide without dummy flange	closed	open, leak

Table 4.9 (*Continued*)

No. (see Figure 4.14)	Designation	Operation state ($\hateq 0$)	Failure state ($\hateq 1$)
26	cut-off armature	open	closed
27	cut-off armature	closed	open, leak
28	cut-off armature without dummy flange	closed	open, leak
29	cut-off armature	open	closed
30	cut-off armature	closed	open, leak
31 T	temperature sensor from TIAS	operational	defective
31 A	alarm in the measuring unit from TIAS	operational	defecive
32	regulating valve in the steam line	operational	defective
P_D	pressure in steam network	$P_D > P_{D\text{perm}}$	$P_D \leqq P_{D\text{perm}}$
W1	tank material at $T < T_{\text{perm}}$	does not fail at $P > P_{B\text{perm}}$	fails
W2	tank material at $T > T_{\text{perm}}$	does not fail at $T > T_{\text{perm}}$	fails
M1	reading of component 19 and reaction	regular reading and reaction	no/wrong reading or no/wrong measures taken
M2	reading of component 11 and reaction	regular reading and reaction	no/wrong reading or no/wrong measures
M3	recognition of alarm by component 15 and reaction	correct recognition and reaction	no alarm recognition or no/wrong measures
M4	recognition of alarm by component 31 and reaction	correct recognition and reaction	no alarm recognition or no/wrong measures

P_D = excess pressure in the steam network; T = heating temperature; P_B = excess pressure in the tank.

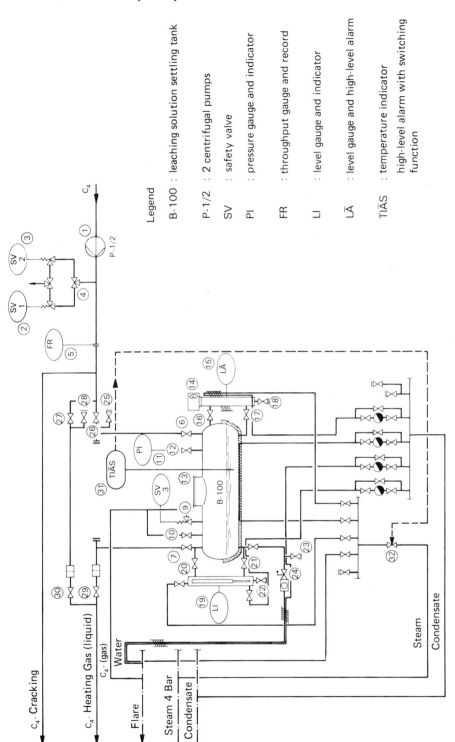

Figure 4.15 PI flow chart of the leaching solution settling tank for butene, corrected according to results obtained from the preliminary hazard analysis, with element designation for determination of the logic structure.

Table 4.9. The components include four measuring instruments with
indicators or alarms by means of which action by service personnel may be
requested if necessary. For this reason human service must be introduced
four times as system components M1 to M4. In addition, the tank material
is formally introduced as components W1 and W2. W1 designates the tank
material at temperature T below permissible tank temperature T_{perm}. W2
designates the tank material at temperatures above permissible operating
temperature. The tank temperature is determined by heating temperature
which in turn depends on pressure P_D in the steam network. For this
reason, pressure P_D is formally treated as a component.

At first there is the general view that, because of its effect, the release of
butene is the undesired event. In the course of further analysis it is,
however, practical not to view the butene release generally. Three
undesired events will be dealt with separately.

UE1: gas release due to internal leakage through valves, slides, etc.
UE2: gas release due to tank rupture at temperatures above the permissible operating temperature and at pressures which exceed the
permissible pressures of the present temperature.
UE3: gas release due to rupture of the tank because of excess above
permissible operating pressure at temperatures below permissible
operating temperature.

For didactic reasons, further procedure will be explained by means of the
undesired event UE2. The two other events will be considered later.

On the basis of the flow chart and component list, it can be assumed
that seven components are possibly involved in the origin of the undesired
incident. They are entered in the upper left field in Table 4.10(a). The
complete decision table would have to include $2^7 = 128$ columns. If it is
noted, however, that occurrence of the undesired incident UE2 is based on
the premise, in every case, that there is excess pressure in the steam
network and that the material then fails, this means that only those
columns are to be considered which include a 1 (one) in the first two lines.
Thus, for the description of the conditions to be observed, the columns
entered in the decision table [Table 4.10(a)] remain, with 1 designating the
failure of a component, or the undesired behavior of a relevant magnitude
and 0 the desired behavior. The system states represented in one column
and leading to the undesired event UE2 are marked by an X. This shows
that the columns marked X differ in pairs such that in the case of
component 9, a 0 or 1 (one) is entered. This means that the component, a
safety valve on the tank, has no effect. Then the first 13 columns must be
considered, while neglecting component 9. From column 8 it is obvious
that the "failure" of components W2, P_D, and 31 T already lead to the
undesired event, irrespective of whether another component does or does
not fail. Therefore columns 1–8 in Table 4.10(a) can be compressed into
column 1 (one) in Table 4.10(b). Columns 9 and 10 differ merely in the

Table 4.10 Decision table for the undesired event UE2 "Gas leak due to tank rupture" in the case of excess permissible operating temperature

(a) Initial table

Component	Column 1	2	3	4	5	6	7	8	9	10	11	12	13	14	15	16	17	18	19	20	21	22	23	24	25	26	27	28	29	30	31	32
W2	1	1	1	1	1	1	1	1	1	1	1	1	1	1	1	1	1	1	1	1	1	1	1	1	1	1	1	1	1	1	1	1
P_D	1	1	1	1	1	1	1	1	1	1	1	1	1	1	1	1	1	1	1	1	1	1	1	1	1	1	1	1	1	1	1	1
9	1	1	1	1	1	1	1	1	1	1	1	1	1	1	1	1	0	0	0	0	0	0	0	0	0	0	0	0	0	0	0	0
31 T	1	1	1	1	1	1	1	1	0	0	0	0	0	0	0	0	1	1	1	1	1	1	1	1	0	0	0	0	0	0	0	0
31 A	1	1	1	1	0	0	0	0	1	1	1	1	0	0	0	0	1	1	1	1	0	0	0	0	1	1	1	1	0	0	0	0
32	1	1	0	0	1	1	0	0	1	1	0	0	1	1	0	0	1	1	0	0	1	1	0	0	1	1	0	0	1	1	0	0
M4	1	0	1	0	1	0	1	0	1	0	1	0	1	0	1	0	1	0	1	0	1	0	1	0	1	0	1	0	1	0	1	0
UE 2	X	X	X	X	X	X	X	X	X	X			X				X	X	X	X	X	X	X	X	X	X			X			

(b) Compressed Table ET2

Component	Column 1	2	3
W2	1	1	1
P_D	1	1	1
9	–	–	–
31 T	1	0	0
31 A	–	1	0
32	–	1	1
M4	–	–	1
UE 2	X	X	X

entry for component M4. The compression provides column 2 in Table 4.10(b). Column 13 remains and is transferred as column 3 into Table 4.10(b). This table shows the compressed decision table ET2 assigned to the undesired event UE2. In a similar manner, examination of the undesired events UE1 and UE3 leads to compressed decision tables. They are shown combined as the complete compressed decision table covering the event "Leak of butene in gas form" in Table 4.11. Here, also, groups of components are combined by an "OR" linkage. Failure, marked by 1 (one), of such a group is given when at least one component fails. Beyond this, only the component failures are marked. A first evaluation of the decision table (Table 4.11) can be accomplished by means of the number of components required for system failure, marked in the table by the number of ones in each column. Analysis provides the following result: unequivocal weak points occur in columns 1 to 3 which come from the decision table ET1 for the undesired incident "gas escape due to leak". Here the failure

Table 4.11 Decision table for the three events UE1 to UE3 "escape of butene in gas form" ($\lor$ = "OR" linkage)

Component	Decision table ET1	ET2	ET3	Comment
18 ∨ 22 ∨ 25	1 – –	– – –	–	Cut-off armatures into atmosphere without dummy flanges
27	– 1 –	– – –	–	
28	– 1 1	– – –	–	Cut-off armatures into atmosphere without dummy flanges
30	– – 1	– – –	–	
W2	– – –	1 1 1	–	
P_D	– – –	1 1 1	–	
31 T	– – –	1 – –	–	Additional technical safety measures as a result of the preliminary hazard analysis
31 A	– – –	– 1 –	–	
32	– – –	– 1 1	–	
M4	– – –	– – 1	–	
W1	– – –	– – –	1	
2/3	– – –	– – –	1	
11 ∨ 12 ∨ M2	– –	– – –	1	PI
14 ∨ 15 ∨ 16 ∨ 17 ∨ 18 ∨ M3	– – –	– – –	1	LA
19 ∨ 20 ∨ 21 ∨ 22 ∨ M1	– – –	– – –	1	LI
7 ∨ 29	– – –	– – –	1	

weak points with decreasing weight

very safe combination

of one or two components already leads to the undesired incident. The components concerned, namely 18, 22, 25, and 28, are cut-off armatures for aeration, i.e., facilitate escape of the operating medium into the atmosphere. Dummy flanges for armatures 18, 22, 25, and 28 are recommended.

Additional weak points are shown in columns 4 to 6. Here, however, three and four component failures, respectively, are required to effect system failure. The effectiveness of measures demanded in the preliminary hazard analysis for regulation of steam heating as reflected in components 31, 32, and M4 comes into play here. Due to increasing steam pressure, component P_D, the tank material, component W2, is exposed to higher temperature stress. Since due to steam-net securing the steam pressure can be maximally 7 bar, the highest temperature stress on the tank is limited to 164°C. Up to this temperature no considerable drop in resistance values of the tank material (Table 4.8) can be expected. Daily check of the butene temperature is recommended.

Column 7, with six required component failures, shows that the system is safely designed with respect to excess tank pressure. This necessity exists because, due to maximum excess pressure of 32.5 bar behind the pumps and permissible operating excess pressure of 23 bar in the tank, the pressure quotient at 1.4 is below the rupture strength of pressure tanks.

Concept Analysis of the Technical Plant after Planning. Preliminary hazard analysis and system structure analysis are intended to facilitate safety evaluation of the planned installation from the basic concept. Comprehensiveness of observation depends on the danger level of the installation. After concluding this work, basic changes in the technical procedure frequently become necessary and must be incorporated into the planning.

Before the planned installation is built, e.g., before individual components are constructed or ordered, the plan must be checked for safety in all its particulars. At that time verification of the incorporation of previously stipulated requirements is indicated.

This verification is referred to as an operability study [4-20], [4-21], technical safety discussion [4-22], or PI flow chart review [4-23] and is implemented according to different principles. H. Lawley [4-20] has established a strict system for the procedure. For the detection of the safety-relevant components, he follows the PI flow chart, and for the detection of hazardous plant conditions, a list of code words. H. Ullrich [4-24] describes a different way of proceeding. He applies a series of checklists in a determined sequence to the planning documents to confirm complete technical safety provisions. Individually such checklists are used to examine the completeness of procedure planning and construction documents as well as measuring and control plans.

The following will be a discussion of the principle of analysis after planning completion. Subsequently the Lawley procedure will be discussed more specifically and then applied to the example of the leaching solution settling tank.

Due to the need to check all safety measures, this part of the system analysis can become very voluminous. The method becomes difficult due to the constant change from the overall review of the plant to the review of individual components. For this reason it is important, in the case of major systems, to place "preliminary hazard analysis" and "system structure analysis" ahead of the evaluation of the planning concept. The advantages of this work plan are the following:

—General overview of the structure of the entire installation with the major hazard points.
—A list of technical safety requirements with respect to the installation.
—Inclusion of technical safety requirements for the plant execution in the preliminary planning stage.

In this way, during detailed consideration of components, the view of the entire installation is preserved. Safety requirements are the basis for the work. Possibly required changes are limited and thereby repeated reworking of subsystems is avoided. Analysis costs remain within acceptable bounds with respect to economics and time.

The following documents must be available for the planning concept analysis:

—process description with particulars on materials, material data, and quantities,
—operational pressures and temperatures,
—PI flow charts,
—list of equipment and data on equipment,
—assembly and position plans,
—technical safety decisions from previous analysis steps,
—a basic concept of operating instructions.

The basic steps in this work are:

—selection of the components according to an outline which must guarantee completeness,
—inquiry into component design data,
—inquiry into requirements with respect to components and technical safety decisions,
—quantitative check of the design with respect to operation definition and possible service failures,
—comparison of design data and actual stress ascertained,
—establishment of a list of required complementary measures,
—check of the effect of these complementary measures on the course of the operation.

In the following, a possible way of proceeding, the hazard and operability study (HAZOP) [4-25], will be described, based on the Lawley's work [4-20] and developed for the area of process technology. The working documents are:

—project papers developed during the planning phase,
—a diagram according to which all installation components are examined one after the other as per Figure 4.16,
—a defined set of guide words for the purpose of ascertaining possible component error functions according to Table 4.12.

Analysis according to the flow diagram of Figure 4.16 is based on the presentation of a technical plant in the form of one or more flow diagrams. It begins with the first vessel (1) and the description of the general intention of a vessel and its lines (2). This is followed first by the examination of all vessel connection lines (3)–(14), then all equipment parts (15)–(19), and finally the vessel (20)–(22).

Examination always follows the same model (5)–(10). After an explanation of the intention, the guide words according to Table 4.12 are applied. Possible deviations will emerge from this procedure. In the case that breakdowns of the operation as defined can be derived from the determined deviations, and if the assumption of these breakdowns is not devoid of sense, these breakdowns will be investigated with respect to cause, consequences, and danger to the environment. Realistic hazard states of the component are recorded. Following the discussion of the entire surrounding field—e.g., a tank—available technical safety measures, related to the hazard conditions found, are checked, and if necessary, additional measures are deduced.

Operability Study Application to the Leaching Solution Settling Tank. The explanation of the working method is completed by applying it to the leaching solution settling tank. Thus, for example, there is an examination of the feed lines from the tank storage to the butene tank and of the tank itself.

The design intention of the line can be described as follows: butene is transferred from the tank storage by means of pumps P-1/2 through the settling tank to the heating system. The delivery quantity and delivery pressure are supposed to remain within specified limits. The guide words for the design intentions, the deviations, and possible causes and consequences, as well as the required measures are shown in Table 4.13. The premise is that examination of documents shows that the release performance of safety valves SV2 and SV3 ((2), (3) in Figure 4.15) is smaller than the maximum delivery of both pumps P-1 and P-2 ((1) in Figure 4.15).

Further analysis of lines and equipment parts of the system "separation butene/washing liquid" shows that evacuation of the washing liquid from the tank to the removal system is the critical component. It is, however,

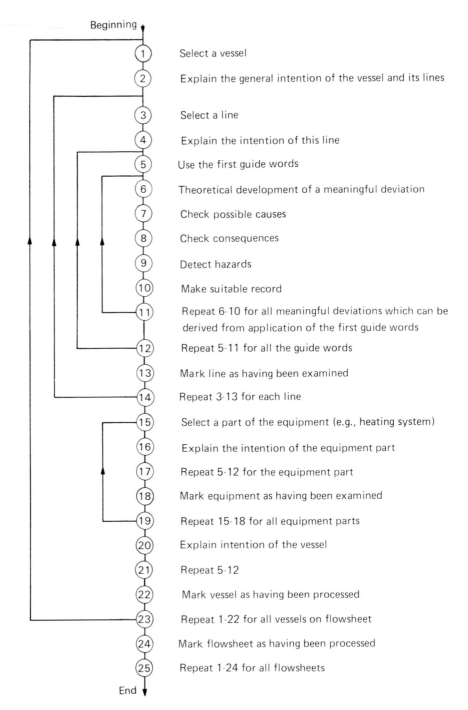

Figure 4.16 Sequence of test steps in the case of a hazard and operability study [4-25] valid for the area of process technology.

Table 4.12 List of guide words for the Hazard and Operability Study [4-25] valid for the area of process technology

Code words	Significance	Comment
NO or NOT (NO or NONE)	Total negation of this function	No part of these functions is exercised, but nothing else happens either.
MORE	Quantitative increase or decrease	This relates to quantities and properties such as mass flow and temperatures,
LESS		but also to functions such as HEATING UP and REACTION.
NOT ONLY BUT ALSO	Qualitative increase	All planned functions and operations are attained. In addition, however, SOMETHING ELSE happens as well.
PARTLY (IN PART)	Qualitative decrease	Only some planned functions are attained, some not.
REVERSAL	The logic opposite of the planned function	This concerns mainly functions, e.g., contrary flow or contrary chemical reaction. It can also be applied to substances, i.e., POISON instead of antidote or D instead of L optical isomers.
OTHER THAN	Complete exchange	Not a single function of those originally established is executed. Something totally different happens.

technically well equipped with a fast-action stop valve, reverse stop, and inspection glass, as well as the satellite heater.

The leaching solution settling tank B-100 in Figure 4.15 remains to be discussed. The design intention can be described as follows: Crude butene undergoes intermediate storage in the tank for the purpose of settling suspended washing liquids. Butene is stored in liquid form under pressure and heated by means of steam to a specific temperature prior to introduction into the heating system. The deviations, causes, and consequences, as well as the required actions resulting from application of the guide words to the design intention and shown in Table 4.14. A check shows that safety valve SV3 [(9) in Figure 4.15] on the tank is adequate for butene vaporization at 22 bar pressure as well as for thermal expansion.

To sum up, the following actions are indicated with respect to the leaching solution settling tank:

— The level gauge and the alarm for butene tank level should be designed for alarm report when the level is below a minimum as well.
— Installation of a position indicator of the control valve of the steam line in the measurement control station.
— Extension of the butene delivery line to below minimum liquid level of the tank.
— Regular check of the line from the tank storage to the butene tank for leaks.

Table 4.13 Results of the butene feed operability study

System: Separation butene/leaching solution   Component: C$_4$ feed pipe

Guide word	Deviation	Possible causes	Consequences	Action required
NO or NOT	No delivery	(1) No butene in tank	No delivery to the heating system (evaluation of consequences there)	(a) Is there information on the quantity stored in the tank storage in the volume indicator?
			Higher degree of corrosion due to leaching solution in system	(b) Possibly an alarm reporting "low" should be installed on the leaching solution settling tank.
		(2) Pump P-1/2 fails	As (1)	Covered by (b)
		(3) Line blocked (cut-off valve closed)	As (1)	Covered by (b)
			Pump running hot	(c) Install cut-off device on pumps covered by (b)
		(4) Line ruptured	As (1)	(d) Regular line inspection
MORE	More delivery	(5) Both pumps deliver at the same time	Butene is released	
			Settling tank is filled above planned level. Not all the leaching solution is separated, heating system exposed to corrosion	(d) Excess over planned level is reported by leaching solution settling tank.
				(e) Prevent simultaneous switching on of pumps covered by (c)
	Higher pressure	(6) Cut-off valve was closed inadvertently	Line pressure is zero delivery pressure	(f) Check line pressure design
				(g) Check safety valve release line covered by (e) and (f)
	Higher pressure and more delivery	(7) Both pumps deliver simultaneously against closed cut-off valve	Line pressure is zero delivery pressure	(h) Are safety valves able to release the entire delivery quantity?
LESS	Reduced delivery	(8) Flange or valve leak and not equipped with blind cover	Butene is released	Covered by (d)

Table 4.14 Results of the leaching solution settling tank B-100 operability study

	System: Separation butene/leaching solution		Component: B-100 tank	
Guide word	Deviation	Possible causes	Consequences	Actions required
NO or NOT	No intermediate storage	(1) Prior to use of butene removal line, the tank was not filled up to minimum level	No leaching solution separation from crude butene possible, increased corrosion of the heating system	(a) Covered by measure (b) (Table 4.13)
	Tank ruptured	(2) Intensive corrosion of tank material	Butene is released, fire, explosion	(b) No measures required since tank material is well suited and anti-corrosion additive is adequate
		(3) Excessive pressure	same as (2)	(c) Check safety valve on the tank
		(4) Excessive temperature	same as (2)	(d) Measure connected with TIAS adequate
		(5) Closed check valves on tank and heating by the steam system	Tank ruptures, butene is released	Covered by (c)
MORE	Intermediate storage at excessively high pressure	(6) TIAS control of the steam heat system does not work	Butene pressure becomes too high, delivery rate by pumps drops. Pressure in the connected heating system increases	(e) Regular maintenance of TIAS is carried out
	Intermediate storage at excessively high temp.			

Table 4.14 (*Continued*)

System: Separation butene/leaching solution   Component: B-100 tank

Guide word	Deviation	Possible causes	Consequences	Actions required
LESS	Intermediate storage at excessively low temp.	(7) Steam system failure. (8) Control valve in steam feed line is closed	Low delivery rate due to excessive resistance. No removal of washing liquid in the case of low outside temperatures	(f) Not required; alarm indicating steam failure is available. (g) Feedback by value position to be established in measurement control station
OTHER THAN	Tank is completely filled with butene	(9) Filling prior to valve opening in the removal line and no LA alarm	Thermal expansion due to heating	(h) Sufficient dimensioning of the safety valve; see (c)
	Large quantities of washing fluid in the container	(10) No removal of the washing liquid	Butene enrichment by entrainment of washing liquid, therefore corrosion in heating system	(i) Not required; as per operating manual the washing liquid is let out at regular intervals
	Static charge of butene in the case of low leaching solution concentration	(11) Incoming stream produces frictional electricity	Electric charge of tank wall, wrong indication by LA	(j) Positioning the feed line below level of liquid

4.4 Quantitative Safety Analysis

Within the overall aim it is the task of quantitative safety analysis to ascertain the frequency or occurrence probability of undesired events leading to incidents. Safety analysis will, in the case of problematic results of qualitative analysis, necessarily inspire the question of whether it should be continued in quantitative form. The question arises in particular when new technical equipment and processes are used. Quantitative safety analysis starts with knowledge of the logic structure of the system to be examined, as has already been ascertained in the course of qualitative analysis. A condition for execution is the presence of sufficient data—information about the behavior of the individual system components and parts. The information must be arranged in such a way that reliability characteristics (failure probabilities, failure rates) and maintenance characteristics (rates of repairs) can be derived. It is only when it is certain that sufficient data are available that quantitative analysis is possible.

Procurement and Evaluation of the Data Base. Determination of reliability characteristics of components and structural parts requires extensive data acquisition and evaluation. Data acquisition can be done in the course of operation or through the systematic investigation of components and structural parts in the laboratory with simulation of operational stress. In the latter case, however, it is necessary to observe conditions necessary for the applicability of test results to actual in-plant behavior. In this connection, test frequencies should be mentioned by way of an example, since in the laboratory test they often have to be higher than actual frequencies during operation. It depends on the individual case whether such circumstances are significant or not. To estimate the failure rate of a component or structural part, it is necessary to examine a large number n_0 of identical test pieces which are still operational at time $t = 0$. After time t, $N(t)$ test pieces should have failed and $n(t)$ test pieces should be intact. Accordingly,

$$n(t) + N(t) = n_0 \qquad (4\text{-}45)$$

will be valid.

In the event that there are, in the time interval Δt, ΔN failures, the failure rate $\lambda(t)$, according to Section 4.2.1, results by approximation from

$$f(t) \approx \frac{\Delta N 1}{\Delta t n_0},$$

$$1 - F(t) \approx \frac{n(t)}{n_0},$$

$$\lambda(t) \approx \frac{1}{n(t)} \frac{\Delta N}{\Delta t}. \qquad (4\text{-}46)$$

The failure probability $F(t)$ dependent on operating time results directly from the data as

$$F(t) \approx \frac{N(t)}{n_0}. \tag{4-47}$$

It is frequently impossible to determine time dependence because of inadequate data and limited available observation periods. In the case of some components, e.g., electronic parts, failure rate time dependence does not exist for practical purposes. Therefore, in such cases, constant failure rates $\lambda(t) = \lambda = \text{const}$ are often used.

Determination of component failure rates can be subject to uncertainties, either because the volume of data for this determination is low or because data from different sources lead to varying results. Experience shows that maximum and minimum values obtained frequently differ by two orders of magnitude. If there is no information about such uncertainties, it would be advisable to view the values 0, 1λ and 10λ as lower and upper control limits, respectively. The effect of these uncertainties on system behavior can be estimated by means of sensitivity analyses.

Due to uncertain data, it is frequently not justified to view data obtained from quantitative analysis on the expected frequency or the occurrence probability of undesired effects in systems as certain absolute quantities. In this case as well, relative comparison of different event progressions or technical constructions frequently provides valuable information. This also holds true when quantitative safety analysis is not applied to entire systems but only to specific safety equipment for the purpose of comparing various implementation alternatives.

Component and System Behavior. Without the inclusion of maintenance and repair measures, the probability $q_i(t)$ that component i fails at time t results on the basis of the distribution function for life according to Section 4.2.1 in dependence on the operating period as

$$q_i(t) = 1 - \exp\left[-\int_0^t \lambda_i(t')\, dt'\right]. \tag{4-48}$$

In the case of constant failure rate, we obtain

$$q_i(t) = 1 - e^{-\lambda_i t}. \tag{4-49}$$

The probability that a system with several components will have failed by time t can be written in the form

$$q_s(t) = f[q_1(t), \ldots, q_n(t)]. \tag{4-50}$$

As long as maintenance measures are not considered, the statement that a system is in the "failed" state after termination of operating period t will be identical with the statement that failure will have occurred by time t. The special form of the function on the right-hand side of Equation (4-50) results from the logic structure of the system concerned, as is shown, for example, in a decision table.

In the case of constant failure rates concerning failure behavior of independent components, the result for a pure "OR" linkage, according to Section 4.2.1, is

$$q_s(t) = 1 - \prod_i [1 - q_i(t)]$$

$$= 1 - \prod_i \exp(- \lambda_i t)$$

$$= 1 - \exp\left[-t \sum_i \lambda_i \right]. \qquad (4\text{-}51)$$

Under the same conditions, in the case of an "AND" linkage, according to Section 4.2.1, the result will be

$$q_s(t) = \prod_i q_i(t)$$

$$= \prod_i (1 - e^{-\lambda_i t}). \qquad (4\text{-}52)$$

If magnitudes or components enter into the system consideration, where undesired behavior of such magnitudes or components is not described by a failure rate but by the event rate for multiple possible events as, perhaps, the appearance of a boundary value excess, then the observation must be varied. This will be shown later in the example of the leaching solution settling tank. Maintenance measures generally exert considerable influence on the probability of the occurrence of undesired system events and therefore neglecting this influence provides results which are too pessimistic. Analysis should, as far as possible, include maintenance measures. In the case of constant inspection intervals and constant failure rates of the components under review, this can be done as, for example, in the following manner. The assumption is that inspection of all components of a system takes place at regular intervals. Failed components are replaced or repaired. The failure rate is assumed constant. The maintenance time is neglected. Immediately upon inspection, at time t_m, component i is intact. The probability that component i has failed at time t between two inspection times t_m and $t_m + 1 = t_m + \Delta t$ is then

$$q_i^I(t) = q_i(t - t_m) \qquad (4\text{-}53)$$

with q_i as in Equation (4-49). This is shown by Figure 4.17 in diagram form.

The probability that a system whose components are intact at the start of an interval between two functions will fail within the inspection interval with length Δt will be

$$q_{S\Delta t}^I = q_s(\Delta t) \qquad (4\text{-}54)$$

with $q_s(\Delta t)$ as in Equation (4-50). As a measure of system safety, the

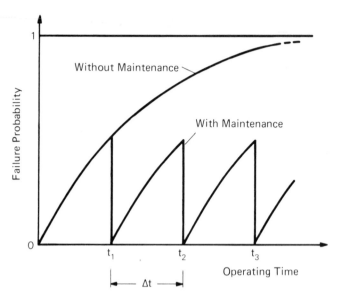

Figure 4.17 Influence of maintenance on component failure probability in the case of constant inspection interval Δt.

probability $q_s^I(t)$ can be viewed as indicating that up to time $t = k\,\Delta t$ the undesired system behavior with possible serious consequences for the plant or the environment has not occurred. The probability that there will not be system failure in any of the k inspection intervals is

$$\begin{aligned} P_S^I &= (1 - q_{S\Delta t}^I)^k \\ &= [1 - q_s(\Delta_t)]^k, \end{aligned} \tag{4-55}$$

and thereby

$$\begin{aligned} q_S^I(t) &= 1 - P_S^I(t) \\ &= 1 - [1 - q_s(\Delta t)]^k \\ &= 1 - [1 - q_s(\Delta t)]^{t/\Delta t}. \end{aligned} \tag{4-56}$$

For $x \ll 1$, the following is generally valid:

$$1 - x \approx e^{-x}. \tag{4-57}$$

When this is applied to Equation (4-56), the result for $q_s(\Delta) \ll 1$ is

$$q_S^I(t) \approx 1 - e^{-q_s(\Delta t)t/\Delta t}. \tag{4-58}$$

If

$$\lambda_S^I = \frac{q_s(\Delta t)}{\Delta t} \tag{4-59}$$

is written, then instead of Equation (4-58), the result is

$$q'_S(t) \approx 1 - e^{-\lambda'_S t}. \tag{4-60}$$

This equation will formally have the form of a life function with constant failure rate. λ'_S can be viewed as the system failure rate in the case of a preset inspection interval. Such a view will of course make sense only when inspection intervals are small in comparison to the entire operation period to be considered. For the "OR" linkage, the following emerges from Equation (4-56):

$$\begin{aligned} q'_S(t) &= 1 - [e^{-\sum \lambda_i \Delta t}]^{t/\Delta t} \\ &= 1 - e^{-\sum \lambda_i t}. \end{aligned} \tag{4-61}$$

In the case of an "OR" linkage, it follows from $q'_S(t) = q_S(t)$ that maintenance in the manner considered is without influence. This is understandable if it is considered that the failure of one component can in itself lead directly to the undesired event and cannot be prevented by subsequent repairs. In real terms, results can be achieved when maintenance is preventive, possibly in such a way that wear phenomena, identified before component failure, are eliminated. Related to the theoretical formulation observed here, this means that preventive maintenance can keep component failure rates at a low level.

For the "AND" linkage, the following emerges from Equation (4-56):

$$q'_S(t) = 1 - \left[1 - \prod_i (1 - e^{-\lambda_i \Delta t})\right]^{t/\Delta t}. \tag{4-62}$$

If $\lambda_i \Delta t \ll 1$, the use of Equation (4-57) can be simplified to

$$q'_S \approx 1 - \left[1 - \prod_i \lambda_i \Delta t\right]^{t/\Delta t}$$

$$\approx 1 - \left[\exp\left(-\prod_i \lambda_i \Delta t\right)\right]^{t/\Delta t}$$

$$\approx 1 - \exp\left[-\frac{t}{\Delta t} \prod_i \lambda_i \Delta_t\right]. \tag{4-63}$$

Analytical representation of the influence of inspection and maintenance measures are in many ways useful for the estimate and show the principles of the handling of these influences.

In working with complex maintenance measures, in particular when inspection intervals vary for different components or vary statistically and repair periods must be taken into account, derivation of the failure probability directly from the system function derived from the logic structure of the system is no longer possible. Complicated analytical observations and computation techniques are required, for example, mathematical simulation methods as mentioned in Section 3.4.

To uncover weak points, it is often inappropriate to ascertain occurrence probabilities q_s from the logic structure via the system function for the undesired system behavior as a whole, but to observe the essentially differing event sequences separately.

In a representation of the logic structure in a compressed decision table, this corresponds to a consideration of individual columns. This also shows when which columns determine the critical state concerning the event sequence dependent on time. This is shown by way of an example in Figure 4.18. Subsequently quantitative safety analysis is applied to the leaching solution settling tack system.

Quantitative Analysis of the Leaching Solution Settling Tank. The point of departure for the quantitative leaching solution settling tank observation is the compressed decision table (Table 4.15). It is identical to Table 4.11 when column 7 is left out of the considerations. Maintenance measures are not taken into account. The slides and cut-off armatures (components 18–30) are, according to [4-32], assigned a failure rate $\lambda_i = 12 \times 10^{-6}/\text{h}$. For the temperature sensor 31T the literature [4-32], [4-33], [4-35] provides a failure rate $\lambda_{31T} = 160 \times 10^{-6}/\text{h}$. For the alarm concerned, 31A, according to [4-33], $\lambda_{31A} = 12.50 \times 10^{-6}/\text{h}$, and for the control valve on the steam line controlled by the temperature sensor, according to [4-34], [4-35], [4-36], a value of $\lambda_{32} = 25 \times 10^{-6}/\text{h}$ was set. In the case of steam pressure

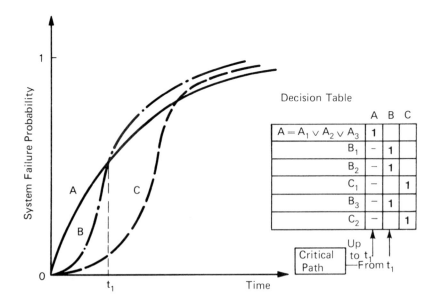

Figure 4.18 Time dependence on critical path.

Table 4.15 Reduced decision table for the system leaching solution settling tank

Component	Column 1	2	3	4	5	6
$18 \vee 22 \vee 25$	1	—	—	—	—	—
27	—	1	—	—	—	—
28	—	1	1	—	—	—
30	—	—	1	—	—	—
W2	—	—	—	1	1	1
P_D	—	—	—	1	1	1
31T	—	—	—	1	—	—
31A	—	—	—	—	1	—
32	—	—	—	—	1	1
M4	—	—	—	—	—	1

P_D, the event frequency λ_p^* is to be inserted for the appearance of excess pressure. On the basis of experience with pressure load variations found, only an excess of P_{Dperm} in 10 years was assumed. This corresponds to an event rate $\lambda_p^* \approx 10 \times 10^{-6}$/h related to an hour. Probability q_{W2} for failure of the tank material W2 in the case that critical temperature T_{perm} is exceeded due to steam excess pressure can be derived from rupture strength. Accordingly, the result is a value of 3×10^{-5} [4-26]. Finally, the probability that the service personnel does not heed or does not react correctly to an alarm released via the temperature measurement or the connected alarm 31A is set at $q_{M4} = 10^{-3}$. This figure is based on [4-33]; there the failure rate is stated as one operating error per average of 10^3 service actions.

The behavior of components 18–30, 31T, 31A, and 32 is described by a constant failure rate, so that according to Equation (4-49), the probability of failure having occurred by time t is given by

$$qi(t) = 1 - e^{-\lambda_i t}. \tag{4-64}$$

If we note that in the decision table the component states recorded in one column are subject to "AND" linkage, the result for the first three columns according to Equations (4-51) and (4-52) will be

$$q_{S1}(t) = e^{-t(\lambda_{18} + \lambda_{22} + \lambda_{25})}, \tag{4-65}$$

$$q_{S2}(t) = (1 - e^{-\lambda_{27} t})(1 - e^{-\lambda_{28} t}), \tag{4-66}$$

$$q_{S3}(t) = (1 - e^{-\lambda_{28} t})(1 - e^{\lambda_{30} t}). \tag{4-67}$$

The system failure assigned to column 4 can be viewed as follows. The condition for system failure is the conjunction of the repeatedly possible pressure excess of steam pressure P_D over the permissible boundary value with failure of component 31T. If $q_{S4}^*(t)$ designates the probability that this

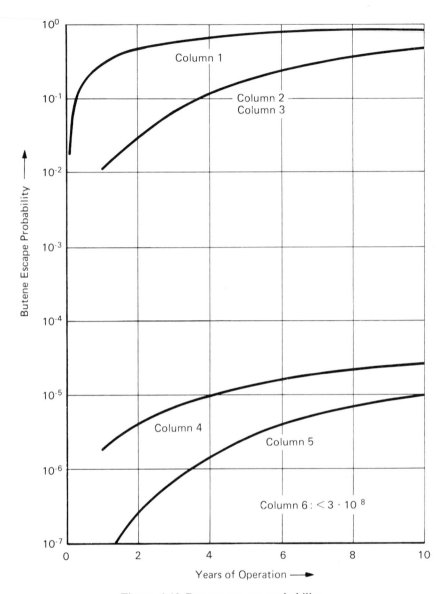

Figure 4.19 Butene escape probability.

coincidence happens at least once up to time t, then for increase Δq_{S4}^* in the time interval t, $t+\Delta t$, the following will be valid:

$$\Delta q_{S4}^* = [1 - q_{S4}^*(t)]\lambda_p^* q_{31T}^*(t) \cdot \Delta t \qquad (4\text{-}68)$$

or at $\Delta t \to 0$

$$\frac{dq_{S4}^*}{dt} \frac{1}{1 - q_{S4}^*(t)} = \lambda_p^* q_{31T}^*(t). \qquad (4\text{-}69)$$

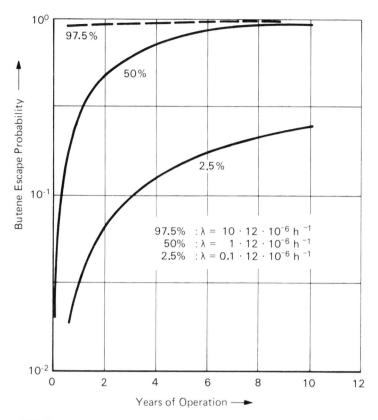

Figure 4.20 Uncertainty limits for system behavior according to column 12 of the decision table in Figure 4.18 on the basis of uncertainty of failure rate λ of components 18, 22, and 25.

The solution of this differential equation is

$$q_{S4}^*(t) = 1 - \exp\left[- \int_0^t \lambda_p^* q_{32T}(t')dt \right].$$ (4-70)

The probability of the system failure occurrence by time t will then be

$$q_{S4}(t) = q_{W2}q_{S4}^*(t).$$ (4-71)

Analogously an exact representation of q_{S5} and q_{S6} is obtained. For an estimate on the safe side, a simple approach will be adequate. The probability $q_p(t)$ that pressure P_D will have exceeded permissible pressure at least once by time t is

$$q_p(t) = 1 - e^{-\lambda_p^* t}.$$ (4-72)

Furthermore, the following is valid:

$$q_{S4}^*(t) < q_p(t)q_{31T}(t)$$ (4-73)

and, in general,

$$1 - e^{-x} < x \qquad \text{for} \quad x > 0. \tag{4-74}$$

Thereby the result is obtained:

$$q_{S4}(t) < q_{W2}q_p(t)(1 - e^{-\lambda_{31T}t} < q_{W2}\lambda_p^* t(1 - e^{-\lambda_{31T}t}) \tag{4-75}$$

and correspondingly

$$q_{S5}(t) < q_{W2}\lambda_p^* t(1 - e^{-\lambda_{31A}t}) \cdot (1 - e^{-\lambda_{32}t}), \tag{4-76}$$

$$q_{S6}(t) < q_{W2}q_{M4}\lambda_p^* t(1 - e^{-\lambda_{32}t}). \tag{4-77}$$

The result of numerical evaluation is shown in Figure 4.19. Particular weak points appear in columns 1–3. They especially include failure of cut-off armatures, in which case escape of the operation medium into the open air is possible. For columns 5 and 6, occurrence probabilities of the undesired event are lower by several orders of magnitude. This relation is not shifted in the event that the input data are not very accurate. The influence of failure rate uncertainty on components in column 1 is shown in Figure 4.20, while the failure rate of components 18, 22, and 25 has been multiplied once by 0.1 and once by 10. Quantitative evaluation in this case clearly confirms the result of qualitative analysis.

The example of the leaching solution settling has shown that for systems with safety importance, quantitative analyses should be superimposed on qualitative observations. If in the future there is more effort made than in the past to procure an adequate data base for component failure rates, this will favor more intensive utilization of quantitative analysis, which is desirable for safety technology.

Chapter 5

The Machine as a Safety Factor

The machine constitutes one of the three major subsystems of the cybernetic total system of man–machine–environment. "Machine," in this connection, denotes a "technical entity" of any type and size (Figure 5.1). It becomes a safety factor solely because of its interaction with man and his environment. The man–machine relationship actually begins during the planning phase of the machine during which the machine's functions and its mode of application are being determined and the necessary conclusions drawn concerning its form and effectiveness. The interrelationship between man and machine continues during the various phases of construction and operation. During the aforesaid phases, man, as a safety factor, has far-reaching possibilities of influencing the machine's properties and characteristics as a safety factor.

A brief look at history reveals that the introduction of new technical methods or the use of new machinery frequently entails a painful, often time consuming, and expensive learning process whereby the results of this process are finally evaluated in a positive manner and are passed on as "experience." However, the danger potentials of modern machines often demand learning processes of quite a different nature. For instance, whenever huge amounts of energy or strongly toxic substances are being used, gaining "experience" from accidents under the "trial and error" method is decidedly more risk prone and costly. Thus it will be more meaningful if the machine, in its capacity as a safety factor, is theoretically explored throughout all phases of planning, construction, and application. Here the machine is regarded as a system subjected to a certain stress and which reacts to that stress with a specific operational behavior. Operational load, in most cases, is therefore not expressed by a single magnitude (e.g., pressure or temperature) alone, but by a number of influencing factors which, moreover, may be interconnected in various ways. The machine's reaction to operational stress expresses itself in its operational behavior. A special characteristic of this operational behavior is its tendency to elude any constant monitoring. In contrast to, e.g., the operational stress factors of pressure or temperature, behavior cannot be measured continuously. In most cases, operational behavior can only be observed in a general way,

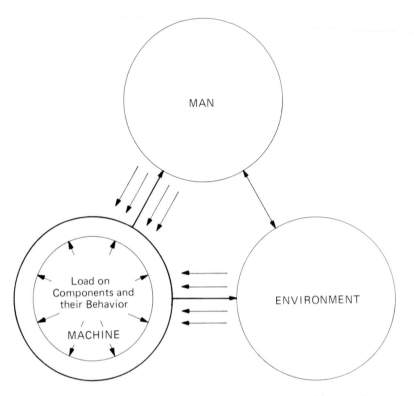

Figure 5.1 The function of the machine within the man–machine–environment–systems.

whereby it is determined whether the machine is capable of fulfilling the requirements demanded of it, fully, in part, or not at all. The point in time when the machine will fail to fulfill the requirements can only be estimated. Therefore, often very careful and time-consuming preparatory examinations will be necessary and numerous observations must be made of its operational behavior and evaluated by means of mathematical-statistical methods. Below we shall try to point out ways and means by which the operational load and the operational behavior of the machine can be determined and evaluated. This determination and evaluation is indispensable for a safety analysis of the type described in Section 4. It will provide access to important safety-related data, e.g., failure rates and/or failure probabilities.

5.1 The Machine under Operational Stress

5.1.1 The Meaning of Operational Stress

A part, a component, or even the entire installation will only fail in such instances where it has been subjected to a stress of intolerable magnitude.

Future stress conditions should therefore be known as accurately as possible during the planning phase and should be taken into consideration accordingly.

While the installation is operational, parts or components, constantly or alternately, are usually either under excessive or insufficient stress. Therefore, after starting the machine, the determination and evaluation of operational stress for the purpose of matching planning and operational data is of the utmost importance. Under operational conditions it is vital that the stress factor be limited in such a way that the design configuration matches the application configuration.

For a systematic survey it is expedient to limit the determination of the possible types of stress in a machine to only a few categories. The above-described method deals with stresses of a primary nature which are directly connected to the operation of the installation and can thus be anticipated with relative ease. These include operational pressures and vibrations.

Other types of stresses are the result of specific operational conditions which cannot be fully taken into consideration during the planning stage. These include the influence of media of different types as well as environmental influence factors.

Pressure and temperature are stresses which, in most cases, can already be determined during the design phase of the machine with a high degree of accuracy. An accurate evaluation of the influence of vibrations, however, will only be possible when the machine is known in all its details. As far as the influence of working media and the impact of the environment is concerned, one may safely say that they can be considered in their entirety during the planning phase only when the type of application of the machine can be predicted with a satisfactory degree of certainty.

Usually the above-mentioned stresses occur simultaneously. However, in most cases, it will not always be possible to find a positive allocation of the stresses to one of the above-mentioned stress categories. Thus, e.g., corrosion can be attributed to the influence of a medium; however, environmental conditions like high humidity in the air or corrosive atmosphere may also lead to corrosion. In specific cases, an evaluation of the operational stress of a machine may constitute an extremely difficult undertaking.

5.1.2 Description and Effects of Operational Stresses

Dynamic Pressure Stresses. Below, methods for the determination and evaluation of pressure and temperature stresses are explained by means of examples. Temperature and pressure stresses have been chosen because these types of stresses are by far the most frequent and are thus of special interest.

Actually, machine parts or components, for reasons of simplification, are often designed for static operation only, although a fluctuating stress pattern may be expected. Therefore, instead of a static stress, pressures whose amplitudes and frequencies are subject to fluctuations will occur. These may be displayed with the help of stress-time functions. Given the above-described conditions, components subjected to dynamic stresses will fail earlier than those under static stresses. The following is meant to show how examinations of components in such cases may be conducted with maximum expediency.

As an example, Figure 5.2 shows the pressure curve at the exit of a pumping station in a pipeline.

According to Buxbaum [5-1] it has proven expedient for the evaluation of stress-time functions to determine the frequency of exceeding predetermined category limits (class limits) for the various stress levels. As an example, this method is represented in Figure 5.3. The entire range over which the operational pressure may move is broken down into classes of a specific width. The class range allocated in each particular case should be chosen in accordance with the course of the frequency distribution to be expected. In case experience is lacking in this field, the starting point should be a class width of 2.5–5.0 bar and should be kept constant across the entire pressure range. The individual classes are allotted the values of the upper pressure limits. A record is kept of how often, during a given period of time, the operational pressure exceeds class limits in the direction from lower to higher values. Often, one day is considered a suitable period of observation time. From the overrun frequencies of the pressure class limits determined in the above-described manner, two cumulative frequen-

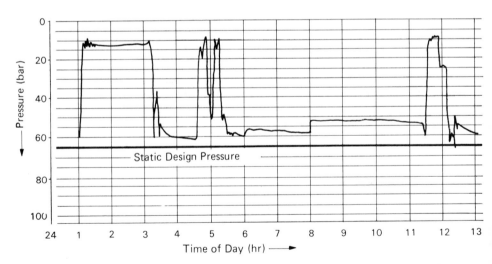

Figure 5.2 Recording strip of the pressure curve at the exit of a pump station.

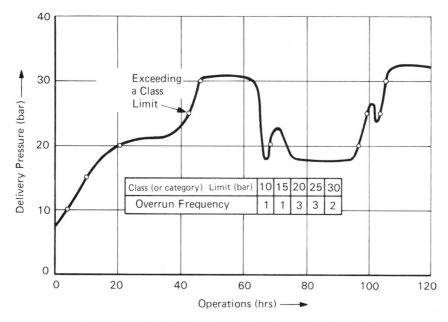

The table within the figure:

Class (or category) Limit (bar)	10	15	20	25	30
Overrun Frequency	1	1	3	3	2

Figure 5.3 Evaluation of fluctuations in operational loads.

cies of the pressure overrun can be determined for each class, depending on whether the overrun frequencies are summed from the lowest or from the highest values. The first is constantly increasing, the other constantly decreasing. Hints as to the above-described method and other counting methods as well can be taken from the voluminous summarizations and explanations by Buxbaum [5-2].

The result of the count is best represented by a simple logarithmic system of coordinates. Figure 5.4 shows the delivery pressure linearly plotted on the ordinate, and the two cumulative frequencies of exceeding the individual class limits are logarithmically plotted on the abscissa (x-coordinate). The two cumulative frequencies are plotted as emanating from the highest or the lowest determined pressure values and are graphically represented up to the point P_m where they intersect. This intersection will be located in the vicinity of that specific class limit which has either been reached or exceeded most frequently. This type of plotting will produce the so-called operational load collective. It allocates to a specific quantitative frequency a specific pressure amplitude P which is located in between the upper and lower load collective (Figure 5.4). The operational load collective is characterized by the following salient points:

—The median pressure of the delivery pressure fluctuation P_m. This is the

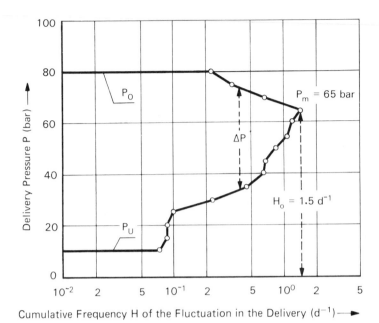

Figure 5.4 Representation of an operational load collective.

pressure which is most frequently attained or exceeded. P_m is allocated to the cumulative frequency H_0.

—The upper pressure limit of the delivery pressure fluctuation P_0. This is the pressure which is frequently determined by the efficiency of the pressure protection device.

—The lower pressure limit of the delivery pressure fluctuations P_u. This is the pressure which has been preset in accordance with operational requirements, e.g., at the entrance to a pumping station, a certain admission pressure must be kept constant.

In its present form, the operational load collective will not permit any conclusion as to the behavior of a correspondingly stressed component.

The result, therefore, is that the behavior of components under dynamic pressure load is determined by so-called "Wöhler tests" (S-N tests). Here, pressure containers during lab tests are subjected to constant pressure amplitudes P of a certain magnitude until they finally fail. If such tests are conducted on identical components using different amplitudes until the component in question fails and the results are plotted in a double-logarithmic system of coordinates having a coordinate pressure amplitude ΔP and a cycle life N, then the result will be a straight line known as the "Wöhler line" (S-N line) which can be represented by the following equation:

$$N = G_1 \cdot \left(\frac{\Delta P}{C_2}\right)^{-k}, \tag{5-1}$$

$$\left.\begin{array}{c} C_1 \\ C_2 \\ k \end{array}\right\} \text{ Constants.}$$

This line displays a course which is component and material dependent and must thus be known beforehand if the behavior of certain components is to be evaluated. Tests have proven that the gradient of the Wöhler line, e.g., for welded pipe, is $k = 4{,}5$. Now the problem is to effect a combination of the information contained in the Wöhler line and the operational load universe; this may be achieved in the following manner.

Assume that there is an operational load collective $\Delta P = \Delta P(H)$ plus a Wöhler line which can be represented by Equation (5-1). The deterioration quotient αS of the pressure amplitudes ΔP of a narrow load collective range between $\Delta P(H)$ and $\Delta P(H + dH)$—according to the linear deterioration hypothesis by Palmgren and Miner [5-3]—is indicated by the frequency dH of the amplitudes within that range and the cumulative number of stress cycles which the latter will sustain, namely $N = C_1(C_2\Delta P(H))^k$:

$$dS = \frac{dH}{N}. \tag{5-2}$$

The total damage or deteriorations caused by the operational load can be ascertained by integrating Equation (5-2) via the amplitudes ΔP contained in it or because of the definite interaction in both directions between amplitudes ΔP and cumulative frequency H—by means of integration via H between the limit values $H = 0$ and $H = H_0$:

$$S = \int_0^{H_0} \frac{dH}{N}. \tag{5-3}$$

If the pressure amplitude ΔP_k, which represents the load collective, is defined as the one which—if applied with a frequency H_0—produces the same degree of deterioration as the universe, then the following equation applies:

$$S = \frac{H_0}{N_K} = \int_0^{H_0} \frac{dH}{N}. \tag{5-4}$$

Here N_k constitutes the endurable number of stress cycles at pressure amplitude P_k as represented by the Wöhler line. Using (5-1) will lead to the following conditional equation for the representative pressure amplitude ΔP_k:

$$(\Delta P_K)^k = \frac{1}{H_0} \int_0^{H_0} (\Delta P(H))^k \, dH. \tag{5-5}$$

In general it will prove useful to approximate the load collective $\Delta P = P(H)$—as demonstrated in Figure 5.5—by a suitable step function (5-5). At this point, the aspects according to which it should be selected can be neglected. Accordingly, the following equation, with the definitions taken from Figure 5.4, will apply:

$$(\Delta P_k)^k = \frac{1}{H_0} \sum_i \int_{H_i}^{H_{i+1}} (\Delta P(H))^k \, dH \qquad (5\text{-}6)$$

$$\approx \frac{1}{H_0} \sum_i (\Delta P_i)^{k h_i}. \qquad (5\text{-}7)$$

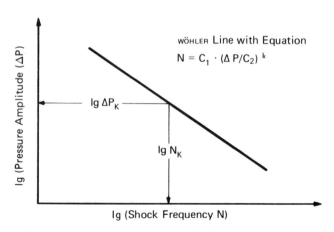

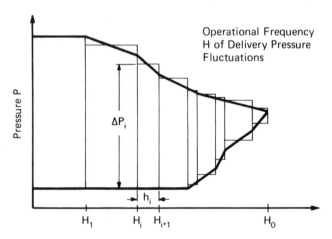

Figure 5.5 Operational load collective and Wöhler line.

Naturally, $H_0 = \sum_i h_i$. Thus,

$$\Delta P_k = \left(\frac{\sum_i h_i \Delta P_i^k}{\sum_i h_i}\right)^{1/k} \tag{5-8}$$

provides the equation which permits the determination of the representative amplitude ΔP_k which will produce the same type and quantity of deterioration in the stressed component as the given operational load collective (5-4).

From practical experience, the representative amplitude ΔP_k is then determined from the operational load collective so that the endurable number of stress cycles N_k can be read from the Wöhler graph (Figure 5.5).

With the above, it will be possible to take into account any dynamic stresses and to forestall any danger of the components failing at a stress magnitude which, as far as its maximal magnitude is concerned, lies distinctly below the statistical design pressure.

Temperature Loads. For steel grades, the connection between material stress and temperature stress as a function of time is often represented as in Figure 5.6. In a double-logarithmic system of coordinates with the axes "stress" and "component life," the interrelationship between stress and time until failure can be represented by a set or family of straight lines, whereby temperature serves as a parameter.

The connection between component life until failure and temperature at constant stress can be represented within a limited range—wherein the straight lines are running parallel to one another and the distance between them shows a linear increase in accordance with the logarithm of temperature—by the interrelationship shown in Figure 5.7. When observing a

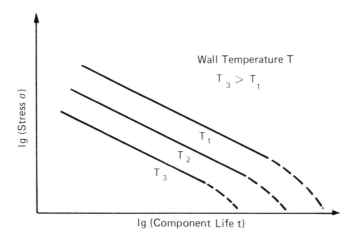

Figure 5.6 Creep stress depending on time (schematic).

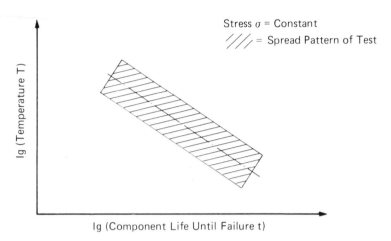

Stress σ = Constant

///// = Spread Pattern of Test

lg (Temperature T)

lg (Component Life Until Failure t)

Figure 5.7 Interrelationship between operational temperature and component life (until failure) (stress = constant).

component which, due to the nature of its application, is subjected to temperature fluctuations which, in turn, influence the behavior of the material, temperature changes may be treated analogously to pressure changes. According to Figure 5.7, the interrelationship between the temperature T and the life t of the material can be expressed by the following equation:

$$\log t_z = A - k \cdot \log T. \tag{5-9}$$

Here A and k are constants which are the results of tests on the material in question and represent a function of stress. This law governing materials according to Equation (5-9) can be compared to Equation (5-1). Since, however, the hypothesis of deterioration is usually also applied to components under creep-to-rupture stress depending on time, the following applies in analogy to Equation (5-3):

$$S = \int_0^{t_g} \frac{dt}{t_z} [T(t)]. \tag{5-10}$$

Here t stands for the current operational life of the component under a temperature $T(t)$ and t_g represents the entire operational time. As in Equation (5-4), here the regression equation

$$S = \frac{t_g}{t_r} = \int_0^{t_g} \frac{dt}{t_z} \tag{5-11}$$

will also apply. t_r is the time during which the representative temperature T_r can be sustained according to Equation (5-9). For the representative temperature magnitude T_r, which characterizes the temperature curve during operation, the following equation applies in analogy to Equations

(5-5) through (5-8):

$$T_r = \left[\frac{\sum (t_i \cdot T_i^k)}{\sum t_i}\right]^{1/k}. \tag{5-12}$$

Here T_i is the temperature of the i temperature step, t_i is the duration of the time period allotted for the i temperature step, and k is the slope of the linear function between the logarithm of temperature and the logarithm of component life whereby the logarithm of operational stress serves as a parameter. The aforesaid magnitude is dependent on the type of material.

In this manner it is possible to determine a representative temperature value for a temperature collective. A prerequisite for this is that the stress on the material be constant. Once the representative temperature T_r has been determined, the failure probability of the component can be estimated with the help of the operation time t_g which is derived from $\sum t_i$. For the representative temperature T_r computed from Equation (5-12), the pertinent mean service life t_z of the material can be determined.

Assuming that the logarithms of the service-life values t determined in tests at constant temperatures present a spread pattern around the mean value log t_z according to a Gaussian error distribution curve, the failure probability W_A of a component subjected during an operational period $t_g = \sum t_i$ to a temperature T_r can be computed from the following equation:

$$W_A(t_g) = \frac{1}{\sqrt{2\pi}} \int_{-\infty}^{u_g} \exp\left(\frac{-u^2}{2}\right) du. \tag{5-13}$$

Here

$$u_g = \frac{\log t_g - \log t_z}{S}. \tag{5-14}$$

The scatter pattern S characterizes the width of the spread band (or zone) depicted in Figure 5.7.

The following, with the help of an example, demonstrates the influence of operationally conditioned temperature fluctuations on the failure behavior of the superheater pipes of a steam generator [5-6]. The following are the specifications of the pipes under test: i.d. $= d = 57$ mm, wall thickness $= s = 13.5$ mm, material 10 CrMo 910. The pipes have been designed for a temperature of 550°C and an interior pressure of 200 bar. Four different modes of operation are being considered which are characterized by the following temperature loads:

1. The operational steam temperature fluctuates below the design temperature, namely between 530 and 505°C.
2. The operational steam temperature fluctuates around the design temperature, between 505 and 545°C.
3. The operational steam temperature remains constant at 350°C, i.e., 5°C above design temperature.

4. The operational steam temperature fluctuates between 530 and 545°C, i.e., above design temperature.

The results of the computations made with the help of Equations (5-9) through (5-13) have been represented in Figure 5.8. In the left half of the picture, the temperature collectives for cases 1 through 4 are shown symbolically. It can be seen quite clearly that the deviations from design temperature exert a considerable influence on the failure behavior of the pipes. Thus in case(1)—operation below design temperature—the failure probability of the individual pipe is lower by a power of 10 than in case(4)—operation above design temperature—and to a large degree independent of the period of time considered.

Combined Pressure and Temperature Loads During the Operation of a Reactor. The reactor considered here is a fluid catalytic cracking reactor [5-7]; its fundamental design is shown in Figure 5.9. The material is 15 Mo 3. Reactors of this type are used in petrochemical plants for the fission of heavy hydrocarbons.

The reactor (at an overpressure load of 2.7 bar and an operating temperature of 510°C) has already exceeded the computed design operation time. All pressures and temperatures during operation were extensively recorded by the utility. Now it must be determined whether continued safe operation is possible. Likewise, the conditions permitting a continued operation must be clearly defined. For this purpose, a measure (magnitude) is derived for the component fatigue from the operational

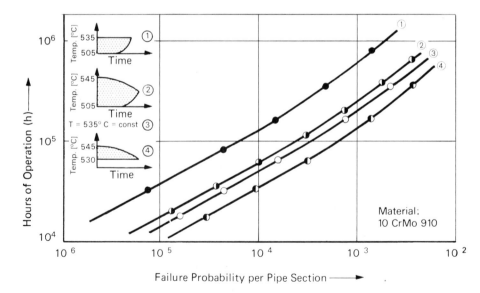

Figure 5.8 Influence of temperature on the failure probability per pipe section.

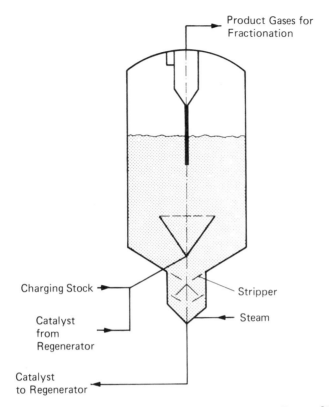

Figure 5.9 Schematic diagram of an FCC reactor according to [5-8].

load through pressure and temperature. If, based on the operational load, it is determined that a continued operation is feasible, the conditions under which a continued operation will be possible will essentially depend on the degree of reliability of an extrapolation of future operational conditions.

During 16 years of operation, design parameters were underrun for 14 years, and only during the last 2 years have prevailing operational conditions approached the design parameters. The operational conditions of those 2 years will be used in the future.

In order to be able to take into account the actual operational conditions, first any available records of the first 14 years, and, second, only the data collected during the past 2 years (since the operational temperature had increased) will be evaluated. Such a separate evaluation will help facilitate the determination of the present component fatigue and the estimation of the further course of component fatigue based on the currently valid operational conditions as well as those planned for the future. Furthermore, it is assumed that operational temperature and wall temperature are identical. From the operational data, one distribution function is deter-

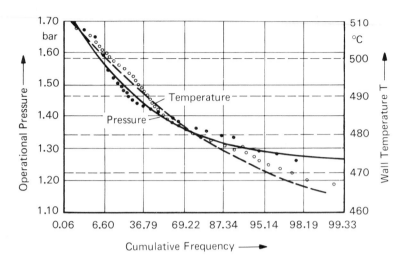

Figure 5.10 Distribution of the cumulative frequencies for pressure and temperature (operational phase I).

mined for pressure and temperature. Thereby an accurate determination of the influence of temperature fluctuations on component fatigue as well as an especially reliable extrapolation into the future will be possible. Due to the fact that the magnitudes of pressure and temperature are independent of one another, the data evaluation will produce the results displayed in Figures 5.10 and 5.11. The individual cumulative frequency values indicate the degree of probability under which a specific pressure or

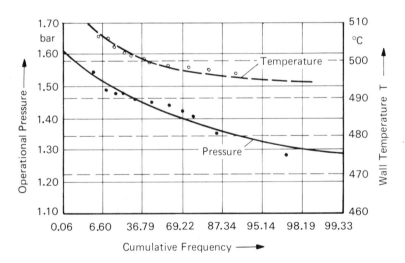

Figure 5.11 Distribution of the cumulative frequencies for pressure and temperature (operational phase II).

temperature magnitude is exceeded or preset. The curves shown here are the result of a regression analysis for these cumulative frequencies. Here the Weibull density function has been selected as a functional approach (Section 4).

$$W(X) = 1 \qquad \text{for} \quad x \leq a$$
$$= \exp\left[-\left(\frac{x-a}{b-a}\right)^y \right] \qquad \text{for} \quad x > a \qquad (b > a,\, y > 0). \quad (5\text{-}15)$$

Here $W(x)$ is the degree of probability for the occurrence of values of a random variable (stochastic variable), i.e., in this particular case, of pressure or temperature, which are greater than or equal to x. The curve parameters a, b, y, which are derived from the regression analysis, have been compiled for the individual operational phases in Table 5.1.

Figure 5.10 shows the distribution function of the wall temperature for the first operational phase. Part of this operational phase is the distribution function of operational pressure which is also represented in Figure 5.10. The expected value of a random variable within a Weibull density distribution is indicated with a tolerable degree of accuracy by parameter b, as long as parameter $y \geq 1$. Parameter b corresponds to a cumulative frequency of approximately 36.8%.

Thus, during the first operational phase, the wall temperature of the vessel showed a mean of only approximately 490°C and an allocated mean pressure of approximately 1.43 bar.

Figure 5.11 describes the operational conditions during the past 2 years (phase II). The wall temperature clearly demonstrates the change in operational conditions. During the first 14 years, the mean value, at 500°C, is approximately 10°C higher. Process control has obviously become more accurate, as can be observed from the insignificant downward variation of temperature, because the asymptote now lies at approximately 493°C compared to 450°C during the first phase.

The distribution function for the occurrence of specific operational pressures has changed in the higher pressure ranges only; in other words, compared to medium pressures, higher pressures are observed less frequently now than during the first operational phase (phase I). This may be

Table 5.1 Curve parameters of the Weibull distributions for pressure and temperature

Curve parameter	Phase I		Phase II	
	Pressure	Temperature	Pressure	Temperature
a	1.25 bar	450°C	1.25 bar	493°C
b	1.43 bar	490°C	1.44 bar	500°C
y	1.9	4.5	3.0	1.55

regarded as a restriction of the range of the possible operational pressure in the upward direction.

Given the distribution functions for pressure and temperature shown in Figures 5.10 and 5.11 in accordance with Equation (5-15), as well as the interrelationship between material stress, temperature, and mean service life of the material under test (15 Mo 3), we now can determine the component fatigue factor S. As a measure for the component fatigue factor S at specific temperature and pressure loads, the relationship between operation time t and computed service life/mean service life t_z is determined under the following stresses, whereby a linear deterioration accumulation is assumed:

$$S = \frac{t}{t_z}. \tag{5-16}$$

The contribution of pressures and temperatures from the (narrower) ranges between P and $P + dP$ or T and $+ dT$ is thus indicated by the following equation:

$$dS = \frac{dt_g(P, P + dP; T, T + dT)}{t_z(P, T)}. \tag{5-17}$$

Here $dt_g(P, P + dP; T, T + dt)$ constitutes that portion of the entire operational service life t_g during which pressure and temperature are to be found in the indicated (narrower) ranges; t_z is the mean service life under these pressures and temperatures which can be seen from Figure 5.12. With the help of the distribution functions $W(P)$ for pressure and $W(T)$ for

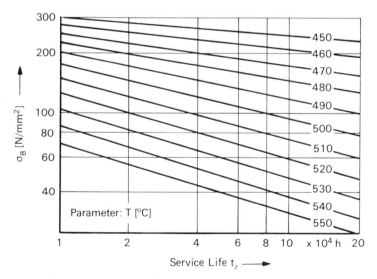

Figure 5.12 Service life diagram for material 15 Mo 3.

temperature—Equation (5-15) in connection with the parameters shown in Table 5.1—dt_g $(P, P + dP; T, T + dt)$ can be expressed as follows:

$$dt_g(P, P + dP; T, T + dT) = [W(P + dP) - W(P)][W(T + dT)$$
$$- W(T)] \cdot t_g$$
$$= W'(P)dP \cdot W'(T) \cdot t_g. \qquad (5\text{-}18)$$

t_g, in this connection, stands for the entire operational time. Therefore,

$$dS = t_g \frac{W'(P)W'(T)}{t_z(P, T)} dPdT. \qquad (5\text{-}19)$$

The totality of the deterioration or damage sustained by the component and caused by the action of these pressures and temperatures is computed by the integrating Equation (5-19) through all pressure and temperature ranges:

$$S = t_g \int_0^\infty \int_0^\infty \frac{W'(P)W'(T)}{t_z(P, T)} dPdT. \qquad (5\text{-}20)$$

In this manner, the component fatigue can first be computed during 14 years of operation and, in a second computation, the fatigue can be determined for the period of 2 more years during the operational phase II. The results of both computations are represented in Figure 5.13. From the accelerated increase in component fatigue S after 125,000 operational hours (roughly 14 years of operation), the influence of the changed operational conditions are readily recognizable.

A further computation with a 15°C safety margin in excess of the temperature measured will yield essentially higher values for the component fatigue. These results, also, are entered in Figure 5.13.

When considering the length of operation time elapsed, the results may be regarded as exceptionally favorable. The computation was completed up to an operation time of 200,000 h. Any further investigation would be futile because of the lack of knowledge of the exact sequence of the material characteristic.

When considering the results of these computations, it should be taken into account that the reactor was operated under conditions which deviate from the design data in a favorable way. Under these circumstances, a positive result of the estimate can be expected, even in cases where the operational temperature has been increased by a safety margin of 15°C in order to compensate for possible faulty measurements.

In Figure 5.13, the results of these computations reveal that, even after 125,000 h of operation, the component fatigue amounts to no more than 0.22 (=22%), considering the fact that the actual temperature was always 15°C higher than the values recorded. Therefore, operating the reactor for an additional 75,000 h under operational conditions which have been planned for the future may be well worth considering. Even after that, a

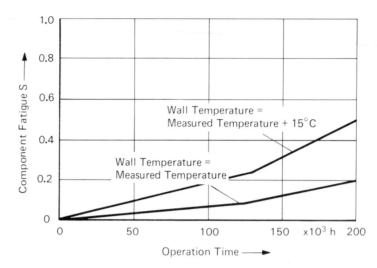

Figure 5.13 Component fatigue curve as a function of operation time.

component fatigue factor of only 0.45 (=45%) should be expected. Such a far-reaching statement clearly demonstrates how very important an exact knowledge of the operational stress in technical systems and components can be.

5.2 The Operational Behavior of the Machine

The operational behavior of a machine is, in fact, a reaction to the sum total of all operational stresses (Figure 5.14). As a rule, the machine is able to withstand these stresses for a limited period of time. During that time, a deterioration process will be in progress, which will eventually lead to the machine's failure. In Figure 5.15, processes of this type are depicted in a simplified manner. As far as fundamental reflections are concerned, the question of which of these laws is actually controlling the increase in fatigue is immaterial. The most important point is the awareness of the fact that, at the moment the machine is started up, the process of material fatigue will begin [5-22]. This way of looking at operational behavior has gained much ground during the past few decades. The consequence of these reflections is the fact that the effects of normal operation and malfunctions (Figure 5.14) cannot be regarded separately. On the contrary, the determination of the extent of fatigue should be effected in the manner shown by the example in Figure 5.15, namely by the open polygon (fatigue curve A).

Operational behavior or performance depends on influence factors, the nature and number of which are essentially determined by the machine's

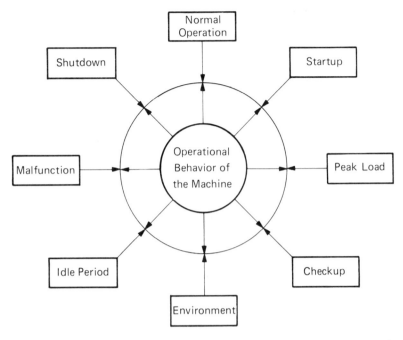

Figure 5.14 The operational behavior of a machine as a reaction to operational load.

design. If we are dealing with simple components, the number of influence factors will be smaller, and it may thus be possible to determine the machine's performance or behavior exactly. However, if a complicated machine is involved, which consists of a large number of components, determining the machine's performance under operational load will become more difficult and can often only be estimated. Apart from the complexity of the machine under investigation, its performance is also determined by random variables which are spread around certain expected values in accordance with specific laws of distribution. Finding these can never be a deterministic, but rather a stochastic, problem. This becomes evident by the example of a simple tensile test during which the material's tensile strength is being determined. According to a great number of tests conducted under identical conditions, tensile strength can only be displayed as a static magnitude.

In the above, we are, for the time being, dealing with the individual component only. In most cases, sufficient information is available on these components so that their performance or operational behavior can be described with a high degree of reliability. Thereupon follows the description of more complex systems whose performance can often be described only by basing it on simplified assumptions.

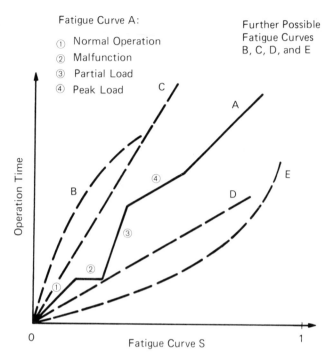

Figure 5.15 Increase in component fatigue under varying loads as a function of time.

5.2.1 The Operational Behavior (Performance) of Components and Individual Parts

In order to determine the operational behavior of a component, it will be necessary to collect the required data either from observations of the operation, from accidents, or from lab tests, in order to be able to determine reliability characteristics such as failure probability or failure rate (malfunction rate). The principles of data processing explained in Section 4 are not limited as to their mode of application; they will apply to any component. If, however, special problems occur in specific components, the above-mentioned methods should be adapted to the prevailing situation.

Below, examples for the determination of operational behavior (performance) are cited, which frequently occur in technical installations, where sufficient data are available.

The Operational Behavior (Performance) of Pipes. Pipes are essential parts of almost any technical installation where substances are being converted, processed, or transported. They are exposed to numerous media of varying states and degrees of aggressiveness. Despite the multitude of possible malfunctions, pipes, in general, are reliable components if manufactured from a suitable material and dimensioned correctly. However, once pipes are defective, accidents cannot be excluded. One especially critical defect is the weakening of the pipe wall in the direction of its axis, because this constitutes a major contribution to a reduction in the pipe's load capacity. Below, such a defect will be used as an example for the determination of reliability characteristics for pipes.

Malfunction Under Static Load. Whenever pipes are subjected to excessive stress by static interior pressure to such a degree that they malfunction, the attained stress level is closely related to the quality of the pipe in question. The static stress level KS is defined as $KS = \sigma_n/\sigma_B$, whereby σ_n stands for the rated stress in the direction of the pipe circumference, while σ_B constitutes the cracking resistance of the pipe material. The pipe quality—one measure for it is the defect magnitude in the pipe section—is described by the nondimensional crack characteristic

$$KR = \frac{(RL \cdot RT)}{s^2}. \qquad (5\text{-}21)$$

Here RL is the length of the crack, RL/s is the effective crack length ($0 \leq RL/s \leq 31.11$), RT is the depth of the crack, and s is the wall thickness.

   With the introduction of the aforementioned nondimensional characteristic KR, a sufficient amount of data have become available for statistical processing, because test results in pipes of a wide variety of different dimensions, materials, and manufacturing quality levels can be observed collectively. If such data ([5-9], [5-10], [5-11], [5-12]) are evaluated using the aforementioned characteristics, the result will be Figure 5.16. The test objects showed wall-thickness/diameter ratios of from $1:100$ to $1:4$ and were manufactured from ferritic steel grades with tensile strengths ranging from 350 to 750 N/mm^2. Figure 5.16 displays the risk quotient KR above the static stress level KS. The averaging curve was determined by means of the functional equation

$$\log(C - KR) = \log C + B \cdot \log(KS), \qquad (5\text{-}22)$$
$$KR = C[1 - (KS)^B].$$

The regression computation yields the values $C = 31.11$; $B = 0.9$. The mean squared deviation of $\log (C - KR)$ in the vertical direction amounts to $S_v = 0.1$. The failure probability W_A of a pipe having a defect corresponding to magnitude KR and subjected to a stress level of the magnitude KS_i can be determined as follows (assuming that $\log (C - KR)$, in Gaussian density distribution, is logarithmically dispersed). The crack characteristic (defect characteristic) $\overline{KR}_j$ is connected with the stress level KS_j through

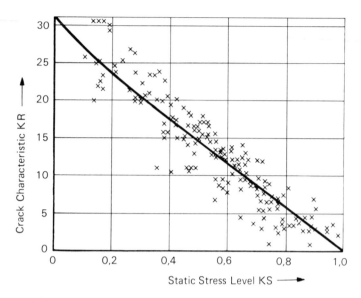

Figure 5.16 Pipe performance under static load [5-21].

Equation (5-21):

$$\overline{KR_j} = C \cdot [1 - (KS_j)^B], \tag{5-23}$$

where $\overline{KR_j}$ denotes that the pipe under a stress of KS_j will usually fail. The failure probability W_a, in the presence of KR_i, should then be determined according to the given premise from

$$W_a = \frac{1}{\sqrt{2\pi}} \int_{-\infty}^{u_0} \exp\left(\frac{-u^2}{2}\right) du \tag{5-24}$$

by means of

$$U_0 = \frac{\log(C - \overline{KR_j}) - \log(C - \overline{KR_i})}{S_v}. \tag{5-25}$$

Equation (5-23) serves to determine the failure probability of a cylindrical vessel. It can be expressed numerically if stress and defect magnitudes are known quantities. This can be explained by the following example. The pipe has a diameter $D = 600$ mm and a wall thickness $s = 8$ mm. The material is RST 370 with a yield strength $\sigma_s = 400$ N/mm² and a break resistance $\sigma_B = 600$ N/mm² (true values). The pipe is to be stressed statically with an operational pressure of $P = 50$ bar. It is assumed that the pipe has a defect which is running parallel to the pipe axis whereby its length is 100 mm and its depth 3 mm. The problem here is to determine the actual failure possibility of that pipe section. From the above data, the stress in the nondefective pipe cross section is

$$\sigma \frac{P \cdot d}{2s} = \frac{50 \cdot 600}{2 \cdot 8} \text{ bar} = 187.5 \text{ N/mm}^2$$

and the static stress level is

$$KS = \frac{\sigma_h}{\sigma_B} = \frac{187.5}{600} = 0.31.$$

According to Equation (5-22), a static stress level of that magnitude requires a mean crack characteristic of

$$\overline{KR} = 31.11[1 - (0.31)^{0.9}] = 20.1.$$

The actual crack characteristic of the pipe section with a crack length $RL = 100$ mm, a crack depth 3 mm, and a wall thickness $s = 8$ mm is

$$\overline{KR}_i = \frac{RL \cdot RT}{s^2} = \frac{100 \cdot 3}{64} = 4.7.$$

If $\overline{KR} = 20$ and $KR_i = 4.7$ are inserted into Equation (5-25), the result is a failure probability of the pipe of

$$W_a = 5 \times 10^{-7}.$$

Thus, among two million pipe sections of that particular grade, one single pipe section can be expected to fail.

This is an insignificant percentage, which means that during startup of the machine, a failure of the pipe section may be discounted. As a prerequisite for a continued safe operation, however, the stress must remain static.

Failure Mode Under Dynamic Load. Whenever a pipe is subjected to a pressure fluctuating between an upper limit P_0 and a lower limit P_u, the pipe will fail after a number of stress cycles N which will essentially depend on the geometrical layout of the defects in the pipe and the magnitude of the fluctuation $\Delta P = P_0 - P_u$ of the pressure.

For the borderline case involving a practically defect-free pipe, a great number of test results are available which are spread in a double-logarithmic system having the coordinates of material stress and number of stress cycles N around the straight line known as the Wöhler (or S-N) curve. The course of the above-mentioned straight line for seam-welded pipe sections—the type we are dealing with—has been examined extensively [5-13]. If the examination is now extended to defective pipe sections, the usual equation for the determination of the interrelationship between stress level and number of cycles can be expanded by one element which takes the defect magnitude into account. The result is a cluster of Wöhler lines (S-N lines) for pipe sections with defects which also include the defect-free pipe as a borderline case. With the magnitude of the defect increasing, the valid Wöhler curve is shifted towards the area of lesser stress cycles. The general formula for this Wöhler curve system reads as follows [5-21]:

$$\log N = -K \cdot \log \sqrt{\frac{\sigma_0^2 - \sigma_u^2}{\sigma_N^2}} + F \cdot \log (C - KR) + G, \qquad (5\text{-}26)$$

where σ_0 and σ_u are the upper and lower limits of nominal stress, respectively; stress $\sigma_N = 1\,\text{N/mm}^2$ is nothing but a nominal value and solely provides the numerical dimension for the argument of the logarithm.

Based on a multitude of test results ([5-14]–[5-20]) for submerged arc-welded pipe sections, the following constants will emerge from a regression computation:

$$K = 3.71; \qquad F = 8.79; \qquad C = 31.11; \qquad G = 0.34.$$

The spread of the logarithm of the actually sustained number of stress cycles around the mean value provided by Equation (5-26) amounts to $S_{\log N} = 0.36$. The family of curves provided by Equation (5-26) with help of these constants, as well as the test results, are represented in Figure 5.17. The crack magnitude KR was selected as the ordinate; the number of stress cycles was defined as the abscissa; the dynamic load KR of the family of curves is the parameter

$$KG = \frac{\sigma_0^2 - \sigma_u^2}{\sigma_N^2}.$$

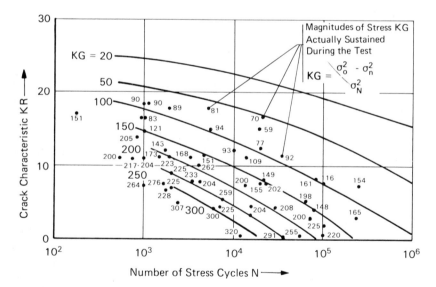

Figure 5.17 Interrelationship between crack characteristic, dynamic load, and endurable number of stress cycles [5-21].

Equation (5-25) can be practically applied as follows. The pipe section to be examined has a specific crack characteristic KR subjected to a stress KG. Equation (5-26) permits the computation of a mean endurable number of stress cycles N_0. However, if the pipe section has been subjected to N number of stress cycles, the failure probability W_a is expressed by

$$W_A = \frac{1}{\sqrt{2\pi}} \cdot \int_\infty^{u_1} \exp\left(\frac{-u^2}{2}\right) du \qquad (5\text{-}27)$$

with

$$u_1 = \frac{\log N - \log N_0}{S_{\log N}},$$

provided the logarithmized number of stress cycles demonstrates a spread pattern in accordance with a Gaussian density distribution.

The following example reveals the type of results which can be obtained with the help of Equation (5-27). The pipe section is again the same one which has been used for the explanation of static behavior. However, now it is assumed that the pipe section is subjected to a dynamic stress with a characteristic $KG = 187.5$. A crack characteristic $KR = 4.7$ has already been determined for the pipe section. During the operation time given, the pipe section is scheduled to sustain $N = 2500$ stress reversals of the stress magnitude KG. The question is: What is the failure probability of the pipe section at the end of the stress period (operation period)?

For this purpose we use Equation (5-26) to compute the mean endurable number of stress reversals N_0 by using $KR_0 = 4.7$ and $KG = 187.5$. The result is a number of stress reversals $N_0 = 23,400$. If the numbers of stress cycles ($N_0 = 23,400$ and $N = 2500$) are inserted into Equation (5-27), the result constitutes the degree of probability of the pipe section failing after the operation period during which the number of stress reversals will reach 2500: $W_a = 3 \times 10^{-5}$.

The Failure Rate of Pipe Sections. Based on Equation (5-26), it will be possible to determine the failure rate of pipes as a function of operation time, stress frequency, and pipe quality. The pipe quality is thereby characterized by the crack characteristic KR. With the help of an example, the failure rate is determined and compared to values from the literature.

Outside Diameter:	$d_a = 10^3$ mm;
Design Pressure:	$P_A = 60$ bar;
Pressure Amplitude:	$P = 10\text{--}60$ bar;
Material:	StE 360.7 TM;
Weld Evaluation Factor:	$v = 1.0$;
Safety Factor:	$S = 1.6$;
Wall Thickness s_v:	$s_v = 13.4$ mm $+ 0.5$ mm.

The result of the computation for the case $KR = 0$, i.e., the originally nondefective pipe, is displayed in Figure 5.18. The ordinate is the operation time and the failure rate is the abscissa. The stress frequency H_0 has been selected as a parameter.

The curves are computed under the assumption that the logarithms of the endurable number of stress reversals demonstrate a spread pattern according to a Gaussian density distribution. The failure rate determined from this shows a tendency to increase along with an increasing number of stress cycles. From a physio-technical point of view, this is understandable, since increasing stress will mean increased wear.

In order to demonstrate how the failure rates computed with the help of Equation (5-26) compare to values obtained from the literature, their specific area is highlighted in Figure 5.18. Any data obtained from the literature are saddled with the disadvantage of having little or nothing to say about the magnitude and frequency of stress as well as the pipe quality. For this reason, these data show a wide spread pattern (see Table 5.2). The only possibility of orientation with regard to the range of application for these data is contained in the "Remarks" column. A comparison of the data qualities stresses the importance of being thoroughly familiar with the type of data to be applied.

In the case of defective pipes, however, the failure rate undergoes a drastic change. For crack characteristic $KR = 5$, the computation was

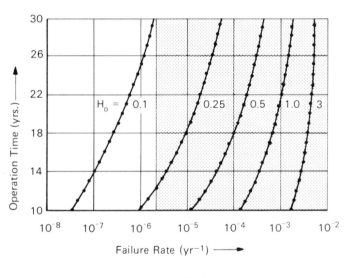

Area of Failure Rates for Pipes (Values from the Literature)

Figure 5.18 Estimate of the failure rate for nondefective welded pipe ($KR = 0$) under varying dynamic loads. Parameter: stress frequency $H_0(Zd^{-1})$.

Table 5.2 Failure rates for pipes and pipe systems (taken from the literature)

Component	Remarks	Failure rates 10^{-6} (h^{-1})	Literature
Pipe		0.2	[5-27]
Pipe system	Observation time 4.5 a	0.19	[5-27]
Pipe system	1352 Reactor years	1.44	[5-27]
Pipe system	9000 Reactor years 399 Defects of all types of defects	5.06	[5-27]
Pipe system	9000 Reactor years 19 Defects Bursting of pipes	0.24	[5-27]
Pipe system	Conventional power plants	11.6	[5-27]
Pipes	Fatigue or leakage	0.2	[5-27]
Pipe system	Cooling circuit	5.6	[5-27]
Pipes	Heat exchanger	9.4	[5-27]
Pipes	Rupture under pressure	0.004	[5-27]
Pipes	Steam system and auxiliary steam system	5.6	[5-27]
Pipes	Industrial application (DM75), rupture	0.0003...1	[5-27]
Pipes	Industrial application (up to DN75), rupture	0.002...5	[5-28]
Pipes	Nuclear application (DN75), rupture	0.0001	[5-28]
Pipes	Nuclear application (up to DN75)	0.002	[5-28]

repeated as an example. The results are displayed in Figure 5.19. Thus it becomes evident that, in the case of defective pipe, the failure rate may absolutely exceed that of nondefective pipes by two or three powers of 10.

Failure Rate and Stress Cycle Resistance. The interrelationship between the failure rate as a reliability characteristic and the conventional stress cycle resistance S_L for components under pressure—defined here as the ratio of mean endurable number of stress cycles N_0 and design stress cycle frequency N—can be obtained from the above in the following manner.

It has already been established that the results of interior cyclic pressure tests [5-14]–[5-20] demonstrate a spread pattern around the curves of mean endurable dynamic load which are similar to a Gaussian density distribution (Figure 5.17). Furthermore, it has been established that the dispersion of the logarithms of the stress cycles is practically independent of the magnitudes of the amplitudes, i.e., independent of *KG*. Under these circumstances, Equation (5-27) permits the computation of the failure probability for any combination of stress cycles N_0 and N. The result of these computations, together with the stress cycle resistance, has been demonstrated in Figure 5.20.

The dispersion $S_{\log N}$ of the individual stress cycles has been selected as a parameter. This representation may be applied in such cases where the spread patterns of the pipes under examination are known. Pipes whose cyclically increasing interior pressures show a behavior in accordance with the indicated Wöhler curves (*S-N* curves) demonstrated a spread pattern

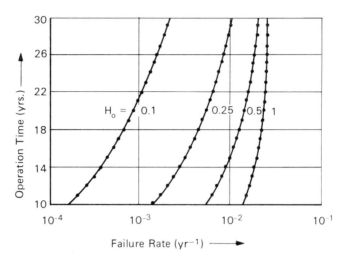

Figure 5.19 Estimate of failure rate for faultily welded pipe ($KR = 5$) under varying dynamic loads. Parameter: stress frequency $H_0(d^{-1})$.

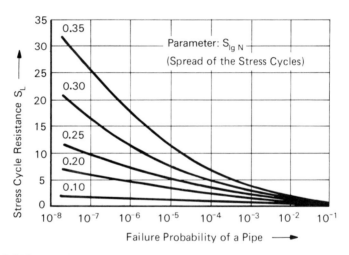

Figure 5.20 Interrelationship between stress cycle resistance and failure probability in a pressure-stressed pipe.

for $S_{\log N}$ which amounted to 0.33 after the results had been evaluated. Based on experience, the computational standard [5-36] considers a cycle reversal number of $S_L = 5$ to be sufficient if the prevalent load conditions are fully known. Therefore, Figure 5.20 tells us that a failure probability of W_A 2×10^{-4} per pipe section obviously constitutes an acceptable magnitude.

The Operational Behavior of Electronic Boards. Boards which are used as plug-in subsystems for electronic controls contain approximately 100 components in the form of transistors, resistors, diodes, and capacitors. The number of components depends on the type of board (AND board, OR gates, clock memory, etc.). Electronic boards are being used increasingly with integrated circuits or microprocessors. Only for matching connections are discrete components still being used. The carrier onto which such electronic components are mounted is a plastic board with a printed circuit. For example, up to 10,000 electronic boards are being used for controlling the operational sequence of a power station. With the help of failure rate and failure effect analyses, as well as failure rates in electronic components obtained from the literature, the failure rate of electronic components can be easily computed. Since, of a total of 10,000 electronic boards used, approximately 10 will fail, failure rates of electronic boards can be easily determined from observations over only a few years. Thus a control of the failure rates of electronic chips computed from theoretical values is possible via the evaluation of the operational behavior of electronic boards. The following description of the operational behavior of electronic boards is meant to compare the failure rate values indicated in the literature—of electronic elements as well as of components—with the failure rates which occur under actual operating conditions.

During numerous examinations, the failure data of over 40,000 electronic boards used in control systems of power plants have been collected and evaluated statistically [5-23], [5-24], [5-25]. As far as thermic stress was concerned, the environmental conditions with regard to thermic load on the structural elements were the same for all boards. The current was supplied by batteries which were recharged from the power network. The structural elements neither displayed overload conditions nor wear.

Table 5.3 displays the determined failure data of the above-mentioned electronic boards. The statistical confidence area was determined with a degree of safety amounting to 95%. The majority of the failures was found in those electronic boards which were used in the input and monitoring areas of the control system. Boards of the input level are those electronic boards marked "contact conversion" in Table 5.3. At the monitoring level, the board types "nonequivalence monitoring," "signalling card," and "display module" are used. The following failure statistics are examples of the types of failures found in boards from one manufacturer:

48% = line voltage applied too high;
19% = defective components;
17% = relays (90% Reed Relays) defective;
7% = short at exit;
5% = surge voltage carry-over;
2% = crack in strip conductor (track);
2% = fine-wire fuse defective.

Table 5.3 Failure rates of electronic cards by one specific manufacturer

Type of card	Failure rate 10^{-6}/h	Confidence area 10^{-6}/h
Nonvalency monitoring	3.2	2.3–4.3
Contact conversion 90	3.8	3.1–4.6
Contact conversion 92	0.55	0.15–1.4
Output amplifier 70	—	up to 2.6
Output amplifier 71	—	up to 11
Control card	1.3	0.8–2
Flasher unit	—	up to 33
Signalling card 90	1.9	0.9–3.4
Signalling card 91	1.3	0.6–2.3
Signal unit	0.38	0.1–1.0
Indicator module	1.4	0.4–2.3
Decision element AND/OR	0.9	0.5–1.4
Time function element, short	0.9	1.2–3.4
Time function element, medium	0.80	0.02–4.4
Time function element, long	0.65	0.03–1.3
Threshold unit	1.4	0.2–5
Decision element AND	0.69	0.2–1.8
Decision element OR 70	0.14	0.03–0.4
Decision element OR 72	1.6	0.6–3.4
Decision element OR 73	—	up to 10
Time function element drive	—	up to 1.7
Relays unit	—	up to 0.5
Dynamic memory	—	up to 164
Manual/automatic switch	3.8	1.2–8.8
Step-function unit	0.47	0.17–1
Storage unit	1.6	0.7–3
Decision unit OR/AND	0.24	0.05–0.7
Control module	—	up to 41
Console light module	0.80	0.02–4.4
Power supply	—	up to 7.5

The failure statistics consider 20 different operators or utilities. As a result of this experiment, it may be safely said that the majority of the boards had been destroyed as a consequence of careless test work and by external influences. Thus the board manufacturers have developed short-circuit-proof electronic boards to replace the old types.

Figure 5.21 demonstrates the failure behavior of 10,000 electronic cards over a certain period of time. From this it becomes evident that the average monthly failure rate of these electronic boards will amount to 8–10. During the shutdown periods of the power plant, however, the failure rate jumps to 16–55. Hidden in these figures are, of course, boards which have failed at an earlier point in time; that is, defective boards which had not yet been detected and were found during tests while the plant was shut down. Taking these uncertainty factors into consideration, the result will be a mean failure rate of 10^{-6}/h for any type of electronic board irrespective of its provenance or manufacture.

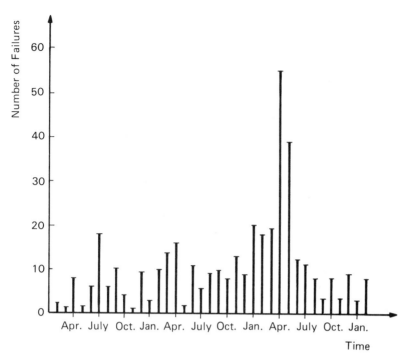

Figure 5.21 Monthly number of failures from 10,000 electronic boards.

The following tabulation (Table 5.4) compares observed and—with the help of values obtained from the literature [5-26]—computed failure rates of electronic boards. The observed failure rates for the first four board types are greater than the computed ones since these boards are being used by first item lists at the input and monitoring levels and are subject to the aforementioned deterioration through external influence factors. However, this comparison of the failure rates still contains an uncertainty factor of

Table 5.4 Comparison of observed and computed failure rates in electronic boards

Type of board rate	Observed failure rate $(10^{-6}/h)$	Computed failure $(10^{-6}/h)$
Nonvalency monitoring	0.5	—
Contact converter	3.8	0.77
Control card	1.3	0.75
Signalling card	1.3	0.66
Recording unit	0.38	1.2
Decision element AND/OR	0.9	0.58
Time function element, short	2.1	0.75
Time function element, medium	0.8	0.7
Decision element OR	0.14	0.18
Step function unit	0.47	0.48
Decision element OR/AND	0.24	0.49

only 6. This is an absolutely acceptable and usual value for failure rates. The results of Table 5.4 prove that it is possible to compute failure rates for electronic boards via failure rates of electronic components obtained from the literature, whereby the latter display sufficient consistency with the rates observed during power plant operation. Of special importance, however, is the fact that—in case of deviations—the computable failure rates, as a rule, are smaller than the rates obtained through observation. This fact should be taken into account by providing corresponding safety margins for computed rates.

5.2.2 Operational Behavior of Installations

The operational behavior or performance of a machine is determined by the interaction between its individual components. In order to fully understand operational performance, it is necessary that the logical interaction of the components is fully known. If a quantitative considera- tion—beyond a qualitative treatment—is deemed necessary, the per- formance of the individual components must be taken into account quantitatively. As was explained in the introductory remarks to Section 5.2, operational behavior or performance is represented as a fatigue process which—at a certain point in time and over the course of a specific operational sequence—will eventually lead to failure. Through observa- tion of the operational behavior we try to determine this particular point in time, and also any other possible events, in order to initiate targeted measures, e.g., changes in design, etc., which are designed to positively influence operational behavior.

The fundamental principles elaborated in Section 4, as well as in Sections 5.1 and 5.2, may be used for the examination of the operational behavior of machines. This will be demonstrated with the help of two simple examples from mechanical engineering. Qualitative as well as quantitative analysis can be applied to these examples. In both cases, the logical structures of the machines considered are determined first, followed by reflections on the determination and elimination of weak points or deficiencies. In quantitative analysis, various methods are employed and discussed briefly.

Operational Behavior of a Gas Dedusting System. Figure 5.22 shows the technical area of a major plant for the production of water gas from coal. The product gas is freed, to a large degree, of suspended dust particles by means of a dust separator and is transported towards the burner of a gas-fired steam boiler. The dust, via an intermediate dust bin, finally falls into the dust trap. For this purpose, valve (1) remains open most of the time while the installation is operational.

Once the dust inside the dust trap reaches a certain level, valve (1) will shut. The dust trap is scoured with nitrogen and the scouring gas is led to

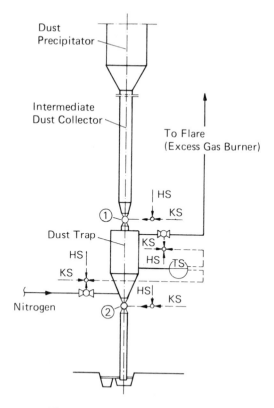

Figure 5.22 Gas dedusting plant.

the flare (excess gas burner). Scouring is necessary in order to prevent the formation of a flammable gas-air mixture inside the trap as well as to forestall any uncontrolled gas leakage from the trap. Subsequent to scouring, valve (2) opens and the dust is removed from the trap from below. Once the trap has been emptied completely, valve (2) is closed again and the filling cycle of the dust trap is resumed.

Here the following problem arises. Within the area of the installation described above, water gas is handled at high temperatures. Whenever the uncooled product gas mixes with air, it will ignite and burn. Therefore, a gas leakage must be prevented at any cost. This is why valves (1) and (2) are interlocked in such a way that (2) can be opened only after (1) has reported the "closed" position. Likewise, valve (1) can be opened only as long as valve (2) reports the "closed" position. It is of the utmost importance that the above-described system is adequately protected against any undesirable gas leakage from the dust removal zone.

This gas dedusting system will not be able to function as designed whenever valve (2), after the dust trap has been emptied, remains open because of a malfunction and the position transmitter (2) reports "Valve (2) closed."

First, the following components will have to be included into the operational behavior analysis: the mechanical design of the valve, the hydraulic control element, the control pulse, the position transmitter of valve (2), the manual control of valve (2), the operation of the manual control element by the operator from the control panel, the acoustical alarm sounded for faulty positioning, and the information on the operational mode of the valve (visual display). Furthermore, a second position indicator (2_z) is introduced as an additional component. The function of the acoustical alarm and the level indicator are dependent on the proper functioning of the position transmitter. An analysis based on the decision table (see Figure 5.23) provides the following information.

1. The functioning of the manual operation, the dust level, the acoustical alarm, and the operation of the valve from the control console are without any significance as far as the system failure is concerned. These components can thus be disregarded from the standpoint of safety unless they are important for system control.
2. The position report (2) is of the utmost importance for the system behavior. The decision chart demonstrates this very clearly, for any failure of the position transmitter is part of any state of the system which leads to the undesirable event.

Since the analysis has revealed that, of the originally considered eight components, only four will determine the system behavior, an attempt to quantitatively estimate the probability for the occurrence of the undesired event suggests itself, whereby the behavior of these four components must

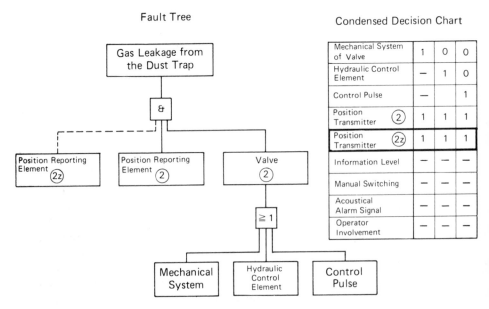

Figure 5.23 Decision chart and fault tree for a gas dedusting plant.

Table 5.5 Failure rates for the computation (interlocking of valves)

Description of component	Failure rate[1] $10^{-1}/h$
Position reporting (position transmitter)	40 (35)
Valve (mechanical)	12.5 (28)
Hydraulic control element	17.5 (28)
Control pulse	90 (35)

[1] Wherever a choice was possible, the mean value of the failure rates was selected.

be taken into consideration. The input data for this computation can be taken from Table 5.5. Figure 5.24 shows the result of this computation in which—based on the failure rates and the exponential function as a density function of the failure probability—the failure rate is determined as the integral of 0 through t at time t.

The above is meant to be an example for the way those states of systems which ultimately will lead to system failure can be determined and which

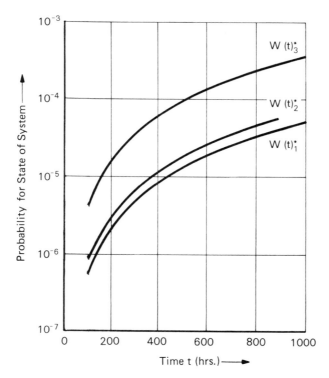

Figure 5.24 Estimating operational behavior. Note: probability factors for system states in accordance with decision chart, Figure 5.23 without additional position transmitter (2_Z).

thus constitute the critical path along which a system failure usually evolves. Figure 5.24 shows the probability curve for the prevailing system states in accordance with the three columns of the decision chart (Figure 5.23) as a function of time between 100 and 1000 hours of operation (without taking into account the additional position report 2_Z). From this it becomes evident that it is the combination of the failure of a control pulse plus the faulty transmission of a position which actually determines the critical path. The curve "state of the system 3" (column 3 of the decision chart)—over the considered period of time from 0 through 1000 h— contributes most heavily to the possibility of gas leakage from the dust trap.

Beyond the above computation, and with the help of a Monte Carlo simulation (see, also, Section 3), the degree to which the individual component contributes to a system failure was determined. The above method also permits taking into account specific maintenance strategies. The life histories of the individual components are created over a period of time from 5×10^4 h and thereby it is also determined whether the system failure can be traced to corresponding component failures at any specific point in time. Here the failure-prone components are recorded in connection with each individual failure. Results demonstrate that position transmission (2) is involved in 100% of all system failures, which is natural because of the system's logical structure (Figure 5.23). The mechanical system of the valve is involved in 12%, the hydraulic control element in 29%, and the control pulse in 59% of all cases of system failure.

According to Figure 5.24, the results of the computation are of a nature which suggest that a decrease of the failure probability is indicated. In the present case, it is safe to suggest that, together with the additional position transmission 2_Z, a redundancy be introduced which may help to improve matters decisively.

Table 5.5 shows a failure rate equal to 40×10^{-6}/h for the position transmitter. If the approximation formula

$$W = 1 - \exp(-\lambda t) \cong \lambda t \qquad \text{for} \quad \lambda t \ll 1 \qquad (5\text{-}28)$$

is employed for estimating the failure probability of the position transmitter, the probability factors $W(t)_1$ through $W(t)_3$ for the system states 1 through 3 (Figure 5.24) will decrease by the factor W because of the additional position transmitter. According to Equation (5-28), this factor assumes magnitudes which, between 100 and 1000 h, lie within the area delimited by the following equation:

$$40 \times 10^{-4} \le W \le 40 \times 10^{-3}. \qquad (5\text{-}29)$$

In evaluating the results of the computation, the absolute values of the failure probabilities should not be stressed too much, but rather their translation by the factor W, because of the introduction of the redundancy.

Operational Behavior Demonstrated in a Standby Power Plant. The plant consists of a diesel engine, the attached generator, and a number of auxiliary systems. Figure 5.25 shows the general plan of the installation. Important auxiliary systems of the plant are the fuel supply system, the lubrication system, the coolant circuits, the generator excitation system, the starter system, the air supply, and the circuitry.

Diesel plants are used in small power plants for the continuous generation of electricity and in standby power plants for the temporary generation of electric power. In the latter case, they must be able to supply, mostly at short notice, important users with power until the temporary breakdown of the regular power supply has been remedied. These plants are located as standby units in airports, radio and TV stations, hospitals, nuclear power plants, large-scale elevator systems, etc.

The following is a general survey of standby power plants. Due to the usually very reliable power supply from the public network, diesel plants in standby power units are very rarely called upon to perform. During blackouts (failure of the power supply), however, the standby power plant is expected to bridge the blackout reliably and with minimal loss of time during startup. Also, this must be achieved automatically. The de-energized or "dead" period between network failure (blackout) and emergency power supply must be as short as possible. Thus the question

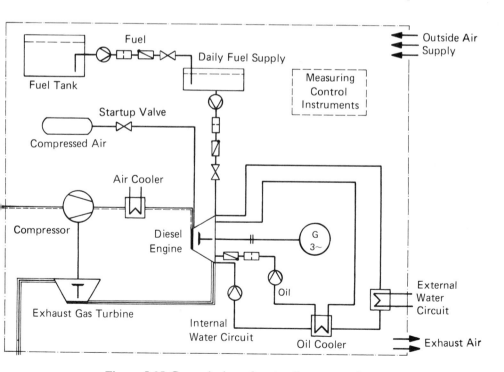

Figure 5.25 General plan of a standby power plant.

"How reliable is the emergency power generation?" is a very important one in the case of blackouts.

The literature ([5-27], [5-28], [5-29]) provides failure rates for diesel standby plants which differ by several magnitudes. However, since the question posed can only be answered with the help of quantitative analysis, a statistical analysis of the operational behavior of diesel plants is indispensable for the determination of reliable failure rates. Figure 5.26 shows how many influence factors may lead to the failure of a diesel plant. This, in turn, requires that the failure rates for standby diesel power plants be determined as to type and mode of application. For this purpose, voluminous data on the operational behavior of standby diesel plants is required. This data can never be collected from the behavior of standby diesel plants under blackout conditions alone, since these blackouts, as mentioned above, are very rare. However, there is another avenue which leads to the acquisition of failure data, namely via monthly trial runs while the public power supply is in full and normal operation.

All utilities have collected values on test runs of standby diesel plants. These have subsequently been evaluated statistically. A first paper deals with the failure statistics of more than 100 standby plants ([5-30], [5-31]). The average test run lasts approximately 1 h. According to an investigation of the average duration of an interruption in the medium voltage network in the Federal Republic of Germany [5-32], there is a high degree of statistical probability that 79% of all interruptions of the power supply last

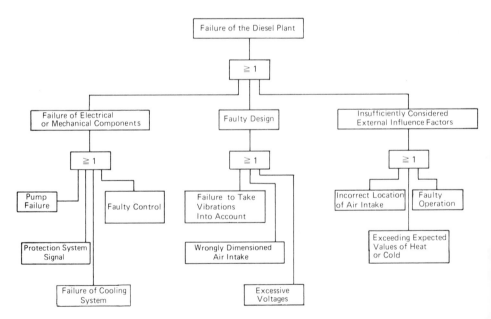

Figure 5.26 Causes of failures in standby diesel power plants.

less than 1 h. Thus the periods of operation of standby diesel plants under normal conditions (test run) and under emergency conditions (blackout) are very similar. Thus it can be expected that a statistical evaluation of the operational behavior of standby power plants will yield data which will be helpful for establishing a prognosis on their behavior under emergency conditions. The available data on failure rates of standby diesel plants during test runs and during actual emergency runs must be evaluated statistically in such a way (see Section 4.2) that mean failure rates and mean repair times can be established. This provides the basis for an answer to the question of the actual reliability of a standby diesel power plant.

First, a fault tree of the standby power plant is established. This fault tree should be quantitatively evaluated by means of the failure rates, the duration of operation, inspection intervals, and repair times. The result will be a statistically safe statement of the failure probability of the standby power plant.

Below, the operational behavior of 100 diesel plants [5-30] is described in depth on the basis of the established failure statistics. The elaborated data are included because similar examinations have not been found in the literature. The failure behavior of standby diesel plants is established by means of the failure probability at startup and the failure rate during operation under load. The evaluation of the failure statistics [5-30], [5-31] demonstrates a double time dependency of the failure rate. On one hand, higher failure rates are observed for all standby plants with short operation times per startup under emergency load conditions, and, on the other hand, small failure rates per startup are found in all plants with a relatively long operation time. The average failure rates for runs under actual emergency conditions are shown in Table 5.6. These are based on dangerous failure categories which usually occur without any prior warning. The operation time per startup is computed as the sum total of all equipment hours recorded on the service recorder (elapsed-time meter) and divided by the number of startups. On the other hand—because of the large amount of early failures—a varying degree of time dependency of the failure rates during startup and those during operation under actual emergency conditions can be observed.

During startup as well as during operation under emergency conditions, constant failure probabilities or failure rates, respectively, have only been

Table 5.6 Comparison of mean operation time per startup and pertinent failure rate

Mean operation time per startup	Failure rate ($\times 10^{-3}$/h)
50	0.1
10	0.7
1	15.0
0.1	30.5

proven in very rare cases. Figure 5.27 demonstrates the fact that constant failure magnitudes can be expected beginning at approximately 250 h of operation and 250 startups.

On the basis of the available data, the influence of the following factors was examined: operation time per startup, engine speed, number of years in operation, engine output, type of starter mechanism, and operator/utility. The influence of operation time per startup and the failure probabilities or failure rates were determined statistically with a high degree of reliability. The available number of 100 units, however, was not large enough to verify others among the numerous mentioned influence factors with a satisfactory degree of confidence. Only over the course of an examination of the type of starter could it be shown that the direct injection of compressed air into the cylinder will decrease the failure rate by approximately the factor 2 as compared to a battery start.

One and the same type of cause was blamed for a great number of failures in standby diesel plants. Failures were observed where the causative factors occurred several times during the first few hours of operation (so-called "systematic failures"). By eliminating this type of failure cause, it was possible to drastically reduce the failure rates.

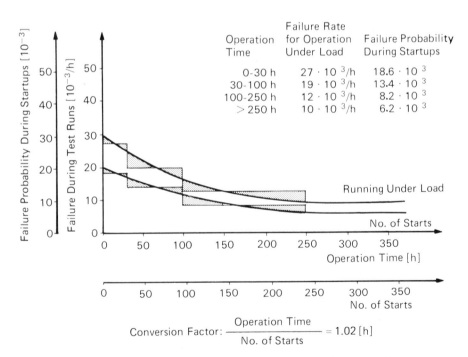

Operation Time	Failure Rate for Operation Under Load	Failure Probability During Startups
0-30 h	$27 \cdot 10^3$/h	$18.6 \cdot 10^3$
30-100 h	$19 \cdot 10^3$/h	$13.4 \cdot 10^3$
100-250 h	$12 \cdot 10^3$/h	$8.2 \cdot 10^3$
> 250 h	$10 \cdot 10^3$/h	$6.2 \cdot 10^3$

Conversion Factor: $\dfrac{\text{Operation Time}}{\text{No. of Starts}} = 1.02\,[\text{h}]$

Figure 5.27 Failure probabilities and failure rate as a function of operation (aggregates of group I: output $> 1000\,\text{kW}$; speed $1000\text{–}1500\,\text{min}^{-1}$).

Besides the above, failures triggered by the simultaneous action of one and the same cause on several units were noted. Due to these "common-mode failures," the advantage of multiple interpretation decreases. We speak of common-mode failures whenever several failures in units under observation within a system occur within a relatively short observation period and are triggered by a single cause of failure. These are caused by the controlling and simultaneous effect of a failure mechanism onto identical parts or components contained in one system. Whenever four standby diesel units with four battery-operated starters fail to start upon request because all four batteries have not been properly maintained, we are dealing with a common-mode failure. Increased reliability in systems by redundant design will not be practical to an unlimited degree because common-mode failures must always be reckoned with. In such cases, units of different designs are used as a preventive measure (diversity principle). Further remedial measures are short test cycles, adequate preventive maintenance, etc. The evaluation of the failure statistics for common-mode failures in the 100 diesel units [5-30] yields a failure probability factor of approximately 10^{-3}. Since the number of common-mode failures was small, that figure is extremely unsafe from a statistical point of view. The failure probability for the simultaneous failure of four units represents a smaller figure (4×10^{-4}) than that for the simultaneous failure of two or three units (3×10^{-3}). For the compilation of a repair statistic, 752 cases have been evaluated. If the standby diesel power plant is regarded as a single unit, 50% of these repairs took 1 to 2 h each and 90% required 55 h.

Repair time means down time, since the repair times include unavoidable waiting times. The longest repair times ranged from one week to three months. For the most important components of the diesel unit, repair times according to Table 5.7 were required. 746 failures could be statistically determined and evaluated for the compilation of a failure statistic [5-30]. These failures are allocated to 17 different component groups and subdivided into failure categories A and B. Failure category A includes failures which do not contribute to system malfunctions. Category B contains those

Table 5.7 Maximal repair times for the major components of a diesel unit

	Maximal repair time	
	50% of all cases	90% of all cases
Engine	25 h	200 h
Charger	20	170
Generator	5	1100
Fuel supply	1.5	20
Compressed air starter system	1	8
Cooling water supply	2	15
Lubrication oil supply	0.8	5
Speed control	4	56

failures which lead to a breakdown of the diesel unit. Allocation to either category A or category B is based on the assumption that blackouts bridged by the generation of emergency current from standby diesel units, as a rule, will last from 1/2 to 12 h. The decision of whether a standby diesel unit has failed or not often constitutes a matter of judgment since the majority of the failures will occur during test runs. Thus such malfunctions and failures are allotted to failure category A, which demand that the system be shut down. An emergency operation over a prolonged period of time, however, would be feasible. Malfunctions of this type would include small leakages in the cooling water circuit. Failures of category A during normal operation do not lead to a breakdown of the standby diesel unit. Failure category B includes failures which make it necessary to shut the unit down after no more than 30 min. Failures of this category will invariably lead to a breakdown of the diesel unit.

Figure 5.28 demonstrates the results of a statistical evaluation of 197 failures in 31 units with an output in excess of 1000 kW. The compilation of a failure cause chart (statistic) is of great importance for two reasons. It can be used for the computation of failure rates for individual components which, in turn, are helpful in establishing prognoses. Figure 5.29, e.g., shows that the diesel engine itself is involved in 3% of the failures. Figure 5.27, for a run under actual load for an elapsed operation time of zero

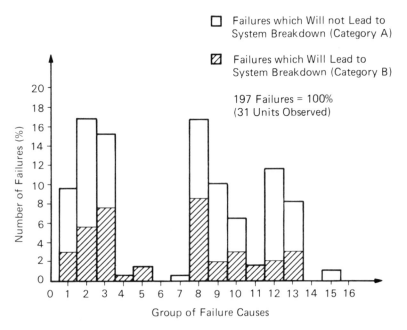

Figure 5.28 Spread of the failure causes (aggregates of group I: output, 1000 kW; speed 1000–1500 min^{-1}) (see Table 5.8).

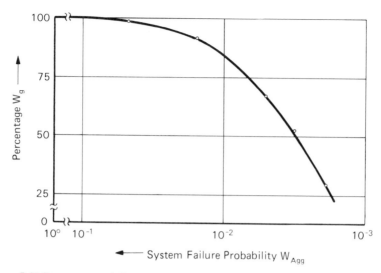

Figure 5.29 Percentage of diesel units involved in the probability of system failure.

hours, shows a failure rate of $30 \times 10^{-3}/h$. In the above example, the diesel engine thus demonstrates a maximal failure rate of $9 \times 10^{-4}/h$.

If no analyses are made with the help of prognostic methods, the failure cause statistic can be used to eliminate weak points in the system. In connection with the repair time statistics, an economically feasible stock of spare parts can be established. The present failure cause statistics demonstrate the fact that the majority of *malfunctions* are caused by defects in auxiliary systems as well as in wiring, air bottles, and filters (Figure 5.28 or Table 5.8, groups 2 and 8, respectively). The majority of *breakdowns* of units was likewise caused by these components as well as through defects in the electronic system (groups 2, 3, and 8). These groups constitute the focal point where possibilities for improvement are concerned. What, now, is the degree of influence that failure rates for standby diesel units have on the failure probability of safety systems? In order to be able to answer this question, a sophisticated fault tree was used as an example and a parameter computation was made for the standby power supply of a nuclear power plant with the help of an analytical computation program [5-34]. The fault tree took into account the combined failure causes of four diesel units, four transformers, 14 engines, and 40 power switches. Within the framework of the parameter computation, only the failure probability W_{Agg} of the diesel units was changed; however, not that of the rest of the components. (For values in the literature consult [5-27].) For the diesel units in question, an operation time under load of 1 h was assumed. The result is the percent involvement W_g of diesel units in the total failure probability of the system.

Table 5.8 Description of failure causes for Figure 5.28

1 Diesel engine: housing, cylinder heads, bearings, pistons, crankshaft, valves, valve timing
2 Auxiliary engine parts of the diesel engine: fuel injection pump with fuel line, cooling water pump, lubricating oil pump, turbocharger
3 Controls, sensors, transmitters, limit switches
4 Generator: housing, bearing, rotor, winding
5 Generator switch
6 Generator: diodes, slip rings, brushes, terminals
7 Exciter unit (mechanical or electrical malfunctions)
8 Lines and tubes, vessels, air bottles, filters
9 External auxiliary parts: compressors, pumps, fans
10 Actuators
11 Starter batteries, starter motor (for electric startup)
12 Electrical control elements, electrical wiring, fuses, voltage regulators, protective switches, etc.
13 Systematic failures: faulty planning, faulty set values, faulty measuring and controlling methods, lack of monitoring by the necessary instrumentation, faulty design, failure to consider environmental conditions
14 Coupling between engine and generator
15 Preheating of cooling water and lubricating oil
16 Other types of malfunctions/failures
17 Failures without localized cause

The following failure rates for the diesel units considered here can be seen from Figure 5.27:

Operation time	Failure rate for running under load	Failure probability Probability during starts
0–30 h	$27 \times 10^{-3}/h$	$18.6 \times 10^{-3}/h$
250 h	$10 \times 10^{-3}/h$	$6.2 \times 10^{-3}/h$

For a single start as well as for 1 h of operation time, the following failure probabilities apply for a diesel unit:

Operation	Failure probability
0–30 h	45.6×10^{-3}
250 h	16.2×10^{-3}

Figure 5.29 shows that the percentage ranges from 88% to 98%. Therefore, diesel engines constitute the system's weak point. The question of whether or not a failure probability percentage of this magnitude can be tolerated in the standby diesel unit must be determined by means of an incident/sequence or risk analysis for the entire nuclear reactor facility.

Chapter 6

Man as a Safety Factor and as an Object of Protection

Within the cybernetic system of a man–machine–environment (MMES) it is man who presents a special challenge to the science of safety. When comparing the time and resources expended through the technical sciences in order to conquer the problems connected with the component "machine", one can only wonder at the comparatively insignificant effort directed toward the elimination of the danger of accidents emanating from or caused by man. The safety of the man–machine–environment–system, however, can only be increased by utilizing the findings of human sciences if the various disciplines agree to close ranks. It must be the goal of these joint efforts to adapt technology to man to a greater degree, because he has proven to be a largely nonvariable system component; however, he has at his command a technology which, to a high degree, is adaptable to the prevailing situations and conditions.

The aim of this chapter is to elaborate on the personal characteristics of man described by the human sciences and their utilization for the planning of a man–machine–environment–system. Sections 6.1, *Medical Aspects*, and 6.2, *Psychological Aspects*, are meant to approach this problem in a detailed way. The human sciences must provide specialized information on as high a measurement level as possible.

For the safeguarding of the entire active system of man–machine–environment, exactly the same principles are applied as have been referred to in *Reliability and Suitability of Machines*. In the case of man, the pertinent procedures can likewise be subdivided into planning, control, and testing. During the planning of a man–machine–environment–system, the anatomical data of the human body and its physiological processes and functions must be taken into account as contributory prerequisites in the same way as the stress characteristics of a material used in the construction of a machine.

When planning such systems, the metropolic processes which take place in the human body must be considered in exactly the same way as the energy transformation in machines. The characteristics of psycho-neural

performance prerequisites in man, whenever used for control purposes
within the man–machine–environment system, must be regarded in the
same light as those technical components of a machine guaranteeing the
controlling functions therein. Of particular interest are the possibilities of
autonomic control in a human being which take place within certain
natural limits subconsciously. Beyond that, the fact that man has been
endowed with the faculty of being aware of his own existence is of special
interest when creating the man–machine–environment–system because,
when considering the variations of this particular characteristic, this would
mean that the limits of human autonomic control can be determined.
Hitherto human characteristics have often been neglected in the planning
of work cycles.

When planning complex man–machine–environment–systems, the natural equipment of the operator, namely man, will have to be taken into
account with some of his specific psychic and physical functions and the
possibilities of compensating for them, and in some cases minimum values
will have to be established. The science called "ergonomics" deals
specifically with the problems of workplace design as far as creating humane
working conditions, the proper performance of the technical system, and
improved safety in the workplace are concerned.

The model in Figure 6.1 shows that the important functions of man
during the operation of modern technical installations chiefly consist of the
absorption of information, the comparison of values either with each other
or with normal values, the making of decisions, and the issuing of control
commands. This basic model must be supplemented to include environmental conditions. These are physical (climate, noise, lighting etc.) as well
as social influence factors. The interaction of the individual functional

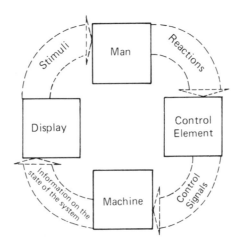

Figure 6.1 Model of feedback control system: man–machine–environment–
system.

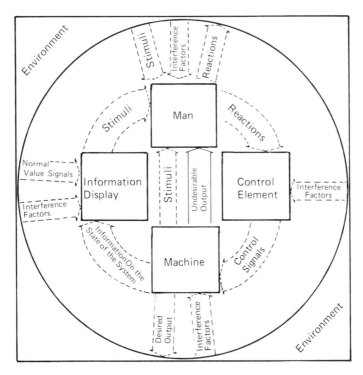

Figure 6.2 Extended feedback control system: man–machine–environment–system.

elements within the man–machine–environment–system is displayed in Figure 6.2.

Man's most important contribution within a modern technical system is his capacity to act as a processor of information. It will thus be necessary to deal with the basic principles which relate to how man receives information, selects, processes, and passes it on. This also includes information on man himself. The individual should not only be able to have a realistic view of himself and react on it but also set goals for himself which will guide his actions. Such goals, introduced into a system which has been planned by others, may have a positive or disruptive influence.

Physical and psycho-neural performance does not remain stable during the course of a day nor during a lifetime. Therefore, with a view to controlling and maintaining normal values of safety, the consequences of stress and strain to which man is subjected and which eventually lead to fatigue, should be taken into account as far as the individual organs of the human body and man as a whole are concerned. The influence of stress and strain which can lead to a decrease in the expected work performance may be equalized by planned work breaks or by recreation. One must distinguish between aspects of work organization and the maintenance of the performance requirements for a normal working life. Not all individuals have the prerequisite minimum equipment to cope with certain functions within the man–machine–environment–system. In this case they will require special training. This is covered in Section 6.4.

Accidents, insofar as they are caused by man, often result from a lack of experience in handling machines and from environmental factors. Therefore, instruction and training play an important integral role when man is introduced to work cycles and their control. Practicing safe behavior and avoiding unsafe acts must be regarded as a special task of safety science and technology. The hitherto unsolved problem of "beginner's risk" must be brought under control.

Risky and unsafe behavior which may lead to accidents, is very often "acquired" behavior which has been more deeply entrenched by "positive" experience. Therefore, on-the-job training which runs parallel with the actual job plays an important role in making certain work cycles safer. It may even require special "subsequent training" or rehabilitation measures whenever the acquired unsafe behavior occurs sporadically in smaller groups or in individuals.

The effectiveness of psychological and pedagogical approaches towards enhanced safety and the protection of man and property from damage through technical production processes are, unfortunately, impeded by political and ideological influences which began with the social sciences in the late 1960s and affected the human sciences in the 1970s. At present this situation has a negative influence on psychology and its application to work safety; this is also the case for ergonomics and occupational medicine.

6.1 Medical Aspects

When planning man–machine–environment–systems, the anatomical data and the physiological functions of man must be regarded as performance prerequisites. They may very well be compared to the stress characteristics of parts and components used in the construction of machinery. Physical performance does not remain stable during the course of the day nor throughout an entire lifetime. This applies to man in his entirety as well as to his organic systems. Fatigue and recreation are performance prerequisites which may play a decisive role in operational safety. Unlike materials used in technology, man's characteristics cannot be expressed in constant values. Bone fractures, e.g., occur at random, depending on the individual and his age, and the skin does not react to heat irradiation by predictable changes. Predictions can only be made stochastically; e.g., for the action of ionizing radiation during an acute accident. Thus, when referring to man, "constant values" are not given in this section. To maintain the uniformity of illustration all diagrams used "unless otherwise indicated" are taken from *Ergonomie* by Rainer W. Löhr [6–11].

6.1.1 The Human Organism and its Functional Capacity

Locomotor System. The skeleton consists of more than 200 bones and cartilages. In adults it determines the size and proportions of the body. The bones are points of action as well as levers for the muscles, and they protect inner organs like the brain, spinal cord, heart, and lungs.

The locomotor system can be regarded as a simple mechanical system consisting of levers and joints (bones, ligaments), energy-producing modules (striated muscles whose tendons transmit the force of the muscle onto the bones), a process control system (central nervous system), and the energy producing unit (power pack) (respiratory and digestive organs, circulation). The circulation also removes metabolic residue from the muscles.

The human skeleton consists of four different types of bones:

—long bones with medullary canals (e.g., the femur; Figure 6.3);
—flat bones with close-knitted trabecular structure (e.g., skull);
—short bones (e.g., carpal bone);
—irregularly formed bones, some with air-filled cavities (visceral cranium).

Because of the trabecular structure of bones, they can withstand stress through pressure or tensile force. The strength of bone substance lies between 110 to 170 N/mm^2 and the elasticity module ranges from 19,000

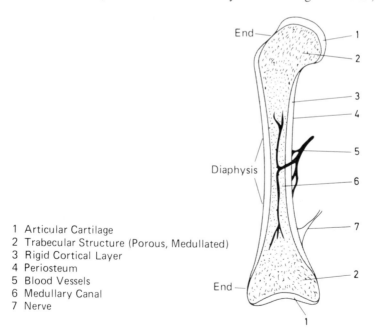

1 Articular Cartilage
2 Trabecular Structure (Porous, Medullated)
3 Rigid Cortical Layer
4 Periosteum
5 Blood Vessels
6 Medullary Canal
7 Nerve

Figure 6.3 Schematic structure of a bone (sectional view of a long bone).

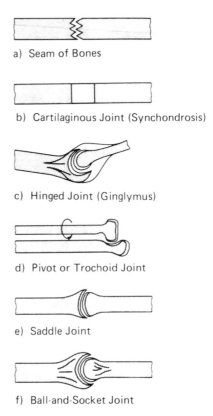

a) Seam of Bones

b) Cartilaginous Joint (Synchondrosis)

c) Hinged Joint (Ginglymus)

d) Pivot or Trochoid Joint

e) Saddle Joint

f) Ball-and-Socket Joint

Figure 6.4 Schematic view of typical joints with different degrees of freedom.

to 26,000 N/mm². At their connecting points, bones are coated with cartilage (for shock absorption, sliding motion) and form joints (Figure 6.4). In accordance with their respective degree of freedom of movement, the joints can be grouped as follows:

—synarthroses (cranium) and synchondrosis (chest) with very limited possibilities of movement;
—hinged joints or ginglymus (elbows and knees) and pivot or trochoid joints (forearm) with one degree of freedom;
—saddle joints (wrist joint at the radius) with two degrees of freedom;
—ball-and-socket joints (shoulder) with three degrees of freedom.

The entire muscular system of man amounts to approximately 40% of body weight. Muscles generate tensile forces only. The following microscopic differentiation is usually made:

—striated muscles (skeletal muscles);
—smooth muscles (flat muscles of inner organs, blood vessels);
—the cardiac muscle.

Only striated muscles are voluntary muscles. They initiate the movements of the body and determine the body's posture. Bones, joints, ligaments, and tendons are involved in each individual movement. Components, structure, and functions of the striated muscles are shown in Figure 6.5. The muscle consists of fine fibers which form groups, bundles, and funicles. The real contractile elements are the myofibrillae within which biochemical energy generates tensile forces. Muscle fibrillae consist of ultrafine muscle fibers with the protein substances actine and myosine which mesh like the bristles of two brushes stuck into one another. The muscle filaments retain their length; contractions are effected by molecular forces which cause the filaments to slide together. Whenever the muscle does not change its length, a holding function will take place. This leads to a rapid tiring of the muscle, since, while in a contracted position, the muscle cannot receive fresh energy. During prolonged muscular exertion, signs of fatigue will appear. These are caused by the accumulation of metabolic products (e.g., lactic acid) which cannot be burned off because of a lack of oxygen or cannot be removed due to interrupted blood supply. Work breaks permit a renewed supply of oxygen and thus a regeneration of the muscle. The muscle power depends on the number of muscle fibers and on the shape of the muscle. Short and thick muscles are able to generate huge forces with small movements. Larger movements are effected by long muscles.

Muscle tension with simultaneous extension entails three different forms of muscular work:

—state of rest;
—dynamic muscular action;
—static muscular action.

During normal muscular movement, static and dynamic action will often overlap. Due to the different degree of stress there is a fundamental difference between these three activities. During dynamic muscular action, the blood requirements of the muscle can be best met. But even during unilateral muscular action of certain muscles (e.g., typing) or of smaller groups of muscles, the performance—as related to the entire organism— does not lead to an increased pulse frequency which would help to improve the blood supply to the active muscles. The consequence is an accelerated localized fatigue. Changes in muscle length can be effected at different speeds and with different muscular forces. The limiting criteria are the maximal speed of the change in length using a minimum of force and the maximum force which can be generated at a low-speed change of length. In practice, the utilization of two-thirds of the maximum muscle force at one-third of the maximum achievable muscle speed will yield the optimal results. During static muscle activity, there will be no movement of limbs. Prolonged contraction against an exterior force will cut off part of the blood supply. Rapid fatigue will be the consequence. Static muscle activity

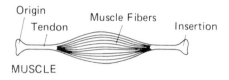

MUSCLE

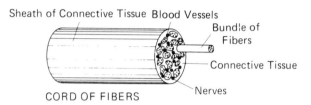

CORD OF FIBERS

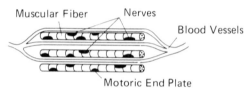

GROUP OF FIBERS

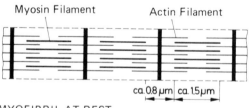

MUSCLE FIBER (20-100 μm ϕ, up to 16 cm long)

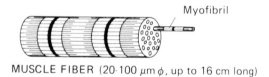

MYOFIBRIL AT REST
(Schematic Longitudinal Section, 1-2 μm ϕ,
Same Length as Muscle Fiber)

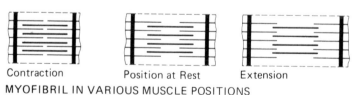

Contraction   Position at Rest   Extension
MYOFIBRIL IN VARIOUS MUSCLE POSITIONS

Figure 6.5 Structure of a striated muscle.

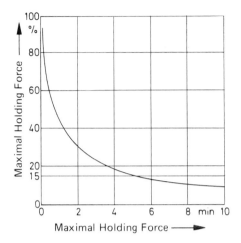

Figure 6.6 Decrease of holding power versus holding time during static muscular activity expressed in percent of the maximum force [6-19].

is proportional to the force applied and to the time interval during which it is held.

Even hypokinetic muscular activity (of the quasistatic type) will cause the muscle to tire quickly. Figure 6.6 demonstrates the decrease of holding power during static muscular activity. According to experimental tests, the percentages of static holding activity during work should remain below 15% of the maximum work load which can be carried out.

If whole groups of muscles are immobilized as, e.g., by means of a plaster cast, muscle force will decrease in the beginning by 25–30% per week. During purposeful muscle training, however, increased muscle power is achieved. The values of muscular stress during exercise must exceed those of the normal daily activity (Figure 6.7). The locomotor system (bones, joints, muscles, and tendons) operates according to a lever

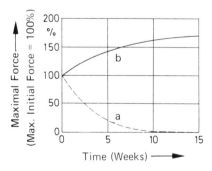

Figure 6.7 Changes in the maximum available muscle force with time (schematic view): (a) after immobilization of the muscle (i.e., by means of plaster cast); (b) after exercising the muscle.

system and permits a wide variety of movements, namely flexing, extension, adduction, abduction, and rotation (see Figure 6.8).

Usually different muscle groups (synergists) are active during a movement. Agonists act in the same direction as the movement, while antagonists act in the opposite direction of the movement. A systematic review of the locomotor system of the human body leads to the following statements: The locomotor system reacts to external influences (stimuli) by mechanical behavior (kinematic, dynamic, or energetic processes). "Stress factors" in this connection, denote a wide variety of influences which are divided into three major groups, namely mechanical, biological, and technical. These

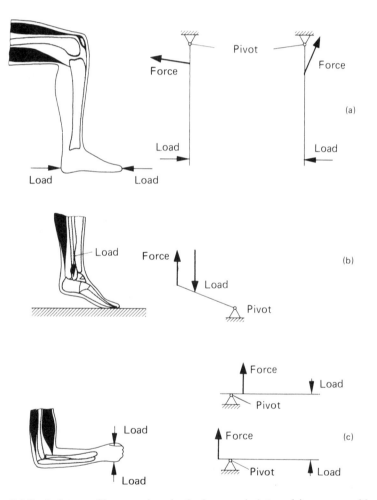

Figure 6.8 Basic forms of lever actions in the human skeleton: (a) one-armed lever with outside application of load (e.g., bending and stretching of the lower leg); (b) one-armed lever with inside application of load (e.g., foot joint on tiptoe); (c) twin-armed lever (e.g., bending and stretching of the lower arm).

are further subdivided according to various aspects, namely mechanical factors—geometric, kinematic, dynamic, and energetic,—biological factors,—anatomical, physiological, and neurophysiological,—and technological factors. Depending on the purpose and goal of the activity, like grasping, reaching, assembling, putting together, etc., the technological factors will be applied. A number of consequences resulting from this affecting construction work are dealt with in Section 6.3.

Metabolism, Energy Exchange, Pulse Frequency. Man—as an open system with a constant metabolism and energy exchange with his environment—needs a constant energy supply in order to be able to maintain his biological functions. Metabolism, the use of energy, and pulse frequency are important indicators for the monitoring and evaluation of man's stress in physical work.

Necessary energy is supplied to the human body through food and oxygen from the air. All chemical processes which take place in the body are called metabolism. This, above all, includes the ingestion and processing of food (carbohydrates, fat, and protein), the digestion and conversion of ingested substances into sugar and fatty acids, amino acids in the gastroenteric tract, the absorption of digested nutrients via the intestines into the bloodstream, the conversion of nutrients in the liver, the combustion of high-energy substances with oxygen, and the formation of low-energy metabolites (carbon dioxide, water, lactic acid, etc.). Blood circulation transports nutrients to both working and storage organs, effects the removal of metabolites to the excretory organs (lung, kidney, etc.), and transmits any waste heat to the skin. During physical work, circulation may play the role of a limiting factor for human performance.

Man's food consists of 50–60% carbohydrates (cereals, potatoes), 30–40% vegetable or animal fats, and 10–12% animal or vegetable albumen. With this variety of food, a mean caloric equivalent of approximately 20.3 kJ per liter of oxygen can be assumed. As well as the above-mentioned nutrients, the human body requires water, vitamins, salts, and minerals. Deficiencies in any of these substances will lead to decreased efficiency or disease. Energy intake and consumption must therefore be balanced. The amount of energy used by man during one day is called energy turnover. Man at rest, with the so-called basic metabolism, guarantees the maintenance of the most important vital functions of the heart, circulation, respiration, and digestion. The basic metabolism depends on age, height, weight, and sex. Figure 6.9 shows the basic metabolic rates for men and women, and Figure 6.10 demonstrates the empirical formula for calculating these rates.

The total metabolic rates according to Table 6.1 depend on the different level of physical exertion during work. Figure 6.11 demonstrates the range of possible metabolic rates in man. An evaluation of human work load based solely on the basic metabolism is possible when mainly dynamic

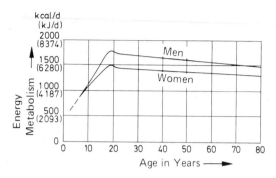

Figure 6.9 Basic metabolic rate for women and men as function of age (schematic view).

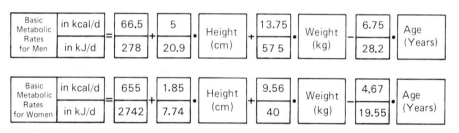

Figure 6.10 Calculation of the basic metabolic rates for adult men and women.

work is carried out by large groups of muscles under normal environmental conditions and without aggravating psychical (mental) stress. During physical work, the muscles require adequate amounts of oxygen, which are provided by blood circulation. For this the heart frequency (pulse frequency) plays an important role in increasing the pumped volume of blood. Therefore the pulse frequency can also be used as a measure of physical stress. The pulse reacts very quickly to physical work stress and to other external influence factors as well.

Table 6.1 Total energy metabolism per day in different occupational groups

Occupation	Metabolic rate kcal/day	kJ/day
Secretary	2400	10050
Goldsmith	2630	11010
Limousine driver	2900	12140
Doctor	3000	12560
Mailman	3410	14280
Baker	3700	15490
Woodcutter	4350	18210
Railroad maintenance worker	4800	20100

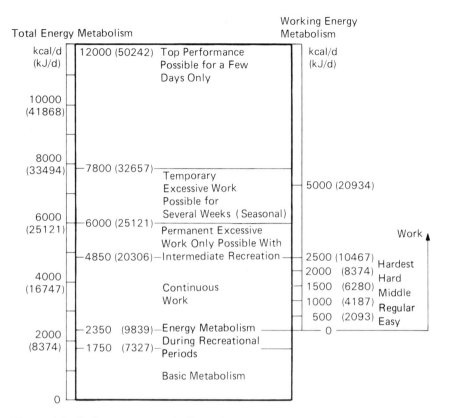

Figure 6.11 Daily energy metabolism of an average male under different work loads.

Physical Activity (Work) and Other External Influence Factors. Usually a distinction is made between pulse rate at rest and pulse rate under physical (work) stress. The work pulse frequency depends on the time and level of work stress as well as the degree of fitness of the person performing the work. Pulse frequency for permanent work is that which at a prolonged uniform level of work will remain fairly constant. Up to that state of equilibrium (steady state) there is almost a linear relationship between pulse frequency and oxygen intake. Beyond the steady state, pulse frequency will increase until complete exhaustion sets in (Figure 6.12). The permanent performance limit for physical work is reached at that point in time when, at a maximum work load, the pulse frequency does not rise any further.

Comparative tests have revealed that the permanent pulse frequency lies approximately 40 beats above the pulse frequency at rest, whereby the latter will differ for each individual. The time required for normalization of the pulse frequency after work has been completed is also of importance. The sum of heart beats during the recovery time (sum of recovery pulses) is

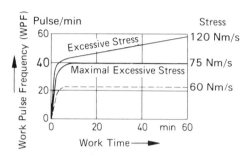

Figure 6.12 Pulse frequency resulting from different levels of work (example: riding a bicycle).

proportional to the degree to which the permanent performance limit has been exceeded (Figure 6.13).

When dynamic work is carried out without any further environmental influencing factors, 40 work pulse beats per minute correspond with an energy metabolism of 17.2 kJ/min (4.1 kcal/min), i.e., 4.3 kJ equal 10 work pulse beats.

If the measured energy consumption is lower than this level, it can be assumed that additional external factors, like high or low temperatures, radiation, noise, additional attention for extra work, or larger shares of static work must have prevailed, all of which require a higher pulse frequency. Metabolic rates and pulse frequencies can be determined by simple measurements. Because of the additional health hazards at workplaces with extreme cold or heat conditions, regular medical checkups are required in many cases [6-28], in order to prevent health impairment. Man

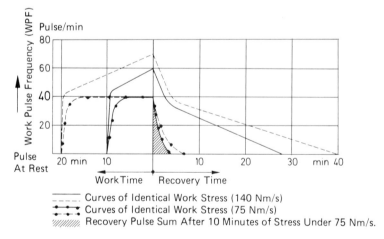

Figure 6.13 Schematic representation of pulse frequency, recovery time, and recovery pulse sum resulting from different levels of work (example: bicycle riding).

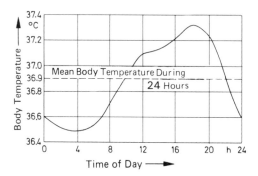

Figure 6.14 Graph of the body temperature during the course of a day.

is a homothermal organism and requires for his vital functions a relatively constant core temperature of 36.9°C which fluctuates only insignificantly during the course of a day (see Figure 6.14).

Major changes in body core temperature can be fatal. Thus the mortality rate at a body core temperature of 43°C is approximately 70%. The same percentage applies to hypothermia at 24°C. Inside the human body, large amounts of heat are produced by chemical combustion processes—one-third by the function of the heart, liver, kidney, and intestines, and two-thirds by the work performed by the skeletal muscles—which are subsequently transmitted to the environment. This is achieved by heat radiation, conduction, and convection as well as by the evaporation of water (sweating). During rapid body cooling, muscular contractions take place (goose pimples) which produce additional heat through chemical processes. Moreover, the blood supply to the skin is reduced in order to minimize the loss of heat from the body. Hypothermia is also often countered by involuntary reflexes. There are no comparable counteracting mechanisms for hyperthermia.

Sensory Organs. Man continuously receives information about the environment as well as about himself, both voluntary and involuntary. These are transmitted by the sensory organs. Special receptors—mechanoreceptors, photoreceptors, thermoreceptors, and chemoreceptors—react to possible stimuli and convert the information into electrical impulses. The frequencies of these impulses are influenced by the intensity of the stimulus and its rate of change. Sensory nerves transmit these impulses to the brain where they are registered and processed as conscious sensations. Whenever constant stimuli act for a prolonged period of time, the impulse frequency of the receptor and the intensity of the sensation will be reduced. This means the receptor has adapted itself.

Receptors preferably react to specific stimuli; e.g., the eye will react to light. However, the effect of electricity or a blow on the eye will also be registered as light. Man possesses the following specific senses: vision (the

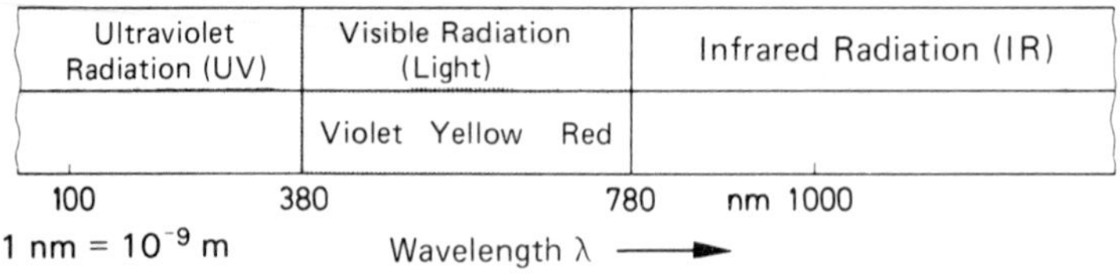

Ultraviolet Radiation (UV)	Visible Radiation (Light)	Infrared Radiation (IR)
	Violet Yellow Red	

```
    100              380                    780      nm 1000
   1 nm = 10⁻⁹ m                  Wavelength λ ⟶
```

Figure 6.15 Section of the electromagnetic spectrum showing the range of visible wavelength.

eye), hearing (the ear), the sense of equilibrium (the internal ear), the olfactory sense (the nose), taste (the mouth and tongue), tactile sense (the skin), temperature sensitivity (the skin and mucous membrane), the sense of pain (skin, mucous membrane, and organs), and the sense for general sensations (e.g., hunger and thirst).

The eye transmits any information on brightness, color, shape, size, and movements in the environment. The eye senses a limited wavelength range as light (Figure 6.15); it is able to differentiate between levels of brightness and a range of colors. Approximately 80% of all information from the environment is received via the eye.

The Eye. The human eye has a structure similar to that of a photographic camera: the pupil, lens, and retina of the eye correspond to the lens opening, lens, and film in a camera (Figure 6.16).

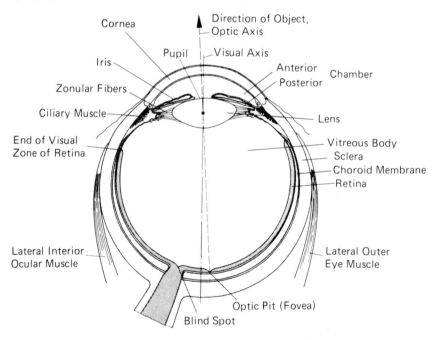

Figure 6.16 Schematic horizontal section through the right eye.

Figure 6.17 Proof of the blind spot in the right eye. If the right eye, with the left eye covered, looks directly at the black cross from a distance of 20–30 cm, the black disk will be eliminated from the picture. At that moment, it is directly reflected at the exit point of the optical nerve.

Inside the retina there are light-sensitive receptors—approximately 6–7 million "thick retinal cones" plus 125 million "slender rods". The optical pit (fovea), the center of the retina, consists of cones only. Towards the outer rim of the retina, the number of cones decreases and that of the rods increases. At the exit point of the optic nerve there are no receptors (blind spot). Normally, this gap in the image is overlapped by the surrounding area and is not perceived at all (Figure 6.17). The receptors are sensitive to a varying degree to energy of different wavelength (Figure 6.18). The cones have a peak sensitivity for green-yellow light which decreases towards blue and red light. They permit daylight vision (photo-optic vision) and color vision. The rods will only react to differences in brightness and are unable to differentiate colors. They are necessary for night vision under poor lighting conditions (scotopic vision). For mesopic vision (i.e., vision

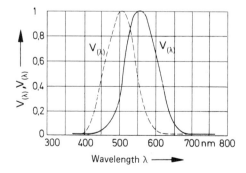

Figure 6.18 Spectral degree of sensitivity to brightness $V_{(\lambda)}$ (daylight vision) and $V'_{(\lambda)}$ (night vision) (according to DIN 5031, P. 3).

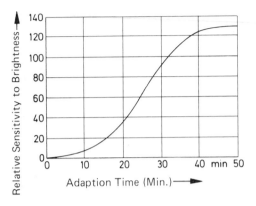

Figure 6.19 Time required in adaptation to darkness as a function of the relative sensitivity to brightness.

between the two extremes of daylight and night vision), e.g., at dawn, cones and rods will interact.

At the center of the iris is a circular orifice (pupil) whose diameter can be varied with the help of a sphincter muscle from between 8 and 2 mm (at an advanced age from between 4 and 2 mm). This mechanism permits the eye to adapt to varying degrees of brightness within a range from $1-10^{10}$. Adaptation to brightness is achieved within seconds, while adaptation to darkness takes much longer, up to one hour (Figure 6.19).

The eye's lens can modify its power of refraction; it will become thicker through its inherent elasticity for viewing close objects. When looking at distant objects, the ciliary muscle will relax and thus tighten the suspensary ligaments of the lens, thereby flattening it. The modification of the lens's refractive power (accommodation) permits the creation of a sharp image on the retina at object distances from approximately 6 cm (as the nearest point of clear vision) to infinity. The ciliary muscle is only relaxed while the eye is viewing distant objects. This is why looking at objects close to the eye always demands prolonged muscle stress and thus leads to eye fatigue after a certain period of time. With advancing age the water content of the lens will decrease and the lens becomes more rigid. The natural consequence is a decrease in accommodation; the result is presbyopia (Figure 6.20).

The eyeball is moved by six different muscles so that any fixed point is always reflected in the optical pit of the retina. Whenever the eye looks straight ahead and 30–40° downward from the horizontal plane, the six muscles are at rest and are in a state of equilibrium. The ability of the eye to perceive separately even very small objects which are placed closely together is called visual acuity. This, of course, also depends on lighting conditions. The perception and recognition of an object is directly connected to certain prerequisites, namely the minimum contrast against the

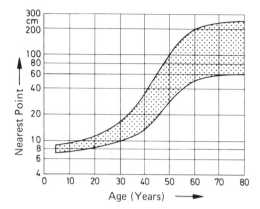

Figure 6.20 Decrease in accommodation capacity with advancing age; the nearest point of clear vision moves farther away from the eye.

environment, the minimum size of the object, the minimum light density of the object to be perceived, the accommodation capability to optical brightness of the eye, and a minimum dwelling time of the object within the field of vision. Whenever the eye is still able to separately resolve two points one minute of arc distant from one another, this is regarded as normal visual acuity 1.0. That is only possible in the optical pit (Figure 6.21).

With continuing receptor stimulation, the sensation of intensity will decrease. Also, light sensitivity will decrease. If one tries to stare at one point for a prolonged period of time, the result will be so-called fixation blindness. Whenever an object must be looked at for a prolonged period of time, the eye will become restless—so-called fixation restlessness. It will make continuous small movements—approximately within the length of one minute of arc. Since, during prolonged acute vision, the eye muscles are being continuously strained, eye fatigue will set in.

Images are reflected on the retina of the right and left eye with minimal difference. This permits stereoscopic or three-dimensional vision. The impression of depth of the object viewed is created in the visual center of the brain. This perception will decrease with increasing distance from the object observed. Human power of imagination, independent from the above, also permits a conditional depth perception with only one eye. This is achieved by means of perspective, the pattern of shadows and light, and the visual effect of color tones. Up to a certain frequency, intermittent light stimuli can be recognized separately or as a sort of flickering movement. The flickering fusion frequency depends on the luminous density and on the size of the retinal area which is being stimulated. It intensifies with increasing luminous density and decreasing stimulated retinal area. Flickering light causes constant fluctuation in the stimulation of the retina and will produce nervousness and early fatigue.

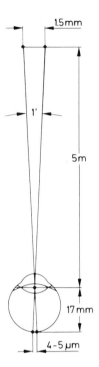

Figure 6.21 Minimum limit of resolution of the eye. Example: two points 1.5 mm from one another and at a distance of 5 mm are barely perceived as two separate points if the visual acuity is 1.0.

Color vision is made possible by cones with different color sensitivity. Cones can react to either short-wave rays (blue), others to light of medium wavelength (green), and still others to long-wave light (red). Nerve connection permits the combination of these three types and leads to the distinction of a wide color spectrum. Approximately 10% of all males are born with defective color perception, 3–4% are red-green blind; red blindness, green blindness, and, less frequently, blue blindness make up the rest. Only 1% of all women have impaired color vision. Women are regarded as carriers of these anomalies; they pass them on to male offspring only. If both partners suffer from impaired color vision, the latter defect may also be found in their daughters.

"Field of vision" is that area which can be viewed when head and eye are both at rest and the viewer is looking straight ahead. In the marginal zones of the field of vision which are reproduced on the rim of the retina where fewer cones are found, only black-and-white vision is possible. Next comes a zone of yellow-blue perception and only in the innermost part the perception of all colors is possible, as demonstrated graphically in Figure 6.22. Blue light and yellow light are thus described as field-of-vision-

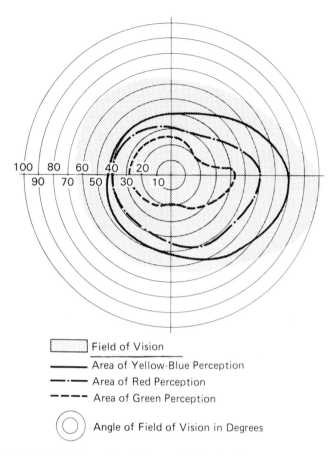

Field of Vision

——— Area of Yellow-Blue Perception

—·— Area of Red Perception

- - - - Area of Green Perception

Angle of Field of Vision in Degrees

Figure 6.22 Field of vision of the right eye displaying the areas of individual color perception.

related warning colors. The special importance of the eye as far as safety in connection with technology is concerned is taken into consideration to a large degree. Thus a driver's license may be issued only to those applicants who can prove they have adequate visual acuity.

The Ear. The ear houses two sensory organs, hearing for the perception of sound waves and the equilibrium mechanism for the perception of accelerations. Figure 6.23 shows a schematic sectional view of the ear.

The external ear (pinna) and the external auditory meatus form the outer ear. The eardrum, a membrane of connective tissue 0.1 mm thick and 8–10 mm in diameter, divides the external ear from the middle ear. The eardrum acts as a resonator with an inherent frequency of approximately 3000 Hz. The middle ear consists of an air-filled cavity, the

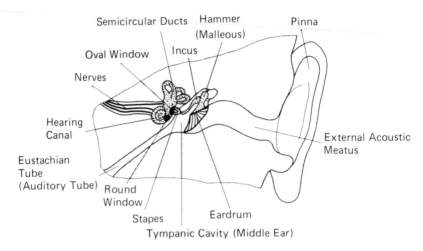

Figure 6.23 Schematic sectional view of the ear.

so-called tympanic cavity, and the group of auditory ossicles consisting of malleus, incus, and stapes. The auditory tube connects the tympanic cavity with the nasopharynx. The chain of auditory ossicles transmits the vibrations of the tympanic membrane (eardrum) onto the so-called cochlea, which is the actual hearing organ. The latter is approximately 32 mm long and on it are arranged the semicircular ducts of the equilibrium mechanism. Inside the cochlea is the organ of Corti with its hair cells which are stimulated by sound waves and transmit electrical impulses to the brain via the cerebral nerves (Figure 6.24).

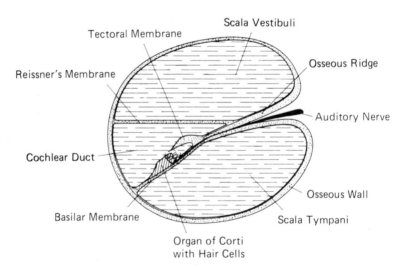

Figure 6.24 Schematic sectional view through the cochlear duct of the inner ear at the lower spiral organ.

The footplate of the stapes is joined to the round window. Since the eardrum and the membrane of the oval window can be stretched taut by means of muscles, a strong damping effect is achieved for frequencies exceeding 1000 Hz. Young people perceive frequencies of between 18,000 and 20,000 Hz. With advancing age, the audibility of the higher frequencies—namely above 12,000 Hz in people over 50—will decrease. This is called presbyakusia.

Besides airborne sound waves, direct sound waves also reach the inner ear via the sound-conducting bones. The mechanical vibration of the stapes footplate leads to the formation of tubular waves in the cochlea. High frequencies are transmitted with higher wave speeds, low frequencies with lower wave speeds. This results in a local dispersion within the Corti organ. At the point where the wavelength of the migratory wave coincides with the depth of channel, a "surf" condition is created. The migratory wave, at this particular spot, is extinguished, but the energy of the vibration accumulates in front of the "surf" and creates a protrusion in the basic membrane. This process, via a permanent excitation, leads to hearing fatigue. If, e.g., a 800-Hz sound acts upon a human ear, the intensity level will be raised. If, after that, a somewhat deeper and quieter sound is offered, this sound is perceived by the fatigued ear as being deeper than it really is. A higher sound will also appear higher than it would normally be perceived by the unfatigued ear.

The human auditory threshold curve and pain tolerance limit are extremely frequency dependent. Figure 6.25 shows a graph of the hearing area with auditory threshold and pain tolerance limit for a test subject with normal hearing. The ear can analyze sounds, tones, and noises according

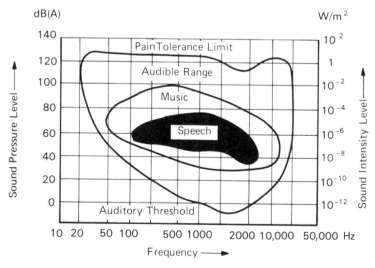

Figure 6.25 Hearing area of a person with normal hearing.

to their frequencies and amplitudes. However, sounds of different phases and identical spectral composition are indistinguishable.

Man's long-distance hearing ability is weak. Over long distances the frequency-dependent damping of the sound is used as auxiliary information. For distances up to 1 m the unequal drop of sound pressure and speed of sound provides the information. The adaptation of hearing to the environment amounts to no more than 1–2 powers of 10. Whenever two sounds differing insignificantly in frequency hit the ear simultaneously, a difference of less than 7 Hz will lead to interference. With greater differences and frequencies, the result is a dissonant tone of constant loudness. In acoustical signal devices, identical sounds should be avoided. The difference should at least exceed 7 Hz. Beat frequencies will only appear with identical loudness of the individual tones. With an increasing sound pressure difference, the louder tone completely smothers the weaker one. This smothering effect of stronger interference sounds over weaker signal sounds decreases with increasing frequency difference between them. When listening with both ears, differences in time are registered for lower frequencies, thereby making directional listening possible. The registered time intervals extend as far down as 1/30,000th of a second. For directional listening at higher frequencies, the intensity drop and the dead angle effect of the human head are used.

Above the cochlea there is the osseous labyrinth with the semicircular ducts, in front of which the bag-like saccule and utricle contain the static sense, namely the sense of equilibrium in their respective vertical and horizontal arrangements (Figure 6.26). By this, gravitational power and many movements in horizontal and vertical directions are perceived. At the entrance point to each of the semicircular ducts, rotary movements and angular accelerations are registered. The anatomy of the semicircular ducts (relative movement of the lymphatic liquid) effects the adaptation to uniform vibrations.

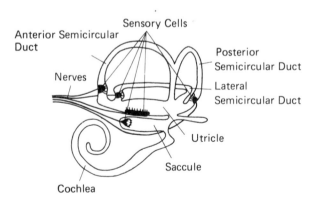

Figure 6.26 Schematic view of the cochlea.

Dermal Receptors and Depth Sensitivity. Experimentally, dermal sensa-tions can be accurately localized. They are the mechanoreceptors, thermo-receptors, and nociceptors. They transmit a wide variety of sensations like pressure, distension, contact, vibration, itch, tickling, cold, and pain. All of the above-mentioned sensations are allocated to the psycho-algesic area in the posterior central gyrus of the cerebral cortex. These stimuli can be localized with different degrees of resolution. The degree of accuracy with which two locally adjacent stimuli can be distinguished is shown by the following parameters:

fingertip = 2 mm;
lips = 4 mm;
sole of the foot = 20 mm;
lower arm = 40 mm;
back = 70 mm.

Mechanical stimulation of the skin is converted into nervous impulses by mechanoreceptors. At first, a distinction between contact and pressure is possible. After prolonged skin irritation through pressure, adaptation will take place. Clothing and glasses, e.g., are no longer registered. Vibrations are classified in a variety of ways according to frequencies. Frequency differences of up to 10% can be registered. The vibration thresholds are frequency independent. Figure 6.27 demonstrates an increased sensitivity at 200 Hz. The mechanical deformation of the skin permits the actual registration of stimuli: Pressure, shear, and bending forces act on the receptor and permit the transmission of information on weight, dimension, shape, physical state, and nature of the surface. Heat and cold receptors separately transmit temperature sensations. Between these two extremes there is the neutral range of indifference. Cold and heat sensation depend on absolute skin temperature, rate of change of the skin temperature, and the area of contact.

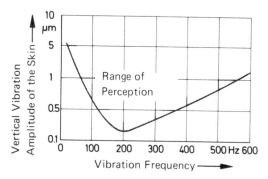

Figure 6.27 Vibration thresholds (vibration amplitudes) of the skin as a function of frequency.

Thermoreceptors of the mucous membranes may also be subjected to irritation by chemicals. They react, e.g., to pepper, paprika, carbon dioxide, and alcohol. Menthol and peppermint act on cold receptors. The thermoreceptors transmit their stimulation to the temperature control center and thus trigger further regulatory processes like being cold and shivering, goose pimples, contractions of blood vessels, or closeness with sweating and dilatation of blood vessels.

Whenever mechanical, thermal, or chemical stimulation of the skin is increased beyond a specific level, pain is felt along with the stimulus-induced perception; nociceptors transmit this type of sensation. The warning signal called pain is generally perceived in two different ways:

— The acute and easily localized surface pain is quickly transmitted to the brain; in most cases it triggers an escape or defensive reaction.
— The dull and less localizable deeper seated pain—because of it's transmission to adjacent areas—is transmitted to the brain at a slower speed; it generally leads to impaired movement.

Although pain receptors never adapt, the feeling of pain can be subdued by other sensory perceptions. In depth sensitivity, i.e., when regulating and controlling movements, tension and tendon receptors in muscles, tendons, and joints play an important part. In safety technology, the gustatory and olfactory senses are of a certain importance as transmitters of information. Man can sniff (breathe in at an accelerated rate) and distinguish more than 1000 types of smells. Smells affect the sensation of comfort to a high degree. The threshold concentration for different fragrances lies between 4×10^{-9} and 5×10^{-14} g/l of air. The adaptive ability of olfactory receptors is strongly developed. Gustatory receptors generally act in unison with olfactory receptors. A stuffy nose also impairs the gustatory sense which, by the way, adapts very easily.

The Nervous System. In order to control logical reactions to the stimuli coming from the environment and the body itself, a central regulation and switching system is needed. Processing of information and control of functions are carried out by the hormonic and the nervous system. Figure 6.28 shows the general cycle of reaction:

— perception of stimuli, stimulation, and transmission of stimuli;
— processing of information: new and stored information are connected which each other leading to conscious or subconscious decisions;
— selecting the reacting organs (transfer of decisions):a reaction is triggered;
— feedback of control values and their effects: real and desired conditions are checked.

The inner organs and their interactions are directed by the hormones of hormonic glands—pituitary gland, thyroid gland, suprarenal bodies, and

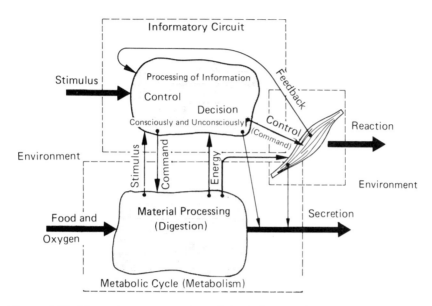

Figure 6.28 Schematic representation of the informatoric and metabolic cycles within the organism.

pancreas. The transmission of information by hormones is slower than that in nerves. Disorders of the hormonic system not only affect their specific organs but also the nervous system as a whole. Anatomically the nervous system can be divided into the following parts:

—the peripheral nervous system (nerve tracts and ganglions);
—central nervous system (brain, spinal cord);
—autonomic nervous system in inner organs, glands, blood vessels, smooth muscles.

The smallest functional unit is a nerve cell, the neuron. Dendrites transmit information into the nerve cell, neurites out of it (Figure 6.29).

Nervous cells (neurons) can be connected by means of synapses which serve as switching units. An adequate stimulation of the neuron produces a nervous impulse through a shift in the ions which will be cancelled out metabolo-energetically after approximately 1 ms: the neuron again becomes receptive to stimuli. The peripheral nervous system consists of nerve pathways in the peripheral area: sensitive nerves transmit stimuli from the environment and the body to the central nervous system (centripetal, afferent pathways). Motoric nerves transmit the commands of the central nervous system to the muscles (centrifugal, efferent pathways).

The number of so-called nodes of Ranvier (Figure 6.29) determines the transmission speed within the nerve pathway which can be measured by means of "myography". Table 6.2 lists a number of transmission speeds in various nerve pathways. In the central nervous system (spinal cord and

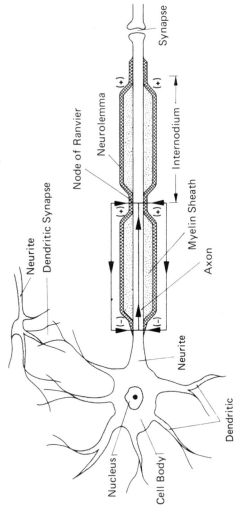

Figure 6.29 View of a nerval cell with schematic demonstration of transmission of stimulation.

Table 6.2 Transmission speeds in various nerve pathways

Fibrous structure	Fiber diameter in μm	Function of the fiber in the human body	Fiber transmission speed in m/s
Thin fiber, without myeloid tissue	1	Slow transmission of impulses from mechano-, thermo-, and noci(re)ceptors; postganglionary vegetative fibers	1
Fiber with thin myelin sheath	1–3	Preganglionary vegetative fibers	10
Fiber with thick myelin sheath	3–20	Fast transmission of pulses from mechano-, thermo-, and noci(re)ceptors	20
		Control pulses to the ends of the muscular spindles (efferent pulses)	40
		Pulses from contact with skin (afferent)	60
		Signal from tendon receptors and muscular spindles (afferent pulses); general motoric pulses	80–120

brain), any information transmitted via nerve pathways is processed, matched against previous information, combined and converted into decisions or stored once more (Figure 6.30). In order to function properly the central nervous system needs specific conditions.

Blood circulation has to be sufficient. In one liter of blood, the brain is supplied each minute with 15–20% of all oxygen required for the entire organism. Thus, lack of oxygen (at great heights, in rarefied air) often affects concentration and attention, judgment, the ability to make decisions, the field of vision, and the perception of color. Determination of the reaction time serves as a functional test for the central nervous system. For optical signals reaction time is longer than for acoustic signals. Reaction times depend on the well being and the individual's determination to perform; they are neither specific nor constant and can be influenced by pharmaceuticals. Fatigue, alcohol, and drugs may prolong reaction time considerably. Within the spinal cord (Figure 6.31), vital reflex functions are triggered and stimulation of the peripheral and autonomic nervous systems are transmitted and switched. The ganglion cells take over the switching function and—without any help from the brain—convert sensitive pulses from the periphery into motoric excitation which, in turn, is transmitted to the muscles (reflexes).

A distinction is made between autoreflexes and extraneous reflexes (Table 6.3). The musculature has a reflectoric muscle tonus which is

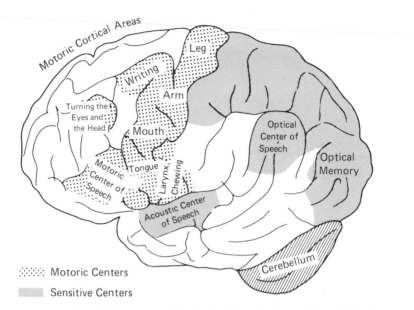

Motoric Cortical Areas

Leg

Writing

Arm

Turning the
Eyes and
the Head

Mouth

Optical
Center of
Speech

Optical
Memory

Motoric
Center of
Speech

Tongue

Larynx, Chewing

Acoustic Center
of Speech

Cerebellum

Motoric Centers

Sensitive Centers

Figure 6.30 Cortical centers of the brain (lateral view of the left half).

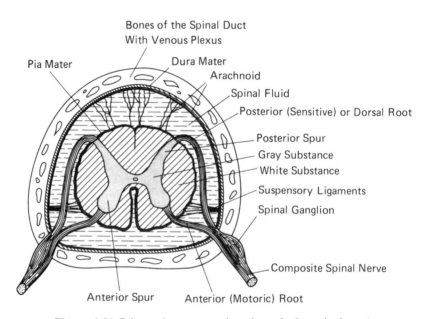

Bones of the Spinal Duct
With Venous Plexus

Pia Mater

Dura Mater

Arachnoid

Spinal Fluid

Posterior (Sensitive) or Dorsal Root

Posterior Spur

Gray Substance

White Substance

Suspensory Ligaments

Spinal Ganglion

Composite Spinal Nerve

Anterior Spur   Anterior (Motoric) Root

Figure 6.31 Schematic cross section through the spinal cord.

Table 6.3 Characteristic criteria of auto- and extraneous reflexes

Autoreflex	Extraneous reflex
Receptor and executive organ are in the same muscle	Receptor and executive organ are in different places
One switching procedure in the spinal cord (monosynaptic)	Several switching procedures, within the spinal cord to different areas (polysynaptic)
Reflex reaction is a single reaction (jerk)	Reflex reaction is a coordinated and complete movement of the muscle
Reflex time constant at ca. 20 ms	Reflex time, duration, and extension are different and depend on intensity of stimulus
May be suppressed by activation of antagonist muscle	May be suppressed by will (cerebral inhibition)
Almost indefatigable, since the muscle recovers between contractions	Fatigable, since muscle contraction is greater, more muscles involved

increased by stimuli through cold or pain and impairs voluntary motoricity. Autoreflexes bring about the normal flow of a movement. They keep the body upright and position the extremities. The movements of speech are also controlled in an autoreflectoric manner. However, extraneous reflexes will also influence, in a coordinating manner, the voluntary activity of the muscles. One example is the deterioration of fine coordination of finger movements at low temperatures. The involuntary bodily functions are controlled via vegetative nerves: the sympathetic system and the parasympathetic system (Table 6.4). The sympathetic system alone effects an increase in performance, the parasympathetic system promotes recuperative processes and thus man's fitness for work.

Stress Characteristics with Predominantly Nonphysical Work. Man absorbs 10^{10} bits of information per second, a figure which corresponds to the

Table 6.4 Examples of important functions of the autonomic nervous system

Influenced part of the body	Effect of sympathetic system	Effect of parasympathetic system
Eye	Dilatation of pupils	Contraction of pupils
Saliva glands	Thickish saliva	Abundant thin saliva
Heart	Accelerated action	Restrained action
Bronchi (smooth muscle)	Dilatation	Constriction
Peristalsis of the stomach and intestine	Inhibitions of movements	Stimulation of movements
Liver	Mobilization of glucose	—
Suprarenal bodies	Excretion of hormones	—
Blood vessels	Contraction	Dilatation

channel capacity of all sensory organs. Vision, e.g., requires approximately up to 10^8 bits of channel capacity, and hearing approximately 10^5 bits. Tests have shown that man is able to consciously process 10^2 bits/s. This means that in the afferent nerve pathways to the brain, there is an information reduction of a ratio of $1:10^8$, which is a meaningful premise for survival. Since the information, on its way to the brain is not lost, but rather processed and used in a multitude of ways, it is possible for man to pass on to his environment an information flow of approximately 10^7 bits/s. (See also Section 6.2.2 and Figure 6.54.)

In connection with the flow of information, the following aspects are of importance for practical application:

—Motoric actions may evolve without any processes of thought in such cases where behavioral patterns have been stored in the pertinent subordinate centers through practice, training, and drill. If these behavioral patterns are being kept functional by corresponding training, then signals from the environment—without penetrating into consciousness—can trigger and run off a voluminous action program in man. During this activity, the memory, with its relatively small conscious apperception and transmitting capacity, is not called upon and remains ready for simultaneous conscious tasks. A classical example for the above is the way an experienced driver functions.

—Immediately prior to consciousness, a switching point must be in existence which ensures that a maximal information flow of 100 bits/s via the individual sensory organs is not exceeded. This flow control always acts alternatively and, as an optimization device, takes prior stored information into account and classifies it according to importance. Conscious concentration on a single sensory channel will displace any other type of information. During concentrated reading, sounds are pushed into the background within a very short period of time and are no longer consciously registered. Here, however, we must consider the fact that man, during the entire duration of the exposure, is subjected to the full flow of information.

Change in Performance as a Function of Time. Man's performance depends on a variety of influence factors. Among them are calculable ones as well as those which cannot be assessed or predicted.

(a) Fitness and Age
Man reaches his maximal state of fitness at the age of 30. Females attain this maximal degree of fitness at the age of 20. With advancing age, there is an increasing decline of fitness (see, e.g., Figure 6.40). As far as mental capacity is concerned, namely concentration and retention of information, there is a similar dependence on age. The performance level of the sensory organs steadily decreases with advancing age. There is a decrease in eye accommodation, a loss of hearing for very high sounds, and a decreased

ability to react. Nevertheless, the total performance of the human organism may remain constant beyond the age of 60. Training and experience make up for the loss of fitness of the individual organs. Taking into account the rapid changes in modern technology, these factors, however, can only be introduced into the compensatory game in a very limited way.

(b) The Influence of Training
Repetitive and identical work processes trigger a training stimulus and will yield improved output. This is valid for all physical functions in general and may, e.g., encompass simple training of muscles or the adaptability to heat or height. Improved output is most noticeable during the first phase of training.

(c) The Influence of Fatigue on Performance
Fatigue, which is an objectively measurable phenomenon in man, and exhaustion, as a subjective sensation, are also called work fatigue and incentive fatigue, respectively. They rank next to biological fatigue which is dependent on daily rhythm. Fatigue always leads to a decrease of ability to perform and function; however, it also leads to a reduction in readiness to perform. The various forms of fatigue always overlap; a sharp delineation is thus impossible, as shown in Figure 6.32.

(d) The Influence of the Rhythm of Daily Life
Man's performance depends on his ability to perform, i.e., on his physical and mental capacity to perform and on his readiness to perform. Motivation, interest, and mood are psychological performance factors, while training, conditioning, daily rhythm, and state of health should be regarded as physiological factors. The influence of the daily rhythm on the physiological readiness to perform and on the faulty performance graph can be seen from Figure 6.33. Figure 6.34 represents the actually achieved

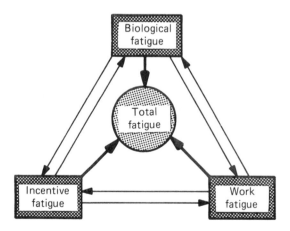

Figure 6.32 Types of fatigue and their interrelationships (schematic).

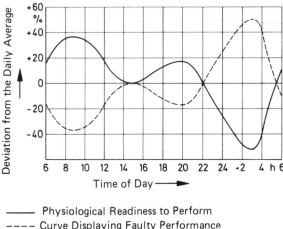

— Physiological Readiness to Perform
- - - - Curve Displaying Faulty Performance

Figure 6.33 Daily dynamics of the physiological readiness to perform (uninfluenced and biologically conditioned rhythm).

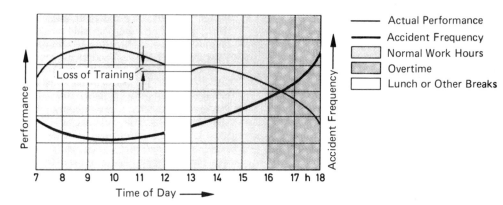

—— Actual Performance
━━ Accident Frequency
▨ Normal Work Hours
▨ Overtime
□ Lunch or Other Breaks

Figure 6.34 Schematic view of actual performance and accident curve.

performance. In general, man has at his disposal so-called capacity reserves which may be mobilized under certain conditions. These also depend on daily rhythm (Figure 6.35).

Anthropometry. Anthropometry, the science of measuring man, concerns itself with the proportions and dimensions of the human body and their functions. The measurements of the human body depend on time. In industrial countries, over the past 100 years, the height of the human body has steadily increased. At the present time, the average German male is 177 cm tall. After the 35th year many dimensions, including height, begin to decrease. Body measurements, besides long-term influence factors,

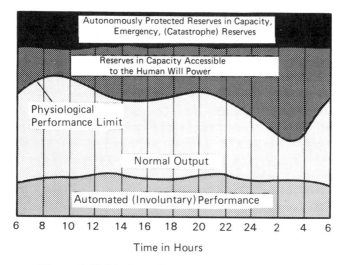

Figure 6.35 Man's performance areas (schematic).

fluctuate during the day as well. The height of man depends on the state of the intervertebral disks and, during the course of one day, may decrease by up to 20 mm.

The average measurements of the body differ according to sex, the various human races, different regions, and, finally, social strata. In our latitudes women, on an average, are 9–10 cm shorter than males. As far as height is concerned, a North–South decrease of approximately 8 cm can be measured from the Scandinavian countries in the North to the Mediterranean in the South. Within the Federal Republic of Germany, this drop—in the same direction—averages approximately 2 cm. Anthropometric parameters can only be used and compared under critical scrutiny and whenever sources of possible error are fully known. The degree of accuracy must be known to the user.

The Centile and Its Application. In general, machinery and appliances cannot be designed for a conglomerate of operators which includes both the tallest and the smallest of men. Thus, for the design and construction of machinery, the "centile" is used, which is based on the scatter patterns of theoretical and natural distributions (Figure 6.36). The 5th centile expresses the idea that for 5% of the number of persons measured, the measurement in question is of a lesser magnitude. The 95th centile means that the measurement in question is greater for 5% of the persons measured (Figure 6.37).

The parameter of the 95th centile will be used in cases where the measurements for openings like doors and hatches or for the width of handgrabs, handles, etc., must be determined. The parameter of the 5th

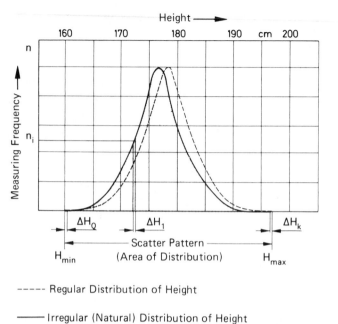

----- Regular Distribution of Height

——— Irregular (Natural) Distribution of Height

Figure 6.36 Measurement frequency of the height of humans, entered into the distribution areas with regular (theoretical) and irregular (natural) distribution; schematic example of a selected population group.

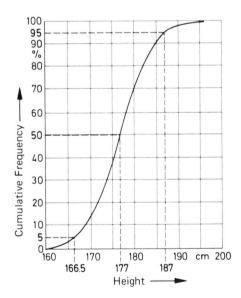

Figure 6.37 Cumulative percentage curve of a selected population group.

centile plays an important part in the arrangement of levers, switches, etc., which the average person must be able to reach. A technical system designed according to the measurements of the 5th and 95th centile can thus be used without any particular problems by 90% of the operators, with a minimum of difficulty in adaptation. The plan, however, is to use the measurements of the 2.5th and 97.5th centiles for work sites and tools.

Methodology of Measuring, Selected Dimensions of the Human Body, and Ranges of Movement. Dimensions and circumferences are usually measured on the surface of the naked human body. Dimensions are usually measured between two protruding bones. Measurements are taken while the person is either in a standing or sitting position. The head thus assumes the "Frankfurt Horizontal Position," i.e., the upper rim of the auditory duct and the lower rim of the bony orbit are situated in one and the same horizontal plane. These "naked measurements" must be supplemented by certain additions for clothes in order to yield design measurements. Figure 6.38 shows anthropometric measurements which are described in Table 6.5. This table contains the centiles of body dimensions.

Parts of these body dimensions are also the angular ranges within which the movements of individual extremities are executed. Thereby, the total range of movement of a part of the body usually consists of a combination of several individual movements. These movement ranges may have a great influence on the design and construction of appliances and controls (area of reach, clearance, safe distance). Any movements demanded of man when handling technology should not be executed in the immediate vicinity of terminal points of the corresponding ranges of movement. Table 6.6 shows the number used for individual angles of movement.

Physical Strength. In human beings, maximal forces are measured. Their percentages serve as a basis for the dimensioning of designs for which a strength-preserving work mode is required. Among these are control levers, cranks, and handwheels. The maximal generated physical strength of man depends on individual influence factors like sex, physical constitution, age, condition, and body symmetry. Figure 6.40 demonstrates muscular force in men and women depending on their respective ages. Through appropriate training, normal muscular strength can be almost doubled (Figure 6.41).

Table 6.7 and Figure 6.42 convey information on the positional strengths of the right arms of men in a standing position. Corresponding values for men sitting can be seen from Figure 6.43 and Table 6.8. The range of maximal deployment of force, however, does not have to be in the center of the total range of movement of a particular joint. Maximal force can only be achieved through the exertion of the utmost will power and over very short periods of time. For general applicability, these forces will have

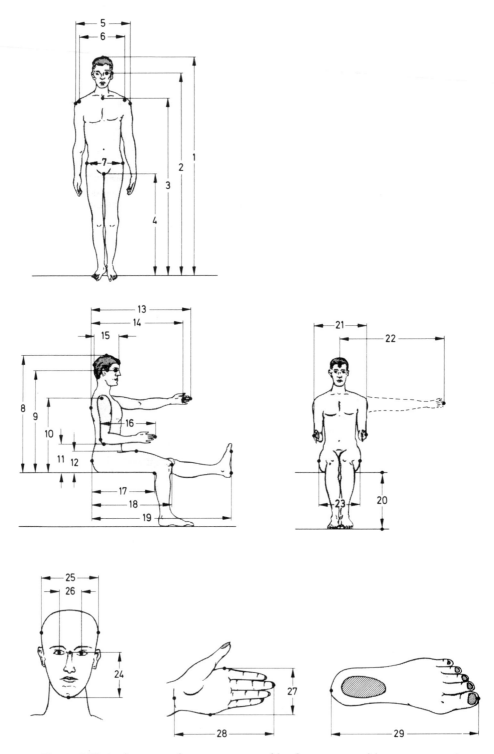

Figure 6.38 Anthropometric measurements (the figures entered here correspond with those in Table 6.5).

Table 6.5 Measurements of the human body (the individual values have been rounded off upward to half a centimeter, values with an asterisk have been estimated in accordance with similar values from the literature, the serial numbers apply to Figure 6.38)

Serial number	Brief description of the object measured	5th, 50th, and 95th centile of height in cm Men			Women		
		5	50	95	5	50	95
1	Height	165.5	176	186.5	156.5	166	175.5
2	Sternal height	135	144	153	127	135.5	143.5
3	Eye level (standing)	153	163.5	174.5	145	154	164
4	Height of step	75.5	82.5	90	69*	74*	81.5*
5	Shoulder width (max.)	41	44.5	48.5	37.5*	40.5*	44.5*
6	Shoulder width (acromion)	36	39.5	42.5	34	36.5	39.5
7	Width of pelvis	26	28.5	31.5	24.5	27	29.5
8	Height (sitting)	86.5	92	97.5	83	87.5	92.5
9	Eye level (sitting)	74	79.5	85.5	70.5	75*	80*
10	Shoulder height (sitting)	57.0	62.0	67	51.5*	56.5*	61.5*
11	Elbow height	19	23	27.5	19*	23*	27.5*
12	Thickness of the femur	12.5	14	16.5	12*	13.5*	16.5*
13	Reach of the arm	78.5*	85	91.5	72.5*	77.5*	83.5*
14	Reach of the arm (functional)	76.5*	80.5*	87*	69*	73.5*	79*
15	Length of head	18	19	20.5	17.5	18.5	19.5
16	Length of the lower arm	44	47.5	50.5	38.5*	42.5*	46.5*
17	Sitting depth	45.5	49.5	54	41.5*	46.5*	50*
18	Length from glutea to knee	55.5	60	64.5	53.5*	57.5*	63.5*
19	Length of glutea including leg	97.0	104	113	84*	98*	112*
20	Height of the seat	40	44	48.5	38*	42.4*	46.5*
21	Outside width of elbows	37	41.5	48.5	31.5*	37.5*	45*
22	Lateral reach	78.5	85	91.5	72.5*	77.5*	83.5*
23	Width of seat	31	34.5	38.5	41.5*	45.5*	50*
24	Height of face (morphological)	10.5	11.5	12.5	10	11.0	12
25	Width of head (max.)	14.5	15.5	16.5	14.0	14.5	15.5
26	Distance between pupils	5.8	6.3	6.8	5.9	6.3	6.7
27	Width of the hand	7.7	8.5	9.2	7.0	7.7	8.4
28	Length of the hand	17	18.5	20	16.5	18	9.5
29	Length of the foot	24.0	26	28	22.5	24.5	26

to be reduced considerably. For short periods of time, 55–60% of the amount of maximum values shown may be permissible. For long-term performances under predominantly dynamic work conditions, 20–25% should never be exceeded. Exclusively static types of work, e.g., holding or grabbing functions, are extremely unfavorable for the groups of muscles under stress.

These parameters of physical force should at most attain 15% of the possible maximum (compare Figure 6.6). Since the frequency of a movement and the required speed of that movement also determine the magnitude of the permissible force, the permissible parameters of physical force must be carefully measured in actual work cycles.

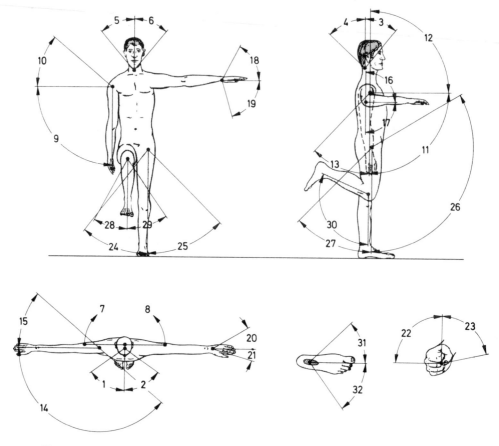

Figure 6.39 Ranges of movement of individual parts of the body (extremities). (The figures correspond with those in Table 6.6.)

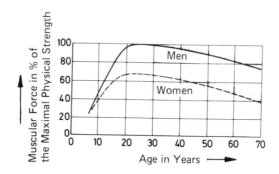

Figure 6.40 Muscular strength in men and women depending on their respective ages, shown as a percentage of the muscular strength (100%) of men.

Table 6.6 Ranges of movement of individual parts of the body (the serial number corresponds with that in Figure 6.39)

Part of body	Joint moved	Direction of movement	No.	Angle of movement in degrees
Head	Spine	Rotation to the right	1	55
		Rotation to the left	2	55
		Flexing	3	40
		Hyperextension	4	50
		Bending sideways	5	40
		Bending sideways	6	40
Shoulder blade	Spine	Rotation to the right	7	40
		Rotation to the left	8	40
Arm	Shoulder joint	Abduction	9	90
		Elevation	10	40
		Flexing	11	90
		Elevation forward	12	90
		Hyperextension	13	45
		Adduction	14	140
		Hyperextension	15	40
		Rotation with abduction		
		(external)	16	90
		(internal)	17	90
Hand	Wrist	Dorsal flexion	18	65
		Palmar flexion	19	75
		Adduction	20	30
		Abduction	21	15
	Trochoid joint	Supination	22	90
		Pronation	23	80
Leg	Hip joint	Adduction	24	40
		Abduction	25	45
		Flexion	26	120
		Hyperextension	27	45
		Rotation during flexion		
		(external)	28	30
		Rotation during flexion		
		(internal)	29	35
Lower leg	Knee joint	Flexion	30	135
Foot	Ankle	Adduction	31	45
		Abduction	32	50

Figure 6.41 Increase of muscular strength through training of varying intensity (schematic).

Table 6.7 Average maximal positional forces in N and torques in Nm of the right arms of standing males. These values correspond to those in Figure 6.42. The highest values in each column have been highlighted

α in degrees	β in degrees	Reach in %	Maximal positional forces in N						Torque in Nm	
			Horizontal		Vertical		Duction (horizontal)		Rotation towards	
			HZ	HD	VZ	VD	AD (tensil force)	AB	inside	outside
−60	0	50	1C2	123	**282**	174	149	111	**22.7**	11.6
		75	127	142	229	159	132	103	20.0	9.0
		100	150	161	181	143	115	95	13.7	6.7
	30	50	113	143	265	160	137	97	21.4	9.0
		75	150	158	221	154	128	92	19.7	7.8
		100	163	172	183	149	121	87	14.7	6.7
	60	50	137	148	215	174	142	85	19.7	9.2
		75	162	153	196	161	124	80	19.1	8.0
		100	175	158	173	151	110	79	13.7	6.5
−30	0	50	127	127	222	157	186	**146**	20.8	15.6
		75	133	133	177	147	151	121	18.1	13.2
		100	139	140	125	134	118	96	11.8	4.9
	30	50	142	151	197	167	182	196	18.5	16.4
		75	151	158	153	147	134	99	17.0	14.5
		100	60	166	116	128	94	92	12.7	5.9
	60	50	167	136	178	177	139	88	17.5	15.7
		75	173	141	154	157	114	82	16.7	14.4
		100	**179**	144	115	142	92	79	12.7	5.9
0	0	50	103	134	151	248	**187**	132	17.9	**18.3**
		75	115	157	117	177	145	111	15.5	15.9
		100	127	179	81	146	105	88	10.8	5.9
	30	50	113	144	147	228	157	103	16.9	16.9
		75	124	169	108	178	123	93	15.2	16.2
		100	139	196	68	126	88	83	9.8	5.9
	60	50	137	162	131	213	126	91	16.1	15.7
		75	150	188	84	175	105	84	15.1	15.7
		100	163	**214**	37	137	84	79	10.8	5.9
+30	0	50	82	73	123	**336**	158	106	16.9	17.5
		75	100	108	105	257	132	98	14.9	16.5
		100	118	143	86	181	108	91	10.8	7.8
	30	50	99	107	122	277	134	96	15.7	17.0
		75	114	146	101	212	120	92	13.7	16.1
		100	130	171	80	143	108	88	9.8	9.8
	60	50	120	147	118	257	111	83	15.0	15.7
		75	136	168	94	201	106	86	12.8	15.7
		100	158	186	69	147	103	89	9.8	8.8

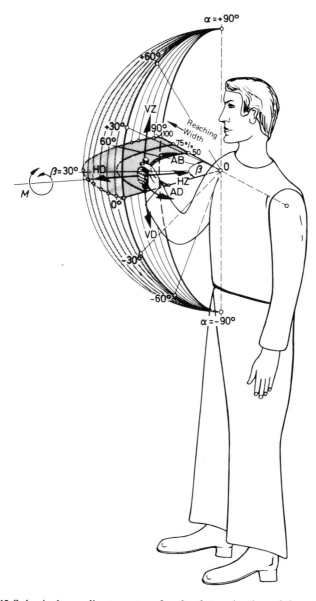

Figure 6.42 Spherical coordinate system for the determination of the point of load application and the systematics of positional forces of the right arm (for the better understanding of Table 6.7): HZ = horizontal tensile force; HD = horizontal pressure; VZ = vertical tensile force; VD = vertical pressure; AD = adduction (pull towards the body); AB = abduction (pull away from the body; M = torque. The accessible space is indicated by the spherical coordinates of range and angle α and β.

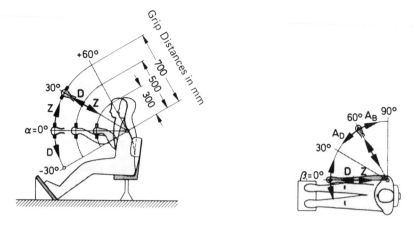

Figure 6.43 Systematics of the positional forces of the right arm of sitting test persons (for the better understanding of Table 6.8).

Table 6.8 Average maximum positional forces in N of the right arm of sitting test persons. The values shown here correspond to those in Figure 6.43. The highest values in each column have been highlighted

α in degrees	β in degrees	Reach in mm	Positional forces in N Back Z	Forward D	Up Z	Down D	Inwards AD	Outwards AB
		300	313	285	**161**	343	180	**127**
	0	500	367	528	116	248	154	106
		700	431	615	122	181	114	84
		300	322	297	157	**384**	180	1111
0	30	500	383	**635**	126	274	147	93
		700	409	556	108	195	112	80
		300	327	276	152	346	165	106
	60	500	379	554	127	260	152	84
		700	398	499	110	179	110	76
		300	300	315	156	275	**195**	107
−30	0	500	424	545	125	263	179	97
		700	**477**	484	105	196	129	83
		300	300	309	159	360	179	112
0	0	500	353	581	121	282	147	91
		700	375	467	107	182	114	78
		300	310	279	155	363	163	109
+30	0	500	351	623	120	252	131	89
		700	381	602	108	190	106	80

6.1.2 The Influence of Environmental Factors on Man's Performance

Environmental factors may influence man's performance level either positively, negatively, or not at all. Among the most important environmental factors in the workplace are lighting and color dynamics, noise and sounds, vibrations, climate, and chemical substances. Environmental influence factors, in general, can be perceived by means of the sensory organs. Some exceptions are chemical substances, invisible irradiation, and inaudible sound.

Light, Lighting, and Color. As mentioned before, approximately 80% of all environmental information is absorbed through the eye. The safety of man when handling technology depends to a very large degree on adequate lighting. Artificial light should always simulate daylight conditions, including the distribution of light and shadow, as well as the reproductive quality of color. Special situations, however, may favor specific frequencies or frequency distributions. In this connection, yellow light should be mentioned as a means of producing strong shadows. The lighting level, which depends on the intensity of light, influences human performance, mood, and the willingness to perform.

Activity, vigilance, and the ability to concentrate are enhanced by stronger lighting. Figures 6.44 and 6.45 show how performance and the feeling of fatigue, and error and accident frequency are dependent on the level of lighting. Figure 6.46 in combination with Table 6.9 explains the

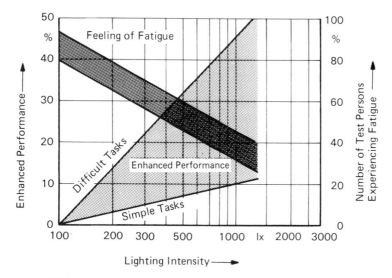

Figure 6.44 Enhanced performance and fatigue as a function of lighting intensity (schematic representation).

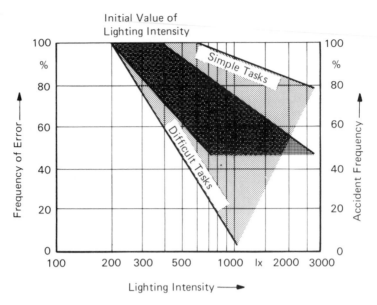

Figure 6.45 Error and accident frequency as a function of lighting intensity.

interaction between rated intensity of lighting, object size, and light density contrast.

Noise. Sound emissions which amount to a molestation or pose a danger to man are perceived as noise and are regarded as negative environmental factors. Noise is a mixture of pressure waves of frequencies from between 18 and 20,000 Hz.

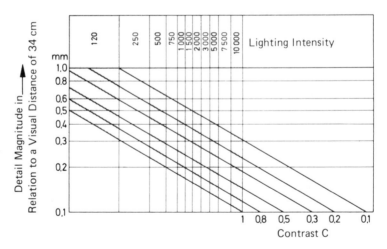

Figure 6.46 Determination of theoretical lighting intensity as a function of object size and light density contrast (compare Table 6.9).

Table 6.9 Example of theoretical light intensities (according to DIN 3035, p.2) (see also Figure 6.46)

Theoretical light intensity E in lx	Visual tasks	Examples
30	Orientation, short stay	Side rooms, unimportant corridors,
120	rough work, large details, and good contrasts	rough rolling in rolling mills, rough cleaning of large parts
250	normal visual tasks, details of medium size, good contrasts	Welding, turning, drilling, milling, locksmith's tasks, meeting rooms
500	More difficult visual tasks, medium details, medium contrasts	fine locksmith's work, assembling of larger parts, normal office tasks
1000	Very difficult visual tasks, small details, insignificant contrasts	Making of textiles, technical drafting, large size office contrasts
2000	Very difficult visual tasks, very small details, insignificant contrasts	Sewing, goldsmith's work, and watchmaking
5000 and above	Exceptionally difficult visual tasks in connection with danger	Lighting for surgical procedures, scientific experiments

The dispersion medium plays a very important part in the development of noise (Figure 6.47). Body noises and liquid noise will spread in solid as well as in liquid substances. The propagation speed of noise in solid substances is approximately 10 times and in liquid media 3–4 times as fast as in normal temperature air.

Sound is indicated as sound level L_p (sound pressure level). The definition Equation (6-1) applies in this case:

$$L_p = 20 \log \frac{P}{P_0} \quad \text{in dB.} \tag{6-1}$$

Here L_p stands for sound level in dB; p is the actual value of sound pressure in N/m^2 (square of the averaged air pressure fluctuations around barometric pressure); $P_0 = 2 \times 10^{-5} \, N/m^2$ reference sound pressure.

The reference sound pressure used here is the sound pressure of a sinus sound with a frequency of 1 kHz; the human ear is barely able to perceive it (hearing threshold). The sound energy penetrating through the area unit per unit of time is described as sound intensity, sound density, or sound level J. The value of sound intensity at the hearing threshold is approximately $J_0 = 10^{-12} \, W/m^2$. This value permits a definition of the sound intensity level with the help of Equation (6-2):

$$L_J = 10 \log \frac{J}{J_0} \quad \text{in dB.} \tag{6-2}$$

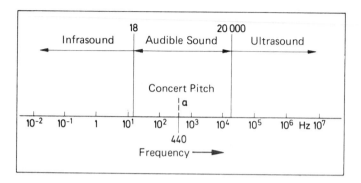

Figure 6.47 Frequency range of sound.

Figure 6.48 shows examples of various sound intensity levels.

The hearing threshold (minimum audibility field = MAF) is decidedly frequency dependent, as demonstrated in Figure 6.25 in Section 6.1.1. Man usually perceives two sounds of differing frequency but with equal sound pressure level, as unequally loud. This is why, for subjective

	Sound Intensity Factor	Sound Intensity in W/m²	Sound Intensity Level in dB (A)		Source of Sound or Noise
Adverse Range	100,000,000,000,000	10^2	140		Jet Engine (25 m Distance)
	10,000,000,000,000		130		Riveting Gun
					Pain Threshold
	1,000,000,000,000	10^0	120		Prop Plane at 50 m Distance
Danger Zone	100,000,000,000		110		Pneumatic Stone Drill
	10,000,000,000	10^{-2}	100		Metal Fabricating Plant
	1,000,000,000		90		Heavy Truck
	100,000,000	10^{-4}	80		Road with Heavy Traffic
	10,000,000		70		Passenger Car
	1,000,000	10^{-6}	60		Normal Conversation (at 1 m Distance)
Safe Range	100,000		50		Subdued Conversation
	10,000	10^{-8}	40	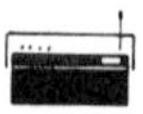	Soft Music
	1000		30		Whispering (at 1 m Distance)
	100	10^{-10}	20		Quiet Apartment in the City
	10		10		Rustling Leaf
	1	10^{-12}	0		Minimum Audible Field (hearing threshold) = MAF

Figure 6.48 Examples of different sound intensity levels. The sound intensity factor reveals the amount of sound by which a specific sound is "louder" than a specific sound close to the MAF. Most sound measurements are conducted approximately 7 m distant from the source of the sound or noise.

perception of sound intensity, a physiological determination of the sound must be carried through. Sound intensity levels L_n are measured in phon. Thus a sound level of 90 phon is understood to be a sound of any frequency spectrum which is perceived with exactly the same loudness as a sinus or pure tone of 1 kHz and a sound pressure level of 90 dB. Tests across the entire sound range will yield curves of equal loudness (Figure 6.49). These curves demonstrate spectral auditory sensitivity. The human ear is especially sensitive to frequencies between 3 and 4 kHz. Noise, to a certain degree, has an activating or stimulating effect on man. Attentiveness, reactivity, and the sensitivity of the sensory organs can be stimulated and sharpened. Once the effect of noise becomes overly strong, the result is a hyperactivation with subsequent reduction of performance. The optimal sound level is usually higher for simple tasks than for complicated ones. During difficult and complicated jobs, noise may have a distracting effect. New, unexpected, and high-intensity noises are transmitted by the switching centers of the nervous system to man's consciousness and interfere with concentration and attentiveness. At sound levels above 85 dB(A), these interferences will increase disproportionately. The result is human error and unsatisfactory performance. The danger of accident increases and reactions of alarm will occur. Simple activities and automatic routine jobs will be negatively affected by sound levels beyond 90 dB(A) only.

Sound may be detrimental in two ways. Via the influencing of the central nervous system, dispersed affectation of various organs may result. The most common form of damage to an organ is defective hearing. Other phenomena are increased general muscle tension (increase of the muscular action potential), decrease of the electric skin resistance, increasing blood pressure, markedly more with impulse sounds (alarm reaction), decrease in skin temperature, dilatation of the pupils, decreased production of

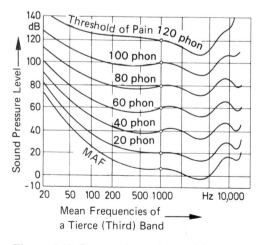

Figure 6.49 Curves of equal sound intensity.

gastric juices and saliva, increased breathing and pulse frequency, constriction of blood vessels of the skin, and disturbed sleep. Defective hearing may overlap age-related presbvakusia (Figure 6.50).

Furthermore, because of the different levels of noise sensitivity in humans, it is extraordinarily difficult to define generally acceptable noise limits which will prevent damage to a person's hearing. Under infrasound (frequency < 16 Hz) individual parts of the human body and organs can be stimulated to produce resonance vibrations. Under prolonged exposure to ultrasound (frequency > 20,000 Hz) the human body may respond with leucopenia. Sound or noise intensity levels above 3 W/cm², when applied for more than 10 min may cause the periosteum to become detached from the bone.

Vibrations. Mechanical vibrations in a frequency range of 0.5–300 Hz, and especially 80 Hz, are transmitted to the human being. Vibrations of over 300 Hz will be dampened by the body tissue; below 1 Hz they will move the entire body. Vibrations have a great variety of effects on man (Figure 6.51). One may distinguish between physical effects as a consequence of resonance frequencies and physiological effects on the functions of specific organs. Irregular vibrations put a heavier stress on man than regular ones. Table 6.10 contains resonance frequencies of various parts of the human body.

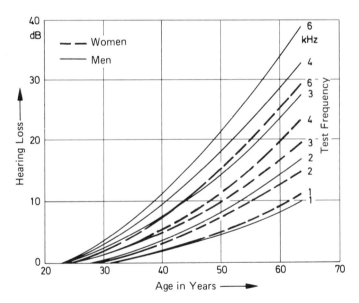

Figure 6.50 Loss of hearing within the various frequency ranges and with advancing age.

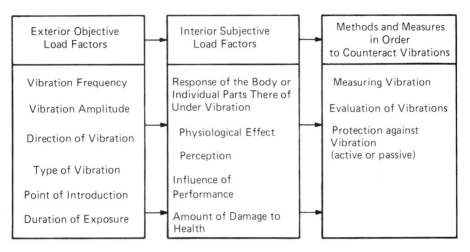

Figure 6.51 Load, stress, and reduction factors of mechanical vibrations on man.

Table 6.10 Resonance frequencies of various parts of the body and organs for vertical vibration excitation

Part of body, organ	Average resonance frequency in Hz
Main resonance of a standing person (shoulders, torso, and pelvis)	4–6
Head	20
Eyeball	80(40–100)
Jaw	6–8
Larynx	8
Bowels and abdominal wall	3
Stomach	4
Secondary pelvic resonance	10–12

The vibration resonance of the eyeballs leads to reduced visual acuity. The human eye can only follow movements of up to 2 Hz. Table 6.11 contains data on critical frequencies of subjectively determined symptoms.

The physiological effects of vibrations on man show a widely scattered pattern; they can be perceived directly and act on the circulation, secretion, digestion, musculature, the sensory organs, and the pulse frequency as well as the psychomotoric system. Under vibration stress, the pulse frequency can increase very strongly. The volume of air intake when breathing and the fibrillation frequency, within the range of resonance frequency of the entire body, show 4–6 Hz as maximum values. Kinetoses, so-called motion sicknesses, as well as air and sea sickness can be generated by influencing the vestibular system, i.e., the equilibrium mechanism.

Table 6.11 Critical frequencies of several subjectively determined symptoms

Type of symptom	Critical frequencies in Hz
General feeling impaired	4–9
Speech disorders through resonance of	
the jaw	13–20
through lack of breath	6–8
Trouble breathing	4–8
Back pains	8–12
Abdominal pains	4–10
Desire to urinate and empty bowels	10–18

The impairment of human performance by vibration can be explained to a great extent by the influencing of the sensory organs and the psychomotoric system. Other symptoms such as uneasiness, pain, etc., on the other hand, have not been explained sufficiently.

Climate. At an earlier stage (Section 6.1.1) details about the necessary constancy of the human body core temperature, the possibility of regulating temperature, and the heat exchange between man and his environment were explained. Climate is the umbrella term for all physical factors influencing the exchange of heat between the body and the environment. Man's well being, health, ability, and willingness to perform depend on climate to a very high degree. Air temperature (dry temperature), air humidity (vapor pressure), movement of the air (air velocity), and heat radiation essentially are the determining factors. In connection with climate, the concept of "comfort" was coined. Subjectively, the latter is judged in a variety of ways, and it is very difficult to create a climate which all human beings consider "comfortable."

The mean human body core temperature fluctuates very little ($\pm 1.0°C$). A compensated heat balance is thus a prerequisite for the ability of the organism to function and stay alive. Since man, when performing physical labor, attains a thermic efficiency of less than 10%, it is safe to assume that in excess of 90% of the energy expended (physical labor) will be in the form of heat which must be transmitted to the environment. A brief overproduction of body heat because of especially intensive muscular work may be absorbed by a temporary increase of the body core temperature. In keeping the body heat balanced, heat conduction, convection, radiation, and perspiration play important parts (Figure 6.52).

Under the influence of heat, blood circulation of the skin increases. Furthermore, higher pulse frequency, thirst, fatigue, fainting, and, in special cases, heat stroke will be the possible consequences, respectively. Air conditioning, protective clothing, and work breaks permit cooling

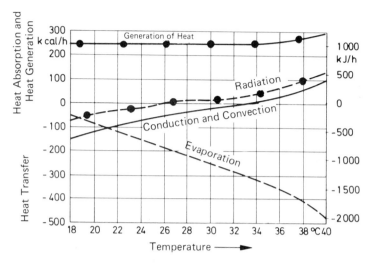

Figure 6.52 Components contributing to heat exchange (for details see text).

down when a person has been exposed to heat; they also allow the body to recuperate and thus resume its activities with renewed vigor and added safety. Eventually, the organism may adapt to heat exposure: it is able to acclimatize itself.

There are many training methods, all of which strive to acquaint the body's mechanisms of heat regulation via endogenous heat generation. Extreme climatic conditions may bring about a drop in performance. However, there is no unequivocal correlation between the influence of the climate and human achievement, because the duration and stress caused by the job in question as well as clothing and suitability of the individual for the type of work play important roles. In addition, the state of health, condition, and motivation exert considerable influence.

Chemical Substances. Chemical substances may act on man in the form of gases, vapors, dust, and liquids. Harmful liquids, in general, are more easily recognized than poisonous gases, vapors, and dust. For the protection of man against noxious substances, so-called MAK-standards (maximum concentration at the work place) as well as MIK-standards (maximum concentration of ambient air) have been determined. Table 6.12 shows a number of MIK and MAK values.

The variety of chemical substances present in the environment leads to a great many different symptoms and impairments of a person's health which, in part, may even be irreversible. Acids, lyes, solvents, etc., often bring about skin damage. Resorption via the skin, respiratory system, and gastroenteric tracts can lead to fatigue, impaired concentration and

Table 6.12 Comparison of several MIK and MAK values

Substance	MIK value in mg/m^3	MAK value in mg/m^3
Sulfur dioxide SO$_2$	0.3	13
Sulfur hydrogen H$_2$S	0.15	15
Nitrogen dioxide NO$_2$	1.0	9
Chlorine Cl$_2$	0.3	1.5

attentiveness, and disturbances in the various organs. Ventilation, exhaust systems, and filtration, as well as other personal protective measures will reduce the influence of chemical substances and thus forestall dangers to a person's health.

6.1.3 The Medical Suitability of Man for Handling Technology

Suitability for handling a technical system, in a general sense of the word, can be understood as the matching of the performance profile of a human being with the requirement profile of the technical system in question. The performance profile is determined by training-related experience and knowledge and by the anatomical, physiological, and psychological characteristics of man. Often there are individual criteria of aptitude which must be taken into account. If, e.g., the actuation of a mechanical valve demands a high degree of muscular power, the latter becomes such a special aptitude criterion. For the evaluation of the suitability of man for a certain type of job, so-called "aptitude checkups" have been introduced. In important cases, such checkups are required by law. Thus, any person working under high air pressure must undergo regular checkups to the extent prescribed by the law. Many regulations contain aptitude criteria for workers handling specific technologies. If there is a suspicion of a health hazard, the employee will be barred from taking on that specific job.

As a prominent example, we cite the preventive medical examinations in occupational medicine for "Driving, Controlling, and Supervisory Activities." The psychological and psychomotoric aspects are explained in Section 6.2. The range of application for aptitude checkups includes employees for the following types of jobs.

Driving Jobs. Driving or operating a passenger car, motorcycle, truck, bus as well as taxicabs, rail vehicles, streetcars, subway trains, ships, and boats on inland waterways (also as pilots), industrial trucks with lifting devices (forklift trucks), industrial trucks without lifting devices, hoists and lifts of high capacities (e.g., traveling cranes, foundry cranes), hoists and lifts for lower capacities (e.g., floor cranes, towing cranes), charging machines and ladle cars, manipulators, dredgers, graders, dumpers, bull-dozers, power shovels, dump trucks, and motor rollers.

Controlling Activities. The control of material, handling of cable-car machinery, controlling jobs with high requirements (e.g., rolling mills), and controlling activities with lower requirements (e.g., transfer lines, mixers).

Supervisory Activities. Supervisory activities of a higher degree of require-ment (e.g., railroad signal boxes), and with lower requirements (e.g., control panels, etc.).

Tables 6.13 and 6.14 contain the minimum requirements for visual acuity, plastic vision, sense of color, field of vision, sense of light, and auditory acuity. Persons who do not conform to these minimum require-ments are considered permanent risks: they are not permitted to hold specific jobs. Beyond the above, a number of additional health impair-ments will classify certain persons as "permanent risks". Among them are:

—temporary loss of consciousness or equilibrium or other physical de-ficiencies of any kind;
—diabetes mellitus with considerable fluctuations of the blood sugar values;
—considerable malfunctions of the glands with internal secretion, espe-cially the thyroid, the epithelial bodies, or the suprarenal bodies;
—chronic alcohol abuse, dependence on narcotics, or other addictions;
—diseases or changes in the heart or circulatory system with accompaning restrictions of the ability to perform or react, changes in blood pressure, of an advanced magnitude, post-infarctial status;
—tendency to renal or gall bladder colics;
—considerable impairment of mobility, loss or decrease in the strength of a limb or extremity necessary for the completion of a specific job;
—diseases or damage to the central or peripheral nervous system with essential dysfunctions, especially organic diseases of the brain or the medulla, dysfunctions as a consequence of brain or skull damage, defective cerebral circulation, psychoses or mental diseases, even after they have subsided, but in the case where a relapse cannot be excluded with a sufficient degree of certainty, abnormal behavior of a consider-able magnitude, mental deficiency;
—disturbances or significant impairment of the maximum stress of the central nervous system.

The exemplary demonstration of an aptitude test shows the problems which must be overcome in the course of integrating man into the man–machine–environment–system (MME–systems).

Table 6.13 Physical requirements for the job group "driving, controlling, and supervisory activities." Requirement categories for characteristics to be tested during initial aptitude tests

Activities	Visual acuity distance	Visual acuity proximity	Plastic vision	Sense of color (color perception)	Field of vision	Light perception	Hearing acuity
Driver's activities							
Limousine driver	2	—	2	3	2	1	3
Motorcycle driver	3	—	—	3	3	2	3
Truck driver	2	—	1	3	3	2	3
Bus/taxi driver	2	2	1	3	3	1	3
Rail vehicle conductor	2	1	—	1	—	—	2
Motorman on a subway train	2	—	2	3	2	2	2
Subway train driver	2	—	—	1	—	—	2
Inland waterway navigation (skipper and pilot)	2	1	2	2	2	—	2
Inland waterway navigation (deckhands)	2	—	2	2	2	—	2
Operators of industrial trucks (with lifting devices)	2	—	2	—	—	—	3
Operators of industrial trucks (without lifting devices)	3	—	—	—	2	—	3
Hoist and lift operator (high requirements)	2	—	1	—	2	2	3
Hoist and lift operator (low requirements)	2	—	2	—	2	—	3
Charging machinery operator and cable car operator	2	—	—	—	2	—	3
Operator of manipulator	2	—	—	—	—	—	3
Drivers of dumpers, loaders, graders, dredges, bulldozers	3	—	—	—	—	—	—
Power shovel drivers (operators)	3	—	—	—	—	—	3
Dump truck operator	3	—	—	—	—	—	3
Street roller operator	3	—	—	—	—	—	—
Controlling activities							
Operator of materials handling system or cable car systems	2	1	—	—	—	—	3
Control console operator	2	1	—	2	2	—	3
Simple control activities (e.g., transfer line)	3	1	—	2	2	—	3
Supervisory activities							
Switch panel operator, instrumentation, measuring instruments	2	1	—	2	—	—	2
Station master railroad, operating the signal box	2	1	—	2	2	—	2

Requirements categories: 1 = Requirement above average.
2 = Requirement average.
3 = Requirement moderate.
— = Of no importance for the job in question.

208

Table 6.14 Physical requirements for the job group "driving, controlling, and supervisory activities." Requirement categories for characteristics to be tested during follow-up examinations

Activities	Visual acuity distance	Visual acuity proximity	Plastic vision	Sense of color (color perception)	Field of vision	Light perception	Hearing acuity
Driver's activities							
Limousine driver	2	—	0	3	2	2	0
Motorcycle driver	3	—	—	3	—	2	0
Truck driver	2	—	0	3	3	2	2
Bus/taxi driver	2	—	1	3	1	2	2
Rail vehicle conductor	2	2	—	2	2	2	2
Motorman on a subway train	2	—	2	3	2	2	2
Subway train driver	2	—	—	2	—	—	2
Inland waterway navigation (skipper and pilot)	2	0	2	2	2	—	2
Inland waterway navigation (deckhands)	2	0	2	2	2	—	2
Operators of industrial trucks (with lifting devices)	2	—	2	—	2	—	0
Operators of industrial trucks (without lifting devices)	3	—	—	—	2	—	0
Hoist and lift operator (high requirements)	2	2	1	—	2	2	0
Hoist and lift operator (low requirements)	2	—	2	—	2	—	0
Charging machinery operator and cable car operator	2	—	—	—	2	—	0
Operator of manipulator	2	—	—	—	—	—	0
Drivers of dumpers, loaders, graders, dredges, bulldozers	3	—	—	—	—	—	—
Power shovel drivers (operators)	3	—	—	—	—	—	0
Dump truck operator	3	—	—	—	—	—	0
Street roller operator	3	—	—	—	—	—	—
Controlling activities							
Operator of materials handling system or cable car systems	2	2	—	—	—	—	0
Control console operator	2	2	—	2	2	—	0
Simple control activities (e.g., transfer line)	3	2	—	2	2	—	0
Supervisory activities							
Switch panel operator, instrumentation, measuring instruments	2	2	—	2	—	—	2
Station master railroad, operating the signal box	2	2	—	2	2	—	2

Requirements categories:
1 = Requirement above average.
2 = Requirement average.
3 = Requirement moderate.
— = Of no importance for the job in question.
0 = This characteristic is not being examined during a follow-up medical examination.

6.2 Psychological Aspects

6.2.1 Minimum Requirements for Psychological Functions

The planner of an MME–system should have an idea of the psychological nature of the working man, the operator in the system. What mental abilities can and must the planning engineer take into account?

This question cannot, as in Section 6.1, be explained by referring to individual organs and their functions, since mental functions cannot be isolated nor can their observable effects be treated as independent variables. Attempts to isolate psychological variables have not led to any solution of the problems facing technical planners. It should, however, be pointed out that the total experience gained in applied psychology has been much less than that gained in physiology, and much of this experience has been obtained in the field of traffic psychology. Thus a certain one-sidedness will become evident in the discussion of this topic. Minimum requirements relating to mental functions have not yet been determined in a manner which would afford a convincing degree of safety for technical planning. This can be readily seen if one looks at the minimum intelligence required of motor vehicle drivers [6-29]. The limit is set at an IQ of 70, as measured by Wechsler's method, which is based on a scale giving a mean value of 100 and having a standard deviation of 15 points. On the basis of this limit, it can be seen that 2% of applicants would be unsuited as drivers of motor vehicles on the basis of an insufficient level of intelligence. Here intelligence is understood as being the manifestation of certain mental traits as measured using psychological techniques. Intelligence, it is true, is an important human characteristic, closely related to success in learning and in work; however, its relationship with criteria of adaptability to vehicular traffic has yet to be proven [6-30].

Similarly, minimum requirements for psychological characteristics have recently been established in industry for driving, controlling, and monitoring activities, which can be tested in a relatively simple manner. Burkhardt and Hahn [6-31] have developed a battery of five tests for the examination of the selected variables (Table 6.15). The tests are carried out in such a way that they can be terminated as soon as it is established that the subject has reached the minimum requirement. This type of psychological diagnostics corresponds to the basic idea that the costs of testing can be kept as low as possible if they do no more than to ascertain that the minimum requirements have been reached.

Psychological tests generally allocate a numerical value to the behavior of a person within the test situation. The reference system for most psychological tests is the manifestation of the tested characteristic in the total population. This is based on the assumption that the psychologically relevant characteristics in the total population conform to the normal distribution. The most frequently used scale for the characterization of

Table 6.15 Psychological requirements for the activity groups: driving, controlling, monitoring. (Source: Committee for Occupational Medicine): Minimum requirements for medical examinations set by the professional association for occupational medicine. Published by the Organization of Professional Associations, Bonn 1980 [6-28]. Requirement level for characteristics to be tested at an initial examination

Activities		Reaction time	Attention	Motor skill	Practical technical understanding	General intelligence
Driving	Car driver	1	2	2	3	2
	Motorbike rider	1	2	2	3	3
	Truck driver	1	2	2	2	3
	Bus/taxi driver/Driving instructor	1	1	2	2	2
	Train driver	2	2	3	2	3
	Tram driver	1	1	2	3	3
	Subway train driver	2	2	3	3	3
	Inland waterways (helmsman and pilot)	–	1	–	2	2
	Inland waterways (deck personnel)	–	3	–	3	3
	Driver of industrial truck (with lifting gear)	2	2	3	3	3
	Driver of industrial truck (without lifting gear)	2	2	3	3	3
	Hoist operator (high demand)	2	2	2	2	3
	Hoist operator (low demand)	3	2	3	3	3
	Charging machine and cable car operator	2	2	3	2	3
	Manipulator operator	2	2	2	2	3
	Dumper driver, earth mover driver	3	3	3	3	3
	Bulldozer driver	3	3	3	3	3
	Skip truck driver	3	3	3	3	3
	Road roller driver	3	3	3	3	3
Controlling	Hauling track and cableway operator	2	2	3	2	3
	Control platform operator	1	1	1	2	2
	Simple control activity	2	2	3	3	3
Monitoring	Control room operator	3	1	–	2	2
	Station master/signal box operator	2	1	–	2	2

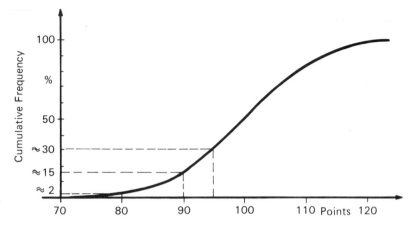

Figure 6.53 Standard distribution of characteristics within the population. Distribution of the characteristic in points (mean = 100 points, standard deviation = 10 points); proportion of the population unable to attain a particular level in the characteristic, in percentages.

behavioral patterns measured in such tests is the standard deviation, to which a mean value of 100 and a standard deviation of 10 points has been allocated. The minimum requirements based on the standard deviation are defined in three levels. Level 1 corresponds to 95 points, by which approximately 30% of the general population would not fulfill the requirement. Level 2 corresponds to 90 points, by which approximately 15% would not fulfill the requirement. Level 3 corresponds to 80 points, by which approximately 2% would not fulfill the requirement (Figure 6.53).

Special requirement profiles have been drawn up according to the five tested variables for individual occupational groups (see Table 6.15). The planning engineer can use this as a tool for personnel selection. For activities other than driving, controlling, and monitoring, however, no minimum requirements for psychological variables have been established. This represents a major area for future research.

6.2.2 Perception and Information Processing

As previously mentioned, man receives information via his sensory organs, principally through the eye, the ear, and the cutaneous senses. Perception is an active process of the organism. The environment is not simply depicted through the sensory organs; on the contrary, selection and interpretation of the incoming data are performed, as has been described within perceptual psychology [6-32].

The so-called "bottleneck" model of perception illustrates the extent of information reduction in perception and the limited ability of man to act as an information processor (see Figure 6.54). Studies of visual perception

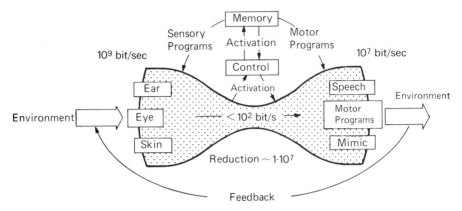

Figure 6.54 The "Bottleneck" model of perception (from Kreidel [6-33]).

have shown that one attempts to extract the structure of the information presented, even where none exists. In the latter case structure is imposed on the information. These findings are the basis of the principle of form building, which although derived from studies of visual perception is equally valid for the other senses, It also forms the basis for design recommendations for the control and monitoring of workplaces, which are discussed in Section 6.3. This principle need not be discussed further at this stage.

Previous experience and knowledge, as well as form, play an important part in the perception and processing of information. This can be seen in an example cited by Senders [6-34]. After an incident in the control room of a U.S. power plant, the question was asked, "Who actually bears the responsibility for human error?" In the control room there was a tempera-ture recorder with two pens, one of which recorded the temperature of cold water in red, the other which recorded the temperature of steam in blue. Since most people associate the color red with hot and the color blue with cold, the operators found it necessary to affix a sign with the message, "Remember!—Red is Cold!" In planning the installation, the designers had failed to consider the humans who were to operate the con-trol consoles. Specific elements of display systems, which should facilitate the perception of information, often relate to population stereotypes. The compatibility between these stereotypes and the design of the information presentation is of vital importance if error-free perception and processing of information is to be achieved. This topic is discussed further in Section 6.3.3.

Error-free information processing is generally more dependent on meaningfulness in the given context than on the maximum decoding performance of the cognitive system. The principle of building forms is valid even for cognitive data processing, whereby the meaning of a signal is often perceived immediately, without any extended search behavior involving a multitude of individual decisions. Human beings, it is true, are generally slower and more error prone than data processing machinery in decoding performance; they are not as good at measuring or counting as the machine. The human is, however, more flexible, does not need such accurate preprogramming, and can process information in uncoded form. Humans can also apply inductive reasoning in the solution of problems, whereas the machine can only "think" deductively, and even then only when specially programmed. Man can generalize experience and thus apply acquired principles to other problems. For information processing during control activities, memory and recall are of decisive importance both for the man and the machine. Humans possess a large long-term storage capacity which is accessible in a variety of ways. The machine on the other hand has a large short-term storage capacity. Man can remember specific facts very fast and at exactly the right moment, while the machine requires precisely coded commands. Machines can recall large amounts of information within a short period; humans can recall large amounts of information neither rapidly nor dependably. Since the machine is unable to anticipate and improvise, man is the better supervisor in critical situations. However, the supervisory activity can only be effective if the work stations are designed according to ergonomic standards and if human limitations and inconsistency are taken into account. In this connection, the scheduling of work and leisure periods is also of importance when considering means to reduce the effects of work-related stress.

6.2.3 Mental Strain and Its Consequences

Mental strain can be explained by means of the arousal model illustrated in Figure 6.55. This model is based on empirical findings concerning the interrelationship between performance and arousal (the Yerkes–Dodson law). The model shows that the mental work load directly influences arousal, and that very low cognitive demands as well as excessively high psychological stress will have a negative effect on efficiency and performance. At a middle level of arousal, performance capacity is at its highest. Doing one's job under these optimum arousal conditions will entail a minimum of mental strain.

An excessively low level of arousal will lead to boredom. This is usually found in repetitive jobs devoid of any change (production lines with short activity cycles) and especially in situations where the following factors come into play:

—Minimum to no more than medium task difficulty (e.g., continuous adding of two digit figures).
—Narrow fields of perception (e.g., concentration on a very narrow work area, where holes are continuously punched at predetermined locations).

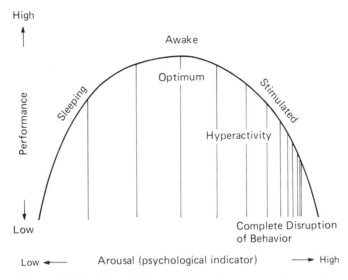

Figure 6.55 Relationship between performance and arousal.

—Prolonged uninterrupted activity (more than about 30 minutes).

Whenever these conditions are present at an isolated work station the probability of boredom is relatively high.

Monotony has effects similar to mental fatigue in that it will produce a decrease in performance quality and quantity, and will be accompanied by sleepiness and demotivation. The characteristic difference between boredom and fatigue lies in the length of the required recuperation period (see Table 6.16). Problems with keeping awake, fatigue, and vigilance are usually found in jobs involving observation, inspection, and monitoring, for example the operator in the control room of a fully automated plant, the radar operator on a ship's bridge, or inspecting the surface quality of sheet metal in a cold-rolling mill. Vigilance studies have shown that after approximately 30 minutes of uninterrupted monitoring, the attentiveness of the observer decreases noticeably. During a prolonged vigilance study of two hours involving 22 seamen, the frequency of missed signals increased rapidly after the first 30 minutes (see Figure 6.56). Performance during vigilance studies is measured as a percentage of the signals detected and correctly responded to per unit of time. If one considers the observer's performance to be a function of the mean number of signals requiring response per unit of time, the graph shown in Figure 6.57 is given. At a rate of 20 signals per hour, approximately 30% of the signals are missed. At a rate of between 200 and 600 signals per hour, the signal detection rate is optimal. At more than 600 signals per hour, where response is required the performance will again decrease (due to over-arousal).

Table 6.16 Comparison between the phenomena of monotony and mental fatigue

Phenomenon	Principal task	Course of activation	Effect on performance	Required recuperation time
Monotony	Continuous activity with relatively low stimulation	Undulating	Impairment of the readiness to perform	Very short, a change of activity is usually sufficient
Mental fatigue	Continuous activity with relatively high stimulation	Falling	Impairment of the ability to perform	Relatively long, can only be countered with prolonged rest periods or sleep

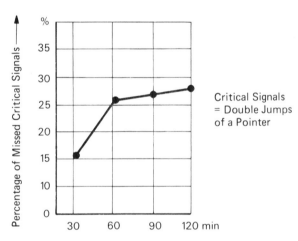

Figure 6.56 Results of a vigilance study: vigilance performance (detection of double jumps of an instrument pointer) as a function of the duration of the monitoring activity.

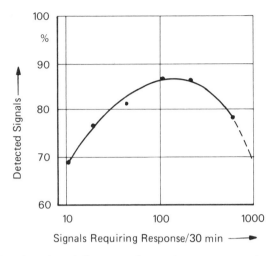

Figure 6.57 Results of a vigilance study: performance as a function of signal frequency (source: Schmidtke [6-20]).

6.2.4 Human Error and Accident Proneness

Human behavior patterns are extremely variable; the performance of each individual has a wide dispersion pattern. Nobody will perform the same activity twice in exactly the same way. A human action is referred to as an error if it exceeds a certain preset tolerance limit. Every human activity thus presents an opportunity for error. The number of errors made is therefore dependent on where the tolerance limit is set.

The possibilities for error are innumerable. According to Murphy's Law, "Anything that can go wrong, will go wrong." Concerning the frequency of human error, Rigby [6-35] can be quoted with authority: "Within the industrial area one may expect that for discrete, work-related activities like reading a five-digit figure, actuating a control element or putting a certain part in its correct place, there will be an error rate of 1 per 1000 to 10,000 operations. Approximately 80 to 90% of these errors will be detected through normal checks and inspections and only 20 to 30% of the undetected errors will entail significant consequences. Thus the probability of a major error occurring during a discrete activity and having grave consequences is often as small as 0.00006 to 0.0000004."

In this context the term "human error" is meaningless because it fails to explain why errors occur. It is even dangerous; since human beings will always make mistakes, the use of such a misnomer means that one must accept that errors arising from a particular source will continually recur. It is also irrational to punish a human for behaving in a human way; that is, for occasionally making mistakes. The ergonomic approach is based on the consideration of human abilities and the prerequisites for reliable work. Work stations, tasks, and procedures, as well as the physical environment should be designed according to human capabilities and limits. In this way it is possible to achieve an increase in the reliability of human performance.

In the past the concept of accident proneness played an important part in connection with the problem of accident prevention, especially for work in areas with unavoidable or environmental hazards. The question here is whether there actually exists an increased probability for the occurrence of accidents to particular individuals. If this were the case it would be possible to identify people with a high individual risk. The distribution of accidents to individuals has been analyzed according to safety research criteria in order to determine whether individual accident frequency is a useful concept for the promotion of safety. This concept led, however, to somewhat excessive expectations. It was thought that occupational safety could be improved simply through the exclusion of "accident prone" individuals. In normal working life the possibility of excluding such individuals is obviously very limited. The validity of studies based on this concept, the majority of which produced results only bordering on significance, often led to wrong prognoses which practicing psychologists could hardly maintain [6-36].

One must not overlook the fact that conventional personnel selection criteria already promote increased safety at the workplace. Thus one can assume that, within the framework of an in-plant personnel safety analysis, the work groups encountered will already have been cleared of individuals having a poor safety prognosis. The existence of individuals who are either pronouncedly accident prone or are habitual offenders has been shown in findings concerning largely unselected groups, for example motor vehicle drivers [6-37]. Two major predictors of accident proneness are a lack of sociocultural integration and a tendency towards excessive or regular consumption of alcohol. These factors are also interconnected; alcoholism can lead to social ostracism, and lack of social integration (e.g. marital troubles) may in turn lead to alcohol abuse and alcoholism. Both phenomena are often related to frequency of job changing, which in turn may be related to safety.

Processes involving social ostracism and developments towards alcoholism do not occur solely in working individuals, but are independent of the individual's work status. Changing a person's habits and developing independence from alcohol is a long and tedious process, and success is associated with great difficulty. For motor vehicle drivers programs for retraining and rehabilitation have been established [6-38]. Because the rehabilitation program within the working environment can be supported with other measures, the chances of success for such programs within a company are much greater than for the more or less sporadic attempts to rehabilitate drivers. The social climate within the company, however, has to be adapted to an appropriate valuation of alcohol consumption. This topic will be discussed further in Section 6.4.

One must not overlook the fact that the achievement of a high degree of personal safety is dependent not only on appropriate design but also on the social climate and an awareness on the part of the individual. Reprimands achieve nothing.

6.2.5 Risk Compensation Through Human Self-Control

The dangers to people inherent in a specific work process within the man–machine–environment–system are often not directly perceived, but merely estimated. In this estimation process considerable errors can occur. Studies in the coal and ore mining industries concerning the relationship between actual and estimated risk have shown that there are only relatively insignificant correlations between the range of objective risks and subjective assessment. This does not mean that the people involved are incapable of differentiating between the various work procedures, but rather, as Burkhardt [6-36] points out, that errors in the estimation of danger are made, which means that the possibility for accident-free work will be low.

Errors in the estimation of risk are attributed by Mort la Brecque [6-39] to the type of faulty thinking which may be considered normal according to psychological principles; "the availability of information determines the results of the thought process. Information which can be recalled easily and which lie in the forefront of consciousness overrides other information which one does not remember or even wishes to re-press."

As far as the estimation of risk is concerned, this means, for example, that the constant emphasis of certain dangers in the media sharpens the consciousness of these—mostly spectacular—dangers, even if they are actually much less frequent than others. A film like "Jaws" exaggerates the actual danger of being attacked by sharks to a grotesquely disproportionate degree.

In connection with road traffic, with its special condition of self-determination of the rate of work, studies by Taylor [6-40] and Wilde [6-41] have indicated that there is an adequate awareness of the risk content in particular situations. Both authors carried out field experiments which showed that drivers regulate their behavior within the MME–system so that the risk they experience fluctuates only between narrow limits. Thus Taylor reported that drivers tend to increase the speed of the vehicle in situations in which low risk is perceived, and to reduce speed if the perceived risk increases again. The perceived risk was determined in these experiments using the psycho-physiological measure of the galvanic skin response. In a similar experiment Galton and Wilde [6-42] showed that the subjective risk was determined on the basis of a verbally formulated risk scale. The test route consisted of 11 sections with heterogeneous traffic conditions, ranging from urban traffic to a four lane expressway. Driving speeds in the various sections varied between 40 and 110 km/h (25 and 65 m.p.h.). Despite differences in driving conditions, the various risk assessments did not differ significantly between the 11 sections. Wilde interprets this as showing that subjective risk assessment is independent of the prevailing traffic conditions. He concludes that drivers "keep variations in situation demands and risk perception as low as possible." According to Wilde, this indicates that "a driver is prepared to accept and tolerate a risk which he perceives as being somehow convenient, suitable, or even useful to him." He discusses a "homeostatically regulated process." The fact that a risk can be perceived as being too low during driving which can be compensated for by increasing the situation demands, particularly through increasing speed, has led to the formulation of the "Risk Compensation Theory" to account for such behavior patterns.

The risk compensation hypothesis derives from the results of field experiments in which drivers performed their tasks under observation. Wilde expresses it as follows: "The level of perceived risk minus the driver's effort to reduce subjective risk—i.e., the degree of caution—is a constant." This constant represents the degree of tolerated risk.

One may assume that the initial psychological analysis and the degree of fatigue during these experiments were relatively constant. The question of whether the individual risk constants remain unchanged over longer periods and under differing environmental conditions therefore remains open. If this was the case, then accidents could not be attributed to the driver being in a hurry or other emotional states independent of the traffic conditions [6-43]. Further, Küppers [6-44] has shown, on the basis of an examination of expressway accidents (see Figure 6.58), that the risk of accidents increases during the course of the day. The risk compensation theory is therefore applicable only within relatively narrow limits. In all situations involving actual risk, the driver simply seems unable to keep the actual risk at a safe level which would exclude accidents. It is doubtful whether theories derived from road traffic studies can be generalized to the man–machine–environment–system of working life. The driver's task differs from many other tasks at work in that driving may often be seen as being rewarding in itself ("One does not travel for the sole purpose of arriving": Goethe), whereby a certain value is placed on overcoming self-created risks.

6.2.6 Ability to Concentrate and Motivation

The degree of concentration on a given task is usually dependent on interest and the satisfaction associated with exercising particular task-oriented skills. This statement is, of course, also valid in working life. Concentration is the active dedication to an object or a task. This dedication consists of the employment of physical or mental forces over a period of time. The expended effort is usually noticed only after the period

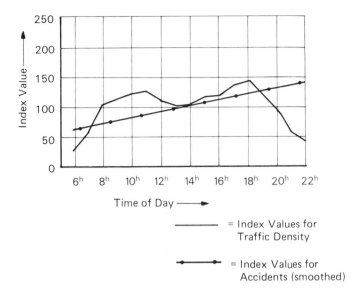

Figure 6.58 Index values (arithmetic mean = 100) of traffic density (cars and trucks) for accidents during the course of a day from data from two expressway sections on normal working days [6-44].

of concentrated activity has come to an end. Concentrated work can be easily achieved when the job holds interest for the working person or the product is of immediate value to him. Poor concentration is not a physical defect which can be overcome with the help of medication, nor is it the lack of some psychological state which one might call "the ability to concentrate." Tests designed to measure the ability to concentrate must therefore always include the provision that high motivation is always present in the test situation. Only then will the test results produce a credible statement. As soon as interest wanes, concentration weakens. When difficulties arise the dedication demanded by others, with which one does not identify, is interrupted. Thus someone can actually be unfit for work while still able to pursue his hobbies. Since safety-oriented behavior is often the most removed from internal inclinations, the effects of fatigue are often accompanied by an increased disregard for safety. This can be measured through an increase in the error rate or the adoption of dangerous behavioral patterns. Lack of concentration is thus a phenomenon which arises early with externally determined requirements and is to be understood as being a sign of excessive mental stress. Failures or reprimands from superiors rapidly put unstable workers under stress, so that the ability to concentrate is further reduced. On the other hand reward and experience of success contribute to an increase in concentration. The key function of motivation for concentrated action can also be determined here. Guidance in error-free and safety-conscious work behavior forms a base on which respect and trust between employees and their supervisors can be built up. Within an externally determined task, a person can discover what he can actually achieve, once he has made that task his own. The aforementioned methods for measuring concentration will therefore only produce credible results if the motivational state of the subject is known.

6.2.7 The Loss of Aptitude-Related Characteristics

In the context of controlling MME–systems and the adherence to set safety standards, the question is often raised as to whether a certain worker is still suitable for a particular technical installation. This question also has implications for personnel assessment and the selection of individuals for particular tasks. However, the problem here is to consider whether the changes which take place in a person during a lifetime may eventually prohibit the employment of that person on the grounds of safety.

In general the physical and psychological bases of action are subject to age and health-related changes which, with advancing age and for specific physical defects, can mean that the person concerned is no longer suited for particular activities or workplaces. Reduced fitness has particularly important implications for driving, controlling, and monitoring activities, as well as for the execution of physical activities which demand a particular degree of elasticity of the body. The prerequisites for the accomplishment of such tasks may already have been lost when the required maturity has been reached.

In driving behavior it is generally accepted that considerable defects can be overcome if they arise during the course of the driver's occupational life, i.e., when they do not interfere with the early training phase. The driver has available a number of possibilities for compensating for his defects through restrictions which he imposes on himself and which he adapts to his existing physical and mental powers. He will, for example, decrease the speed in accordance with his indisposition, thus slowing down the succession of stimuli and giving him more time to react.

Despite the slowing down of the physiological and psychological reactive processes during the course of a lifetime, it is impossible to demonstrate an increase in the risk factor for older drivers if the beginner's risk is used as a yardstick for the magnitude of risk. This is illustrated in Figure 6.59 which shows the indices for degree of accident risk for car drivers in individual age groups. The compensation of the individual accident risk by selecting more prudent speeds is generally achieved by the average driver only after going through a practice phase characterized by an especially high average risk. The increase of a driver's risk at an advanced age is insignificant in comparison.

At this point it can be seen what the risk compensation theory should express; it is possible to compensate for risks arising from human defects through the will to achieve safety. Indeed many elderly people are very careful to conserve the faculties still left to them and thus remain fully

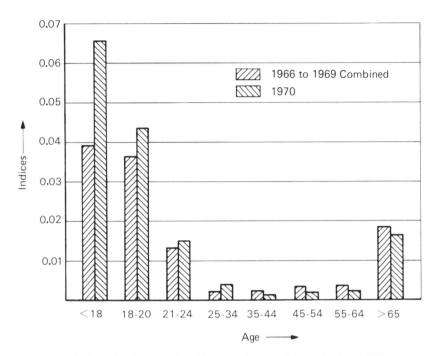

Figure 6.59 Indices for degree of accident risk for car drivers for 1966–1969 and for 1970 (from Heller [6-45]).

capable of participating in many work processes, even though the strength they possessed when younger is no longer available and their physical functions are slower or restricted. Only he who goes beyond his capabilities is courting danger. This is, however, a characteristic of youthful activity during work or in traffic, rather than of old age.

This does not mean, however, that everyone succeeds in setting the appropriate level for their physical and mental state. There are many people who, in a wide variety of situations, misjudge their capabilities and take too much upon themselves. Such a behavioral tendency may escalate to pathological proportions and is then referred to as "behavioral idiocy," which means that the individual would not appear to be so unbalanced if he would refrain from taking so much upon himself. In order to assess a person's suitability for a particular workplace, both before hiring and during the course of his working life, it is of the utmost importance to evaluate his capability for self-assessment and self-control. Many workplaces have in common with traffic the fact that the individual can regulate the pace of his own actions within specific limits. Thereby he is in a position to compensate for physical defects affecting his speed of reaction, be they of a passing or a permanent nature, by selecting a slower succession of stimuli in order to achieve a safer behavioral pattern. The same is true for mental processes as well as for physical actions. For work processes which are not self-paced, the approach elaborated from the risk compensation theory for personnel assessment does not apply. Work safety problems in this area remain unresolved. There are still no criteria available for the determination of the necessary reaction or manipulation speeds at individual work places. In the long term, however, the necessity for the determination of minimum parameters of psychological capability for particular risk-laden activities can no longer be avoided.

6.3 Ergonomics

Ergonomics is concerned with the scientifically based design of work. The aim is to increase the output and efficiency of the work system, to create optimal and humane working conditions, and to improve safety at work.

Because humans often constitute a special problem within the man–machine–environment–system, it is of vital importance that human capabilities and limits be taken into account in every phase of system development. This is the task of systems ergonomics, which may be regarded as fitting the technology to the man.

Beyond this, ergonomics deals with further topics. A suitable representation of all functions of the MME–system between the technical components and the human components must be achieved. This is important, for example, for the determination of the degree of automation. In addition, the relationship between safety at the workplace and the safety of the plant must be emphasized. In many areas of modern technology, the incorporation of systems ergonomics has become imperative. The landing of a manned spacecraft on the moon without a major contribution of systems ergonomics to the Apollo program would have been impossible.

6.3.1 Application of Established Knowledge

The ergonomist has access to the scientific findings on the capabilities and limitations of man from anatomy, physiology, and psychology. Many ergonomic problems can be solved on this basis without the necessity for utilizing empirical methods. This approach can be illustrated by means of an anthropometric example.

Figure 6.60 shows clearly what happens if human body dimensions are not considered when designing pieces of equipment. In the upper half of the illustration, a man of average build stands in front of the controls of a commercially available lathe. It is readily apparent that the controls have been placed very unfavorably. In order to reach all the controls comfortably the operator would have to have the body dimensions shown in the lower half of the illustration, i.e., a height of 1372 mm, a shoulder width of 610 mm, and a lateral reach of 1174 mm.

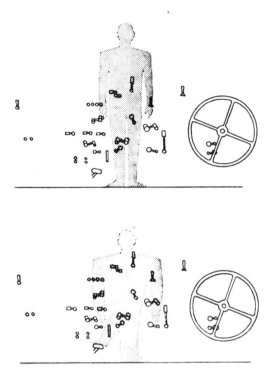

Figure 6.60 Example of the unfavorable placement of controls on a lathe.

The first step in fitting the machine to the man consists of ensuring that the dimensions of the workplace, tools, and controls correspond to the body dimensions of the operator (see Section 6.1.1). For this purpose data files on human body dimensions and physical strength are available. In the compilation and utilization of anthropotechnic data files, the following must be taken into consideration:

—Not only is the "average dimension" important, but so is the "dispersion range."
—The arithmetic mean of a large number of measurements is taken as the "average dimension", e.g., height, length of the lower arm, distance between the eyes, leg strength when sitting, etc.
—The distribution is usually represented by the 5th and 95th percentiles (see Section 6.1.1).
—The anthropometric measurements for men and women differ. When designing seats, tools, or machines, attention should be paid to whether the workplace is to be occupied by men or women.

On the basis of anthropometric data, recommendations can be made concerning the dimensions of workplaces. Figure 6.61, for example, shows the recommended heights for a person sitting down and carrying out a number of different activities.

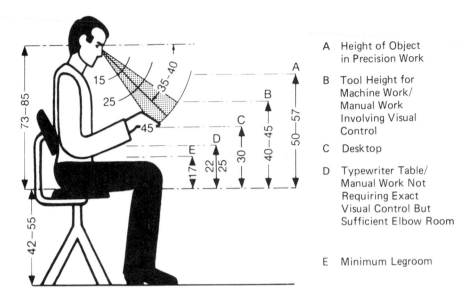

A   Height of Object
    in Precision Work

B   Tool Height for
    Machine Work/
    Manual Work
    Involving Visual
    Control

C   Desktop

D   Typewriter Table/
    Manual Work Not
    Requiring Exact
    Visual Control But
    Sufficient Elbow Room

E   Minimum Legroom

Figure 6.61 Work heights in a sitting position in cm (from Stier [6-46]).

At tabletop height all parts, tools, controls, and material containers should be arranged within the reach envelope. The dimensions of the reach envelope are determined by the length and flexibility of the arms. The optimal reach envelope is shown in Figure 6.62.

In direct relationship to the reach envelope is the requirement for the arrangement of displays which must be placed within the usable portion of

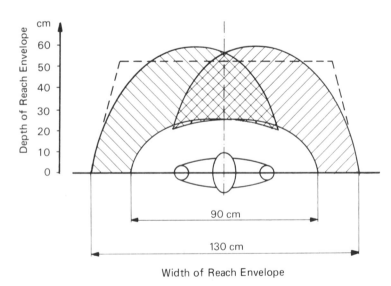

Figure 6.62 Optimal reach envelope at tabletop height (from Schmidtke [6-47]).

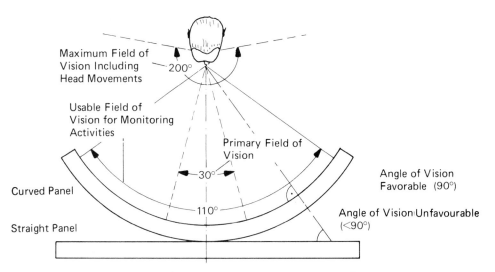

Figure 6.63 Visual conditions for straight and curved displays (from Neumann and Timpe [6-48]).

the field of vision (Figure 6.63). It is true that a horizontal curved arrangement of the displays is more difficult to achieve. However, such an arrangement offers the distinct advantages of a larger useful field of vision and equal distances from the eye. This will reduce the demands on accommodation and convergence changes, and the possibility for arranging the controls within the optimal reach envelope is increased.

6.3.2 The Use of Empirical Methods

In order to determine the ergonomic state of existing systems and to evaluate ergonomic design suggestions, empirical methods are used. This is done especially when established scientific knowledge is lacking. Several important methods for determining the ergonomic state are available.

Interviews. Discussions are held with the operators themselves, as well as with supervisors and technical personnel, concerning the content of the work and the problems arising from it. These discussions should be linked to general observations at the workplace. In special cases, structured interviews using a fixed schedule are used.

Questionnaires and Checklists. These methods correspond to a rigidly structured interview. With a checklist the data collection is independent of the questioning of a single worker. There are a number of standardized ergonomic questionnaires and checklists available, e.g., the FAA (questionnaire for ergonomic analysis), the AET (ergonomic data collection

Table 6.17 Excerpt from Köck and Odehnal's "Ergonomic Checklist" [6-49]

1.4	**Display instruments**
1.4.1	Are the most important and most frequently read instruments arranged in the most favorable locations within the field of vision?
1.4.2	Are the display instruments clearly discernable from the background?
1.4.3	Are excessive differences in brightness between displays and background avoided?
1.4.4	Do the scale divisions and the accuracy of the instrument correspond to the required reading accuracy?
1.4.5	Are pointers so designed and arranged that readings can easily be made? If not, are the causes:

    [49] shadows   [51] pointers covering numerals
    [50] too short pointers   [52] other

1.4.6	Have provisions been made to ensure that fast and irregular movements of the pointers cannot impair the reading of the instruments?
1.4.7	Are displays and related controls arranged closely together?
1.4.8	Are functionally different groups of instruments easy to distinguish through differences in shape or color?
1.4.9	Are instrument malfunctions caused by disturbances indicated in an effective manner?
1.4.10	Are digital displays used where accurate readings or the presetting of control values are required?
1.4.11	Are colored area displays used for voluminous and fast readings?

procedure for task analysis), and the "Ergonomic Checklist." Table 6.17 shows an excerpt from the "Ergonomic Checklist."

Observation. When determining the actual state, especially with regard to details of the work procedure, observations on site can hardly be dispensed with. Care must be taken, however, to ensure that such observation will in no way disturb or influence the work process itself. Combinations of informal observations and discussions at the workplace can be very helpful for obtaining the desired information. In certain cases more systematic methods are employed where procedures are recorded using film cameras and other equipment.

For data collection in systems ergonomics the time and motion methods of work science are often excessively detailed. More global methods are much more applicable; for example, activity sampling and link analyses. In activity sampling the activities are recorded at preset times. Thus, making a continuous record becomes superfluous. An example of this is given in Section 6.3.7. In a link analysis the connections (information flow, movements, etc.) between men and machines are quantified. On such a basis an arrangement of the individual elements can be developed, as shown in Figure 6.64.

In order to evaluate an arrangement of displays, a link analysis can be carried out with the aid of recordings of head and eye movements. In Figure 6.65 the percentages of connections between displays during an instrument landing are shown (values under 2% have not been included).

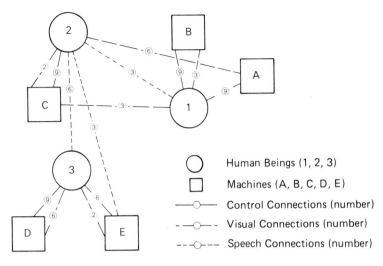

(a) Original Arrangement

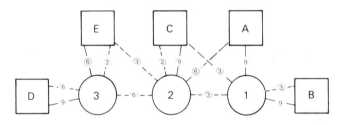

(b) Improved Arrangement

Figure 6.64 Comparison of two arrangements on the basis of a link analysis (from McCormick [6-51]).

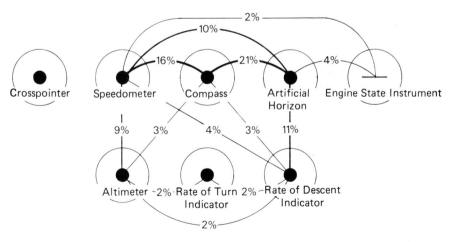

Figure 6.65 Results of a link analysis for an instrument landing system (from McCormick [6-51]).

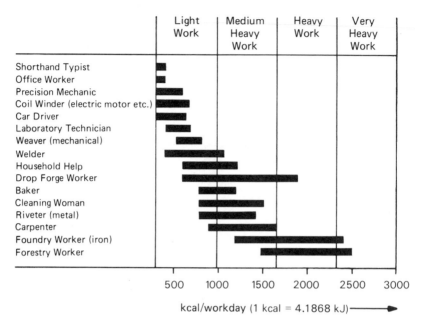

Figure 6.66 Comparison of the effort required for various jobs.

This evaluation demonstrates that the arrangement of instruments corresponds well to the sequence of use.

Strain Analyses. As mentioned earlier, the stress and strain experienced during physical exertion can be determined by measuring oxygen consumption or heartbeat frequency. By measuring oxygen consumption, the energy consumption and thus the physical stress can be directly derived. Figure 6.66 shows the work-related energy consumption per work day (effort) for a variety of jobs.

Accident Analyses. Among the written documentation used for obtaining ergonomic data are accident statistics and reports. It should be mentioned, however, that the content of such documentation is somewhat unreliable, especially where the description of human error is concerned. Since accidents occur relatively rarely, an alternative approach is to study "critical incidents" instead of accidents. These are intermediate events which do not result in accidents, e.g., the faulty reading of an airplane instrument. The required information is obtained by means of interviews or questionnaires, whereby it is of vital importance to convince the respondents of the confidentiality of their statements.

Knowledge concerning the present state of a system can serve as a basis for ergonomic improvement, and beyond this for the ergonomically optimal design of a new applied system. If during the evaluation of a design

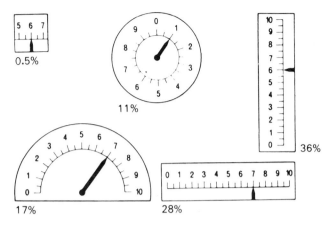

Figure 6.67 Percentage of faulty readings in an experiment comparing scales of different forms (from Grandjean [6-6]).

proposal the established information available proves to be inadequate, a practical test of the proposal will be made. For larger systems, a practical test before startup is, in any case, usually required. Some of the most important methods used for ergonomic evaluation are briefly described below.

Experimental Evaluation. In order to compare the design proposal with the previous design or with other alternatives, a specific experiment will be carried out. The evaluation criteria are measures of system performance; for example, the amount of time required to complete a certain procedure or the number of errors made. Figure 6.67 shows the results of an experimental evaluation of five scale types, in which the criterion was the number of erroneous readings.

Construction of a Mock-up. A mock-up is a static test structure which accurately represents the spatial arrangement of the proposed design. The mock-up does not need to be functional; displays may be drawn on cardboard and controls may be made from wood or other easily worked materials. The mock-up is used to test the spatial arrangement; for example, to make sure that all displays and controls lie within the field of vision and the reach envelope, respectively. The advantage of a mock-up is that problems as well as appropriate solutions may become evident, which would not be detected if the examination was limited to drawings or other documents. Figure 6.68 shows the mock-up of the cab of a street sweeping vehicle which was built for the optimization of the driver's seat and vision.

Simulation. In simulation a dynamic test structure is used which may be regarded as an extended mock-up. In this manner the operation of a

Figure 6.68 Mock-up of the cab of a street sweeping vehicle.

system can be evaluated under realistic conditions. Simulators are expensive, but they permit an extensive ergonomic assessment. They can also be used for training such personnel as nuclear power station operators, pilots, and naval officers.

6.3.3 Design and Arrangement of Displays and Controls

Displays. Information concerning the state of a technical installation is primarily provided to the operator via displays. To adapt these displays to human requirements is of vital importance. The first consideration is to ensure that the most suitable type of display is used for the intended application. Table 6.18 shows a comparison of three important display types with regard to their suitability for specific purposes.

When designing a display, a number of points should be particularly considered. The zero point of a circular scale, or the beginning of the scale, should be arranged on the left or at the top. The gradation of a scale on a dial should increase clockwise and on a straight scale from left to right or from bottom to top. A normal scale has optimally 20 gradations. The interpolation interval should be of 5, 2, or 1 unit(s). For the numbering

Table 6.18 Advantages and disadvantages of various types of optical display

Application	Digital display	Fixed pointer	Fixed scale
1. Quantitative reading	*good* short reading time and minimum errors in numerical value acquisition	*moderate*	*moderate*
2. Quantitative reading and comparison	*unfavorable* numbers have to be read and position changes escape notice	*unfavorable* direction and magnitude of deviation hard to assess without actually reaching the scale value	*good* pointer position readily noted, scale values do not need to be read off. Changes in position quickly registered
3. Setting values	*good* exact monitoring of the numerical setting, but relationship with control movement less direct than moving pointer. Difficult to read with rapid settings	*moderate* equivocal relationship with control movement. No change of position of pointer which would assist monitoring. Difficult to read with rapid settings	*good* obvious relationship between movement of pointer and control element. Changes in pointer position facilitate monitoring. Quick setting possible
4. Manual control	*unfavorable* for monitoring tasks no change in position. Relationship to control not easily comprehended. Fast changes difficult to read.	*moderate* for monitoring tasks no change in position. Relationship to control comprehended. Fast changes difficult to read.	*good* pointer setting easily monitored and controlled. Easily comprehended relationship with control movement.

the following rules apply:

Very good	0 —	1 —	2 —	3 —	4 —	5...
Good	0 —	2 —	4 —	6 —	8	....
	0 —	5 —	10 —	15 —	20	......
Poor	0 —	8 —	16 —	24		....
	0 —	4 —	8 —	12		....
	0 —	9 —	18 —	27		....
	0 —	6 —	12 —	18		....
	0 —	3 —	6 —	9		....
	0 —	7 —	14 —	21		....
	0 —	2.5 —	5 —	7.5		....
Unusable	0 —	1.25 —	2.50 —	3.75		....

increasingly
unfavorable

Pointers should not cover up the numerals. The script used should be clear and easily read (sans serif). If the largest expected reading distance is 1 mm, the minimum dimensions can be computed as follows:

Lengths of the main stroke $= \dfrac{a}{14.4}$ mm

Thickness of the scale gradation lines $= \dfrac{a}{5000}$ mm

Height of the fine gradations $= \dfrac{a}{200}$ mm

Height of the main gradations $= \dfrac{a}{90}$ mm

Distance between fine gradations $= \dfrac{a}{600}$ mm

Distance between main gradations $= \dfrac{a}{50}$ mm.

In practical applications, these minimum dimensions are usually far exceeded in order to ensure sufficient reading accuracy even under nonoptimal conditions such as low lighting levels, vibration, time pressure, etc. In order to illustrate the guidelines given here for the design of display scales, a number of incorrect and correct scale gradations are shown in Figure 6.69.

Controls. The selection of suitable controls also depends on the area of application. Table 6.19 shows a comparison of the most frequent types of control with a view to their suitability for various purposes.

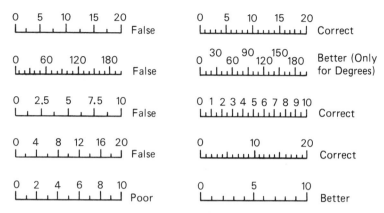

Figure 6.69 Examples of correct and incorrect scale gradations (from Grandjean [6-6]).

Table 6.19 Areas of application for various types of control

Type of control	Rapid setting	Suitability for Accurate setting	Transmission of forces	Setting in a wide range
Crank				
large	good	poor	unsuitable	good
small	poor	unsuitable	good	good
Wheel	poor	good	moderate/poor	moderate
Knob	unsuitable	moderate	unsuitable	moderate
Lever				
horizontal	good	poor	poor	poor
vertical (longitudinal)	good	moderate	short: poor/ long: moderate	poor
vertical (transverse)	moderate	moderate	moderate	unsuitable
Joystick	good	moderate	poor	poor
Foot pedal	good	poor	good	unsuitable
Pushbutton	good	unsuitable	unsuitable	unsuitable
Clamp/locking handle	good	good	unsuitable	unsuitable
Toggle switch	good	good	poor	unsuitable

Arrangement of Displays and Controls: Compatibility. Most people have certain expectations and ideas about the properties and interrelationships of stimuli and reactions, which are referred to as "population stereotypes." The design, arrangement, and allocation of displays and controls should be compatible with these stereotypes. Fulfilling this requirement will lead to time saving and fewer errors being made, because the necessity for coordination and decoding is decreased, reducing the number of steps

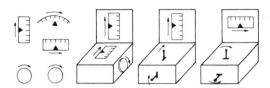

Figure 6.70 Examples of the expected relationships between the movements of controls and displays.

required to process the information. Three types of stereotypes can be differentiated.

Conceptual Compatibility. People associate, for example, the color green (as in a traffic light) with the idea of safety, absence of danger, etc.

Movement Compatibility. People expect a clockwise turn on an electrical appliance to produce an increase in power, loudness, etc. They expect water to flow if a tap is turned anti-clockwise. A car is expected to turn right if the steering wheel is moved clockwise. Moreover there are particular expectations concerning the relationship between the movements of controls and displays (see Figure 6.70).

An example of poor movement compatibility is given by the ship's compass. If a ship alters its course to starboard, the compass scale moves to the left. In order to correct any deviations, one tends to turn the wheel to the right, which tends to increase the deviation from the course. In order to counteract this natural tendency, the course line can be depicted in the shape of a ship, as shown in Figure 6.71. Now a course deviation to starboard will be perceived as a movement of the ship-form to the right, so the deviation is corrected by turning the wheel to the left.

Spatial or Positional Compatibility. Most people assume that when controls and displays are arranged in separate groups, elements in the same relative

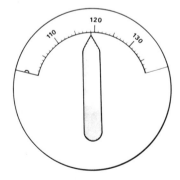

Figure 6.71 A "compatible" ship's compass (see text for explanation).

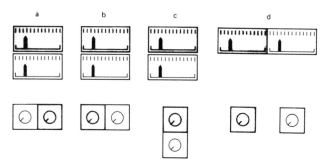

Figure 6.72 Various forms of incompatible and compatible arrangements of displays and controls.

location belong together, e.g., top left belongs with top left, middle belongs with middle, etc. Figure 6.72 shows some examples of poor and satisfactory positional compatibility. In this case the operator of a nuclear power station, because of lack of space, planned a layout with adjacent displays as shown in Figure 6.72a. This arrangement corresponds to no population stereotype, so relatively frequent confusion in the use of the controls would be expected. If no other possibility exists, Figure 6.72b would be preferred, although not recommended. Figure 6.72c is preferable, and Figure 6.72d is clearer still. In terms of safety it should be noted that training can overcome incompatible associations. However, in the event of stress or the sudden occurrence of a dangerous situation, the "natural" expectations and behavior patterns are likely to dominate.

Guidelines for the Layout of Displays and Controls. The layout of displays and controls within the operator's field of vision and reach envelope is determined by the following principles.

Importance. The displays and controls most important for the operation of a technical system, especially those fulfilling a safety function, should be placed where they are most visible and most accessible.

Frequency of Use. The most frequently used displays and controls should be placed in the most visible and accessible locations.

Function. Displays and controls which have a functional relationship should be placed close together.

Sequence of Operation. The arrangement of displays and controls should correspond to the sequence of use in normal operation.

The aim should be to conform to as many of these guidelines as possible. In situations where the guidelines conflict, they should be weighted

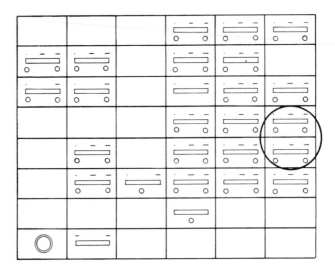

Figure 6.73 Location of an operational error (encircled) on a typical control room console.

according to their relevance for the system in question. Figure 6.73 gives an example of a poor instrument layout. During the course of an incident an operator confused the two encircled modules. This type of error is likely to occur very frequently, because in this type of arrangement human error is virtually built in. Similar instruments with differing functions are bunched together with almost no semblance of order. They can only be identified by means of a label, which is located above the controls and is in largely coded form.

Deviations from the normal state of operation can be detected faster and more dependably when the displays are arranged so that during normal operation the pointers all point in the same direction. Thus the 28 displays in the right half of Figure 6.74 can be checked for deviations just as rapidly and as accurately as the four displays in the left half.

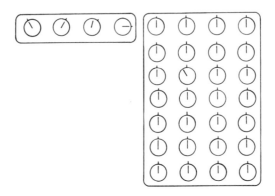

Figure 6.74 Example of a suitable arrangement of many displays (right), which can be checked as rapidly and as accurately as the arrangement of a few displays (left).

6.3.4 Monotony and Vigilance Problems

In Section 6.2.2 "mental strain" was discussed. Some practical recommendations were made concerning situations in which boredom or lack of attention can occur.

A low level of arousal can itself produce the danger of monotony. Unchanging or repetitive activities, such as working on a production line with short activity cycles, typify this situation. In practice, the following measures can be taken to counter monotony:

—Replace conveyor assembly work by group work with autonomous allocation of tasks to individuals within the group.
—Job rotation (the individual moves from time to time to other work stations).
—Avoid spatial isolation.
—Institute short breaks involving physical movement.
—As far as compatible with the work flow, allow the individual to regulate pauses during the work phase between official breaks.
—Avoid monotonous environmental conditions (changing the setting of the air conditioning during the day in accordance with climatic changes, playing stimulating music for ten to twenty minutes at frequent intervals during a shift).

For activities requiring a high degree of attention because of weak and rarely occurring signals, a reduced degree of vigilance as a consequence of monotony and fatigue presents a particular problem.

The following recommendations for practical application can be derived from vigilance studies:

—During pure monitoring and checking activities, the uninterrupted period of activity for the individual should not exceed 30 minutes. This requirement can be met, for example, through job rotation or by integrating the monitoring activity into other activities. In the latter case the individual will no longer be engaged in a pure monitoring activity but in a mixed activity.
—Critical signals should be presented in such a way that the contrast between them and their surroundings through color, shape, brightness, or by their dynamics (movement, blinking) is as sharp as possible.
—In the system conception it should be ensured that no dangerous situations can develop if a signal is missed. The system should, by means of diverse technical monitoring and protective devices, be brought automatically to a safe condition if the monitoring operator fails to react to a critical signal within a predetermined period of time.

6.3.5 Factors of the Physical Work Environment

The factors of the physical work environment—microclimate, noise, mechanical vibration, lighting, radioactive radiation, etc.—often have a direct effect on the safety and health of the working person. They are

Table 6.20 Overview of emission noise levels (evaluative levels) at the workplace according to German law ([6-52]). (1) Standard on noise prevention at the workplace [6-53]; (2) accident prevention regulation—noise [6-54]; (3) regulation governing the workplace [6-55]

Permissible emission level (dB (A))	Commentary on the application	Regulation
up to 55	mostly mental activity/breaks, waiting rooms, dispensaries	(3)
up to 70	simple or mostly mechanical office work and comparable activities	(3)
up to 85	all other activities in new workplaces; not in workplaces constructed before 1/5/76 if major changes are required.	(3)
up to 90	in general, all workplaces/in workplaces constructed before 1/5/76; in new workplaces if technical, organizational, or economic reasons prohibit the application of the 85 dB (A) limit	(1) (3)
over 90	only with special government permit	(3)
over 85	personal noise protection devices must be provided	(2)
90 and above	hearing tests/personal noise protection devices must be worn, preventive checkups, workplaces must be marked as "noise areas"	(1) (2)

therefore of great importance. When certain limits are exceeded, some of which are set by law, suitable protective measures must be taken, e.g., wearing heat protective clothing or using personal noise protection devices. Table 6.20, for example, gives an overview of permissible noise emission levels at the workplace and possible countermeasures.

From the systems ergonomics point of view environmental factors are also important in that they influence the performance of the operator, and thus the efficiency of the entire system. This has an impact on the safety of the system. The influence of environmental factors on the performance of human beings will be illustrated in three examples.

Example One—Climate
The curve in Figure 6.75 shows the temperature limits according to exposure time at which performance decrement will occur. The curve is based on a large number of studies. The actual temperature for work involving mostly mental activity should not exceed 25°C. Higher temperatures impair not only personal well-being but also bring about a distinct decrease in performance which in turn is manifested in a higher error rate.

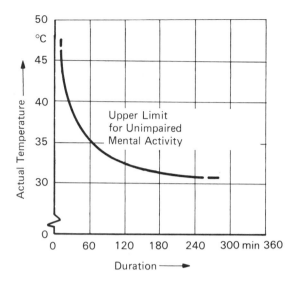

Figure 6.75 Temperature curve showing the limits over time for mental activity (from McCormick [6-50]).

Example Two—Noise

An important effect of noise emissions in a man–machine–environment–system is the disturbance in verbal communication. When full understanding of sentences is required, the interference sound level for normal messages should be no higher than 10 dB (A) below the speech sound level. For unusual or difficult messages this difference should be increased to 20 dB (A). The maximum distance between speaker and listener which still ensures satisfactory communication can be derived from the speech interference level (SIL). The SIL is the arithmetic mean of the octave sound level in the four octaves with the mean frequencies of 0.5/1/2/4 kHz. Table 6.21 shows the maximum distance between speaker and listener for various SILs.

Example Three—Lighting

The intensity of lighting can fundamentally influence work performance, especially when the work places demands on visual acuity. Better lighting can lead not only to improved performance but also to a reduction in the error rate (see Figure 6.76). The results, obtained in a cotton spinning mill, show that an increase in the lighting intensity from 170 to 750 Lux led to an increase in production from 100 to 110.5%, while the expenditure for rejects was reduced from 100 to 60.4%.

Table 6.21 Maximum distances for satisfactory verbal communication with various speech interference levels (SILs) (from [6-56])

SIL in dB	max. distance in meters Mode of speech	
	normal	raised
35	7.5	15
40	4.2	8.4
45	2.3	4.6
50	1.3	2.4
55	0.75	1.5
60	0.42	0.84
65	0.25	0.5
70	0.13	0.26

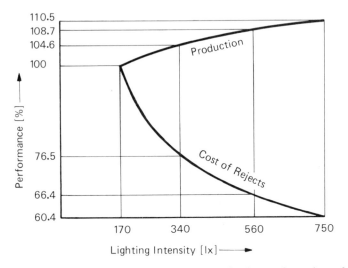

Figure 6.76 The effect of lighting intensity on production and number of rejects (from Grandjean [6-6]).

6.3.6 Avoiding Switching Errors—Illustrated by a Practical Example

An ergonomic study was carried out to investigate the occurrence of switching errors in the operation of the power distribution network of an electrical utility. The main purpose of the study was to determine the cause of these faulty switching operations and to make ergonomically sound recommendations designed to reduce the frequency of the switching errors. In order to obtain the necessary information, extensive discussions were held with the responsible department heads of the utility; the foremen of the switching installation were accompanied on their rounds by ergonomists and interviewed; and finally, written records concerning the nature of the operation and the switching errors in question were critically analyzed.

Evaluation of the information obtained showed that there was no detectable lack of motivation or safety consciousness on the part of those responsible. The causes of the switching errors lay in technical and ergonomic shortcomings in the installation, for example, in differences between individual types of installation, which required a great variety of operational procedures.

It was possible to arrange the apparently appropriate measures for the prevention of switching errors in a clear order of importance. This order, shown in Table 6.22, confirms that technical–ergonomic design principles are more effective than, for example, motivational or disciplinary measures. This is also why the holding of a psychology seminar, as had originally been proposed, was advised against. Instead, concrete proposals for ergonomic improvements were put forward. In order to avoid confusion in operation, future plans should be to achieve as great a uniformity as possible in the type of installation used.

Table 6.22 Rank ordering of measures for the prevention of switching errors.

Type of measure	Example	Installation costs	Dependence on human reliability	Possibility to check in practice	Effectiveness
Technical	Interlocking	very high	none	no problem	very high
Ergonomic	Uniform coding	high	low	no problem	high
Organizational (work aids)	Checklists	high	moderate	relatively simple	relatively high
Organizational (verbal information)	Safety training	high	relatively high	mostly difficult	medium
Organizational (written information)	Entry in S-Book	high	relatively high	mostly difficult	moderate
Organizational (regulations)	Instructions	moderate	relatively high	mostly difficult	moderate
Motivational	Psychology seminar	moderate	very high	almost impossible	low
Disciplinary	Reprimand	low	—	mostly difficult	low
Laissez-faire	Individual procedure	very low	very high	almost impossible	very low

6.3.7 Ergonomic Design of the Cab of a Street Sweeping Vehicle

The workplace "cab of a street sweeping vehicle" has a number of special problems associated with it. These include the problem of sweeping while driving in the midst of traffic as well as the unfavorable environmental conditions of noise, dust, and external weather. When reversing, the limited visibility from the cab presents a particularly important safety problem.

In a study of the working conditions in urban cleaning jobs [6-57], in addition to measuring environmental conditions in street sweeper cabs, an analysis of the driver's activities was made. This was done using the activity sampling method. The observer noted the driver's direction of vision every four seconds following the emission of a signal via an earphone. During the activity sampling period all operations of the controls of the sweeping equipment as well as all driving maneuvers were also registered.

Figure 6.77 shows the distribution of the driver's lines of sight in two examined types of street sweeping vehicle. It can be seen that the observation of the circular brush in sweeping vehicles with a truck chassis takes up more than twice as much time as with the three-wheeled type of sweeping vehicle. This is attributable to the fact that on the conventional type of vehicle the circular brush is positioned behind the driver's cab and is observed either in the outside mirror or through the window (see Figure 6.78). On the three-wheeled vehicle the circular brish is arranged in front of the cab and can be observed easily and directly through a special

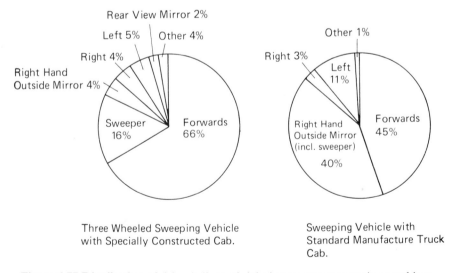

Figure 6.77 Distribution of driver's lines of sight in two street sweeping machines.

Figure 6.78 Street sweeping vehicle with truck chassis and standard manufacture truck cab (Kuka type 246). The illustration shows a prototype from the Berlin Sanitation Department equipped with air conditioning and television cameras.

window, which leaves the driver more time to look directly ahead (see Figure 6.79). This has an undoubtedly positive effect on safety in traffic.

Despite the opinion of the utility that the schedules of the vehicles had been so drawn up that reversing the vehicle would be unnecessary, it was frequently noted that reversing took place (up to 7 times within 30 minutes). This emphasizes the need in future designs for improved rear vision, which could be achieved with the use of television cameras. It was further noted that almost without exception the windows are kept open while driving. Figure 6.80 shows the factors which contribute to this.

Driving with the windows open raises the stress caused by noise and dust, since the brush units are mounted on the right hand side of the vehicle. In order to reduce the levels of noise and dust, measures must be introduced which allow the windows to be kept closed. Such measures would include equipping the vehicle with air conditioning and with television cameras to improve visibility.

For the development of a new street sweeping vehicle a mock-up was built (see Section 6.3.2). By means of this mock-up the spatial arrangement of the cab's interior and the visibility could be optimized.

Figure 6.79 Three-wheeled street sweeping vehicle with specially constructed cab (Faum AK 320 HB).

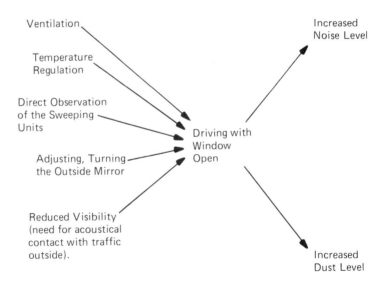

Figure 6.80 Interaction of the most important factors within the problem complex "street sweeping vehicle cab."

6.4 Education and Training

In order to avoid dangers within the man–machine–environment–system, the efforts directed towards the promotion of safety-related behavior or towards changes in human behavior aimed at preventing accidents should be considered. For these purposes the principles of learning are applied, about which some basic statements need to be made.

For the purpose of avoiding some of the dangers lurking in the natural environment, man has been equipped with reflexes designed to forestall any injury. An example is the closing of the eyelid whenever an airborne particle approaches the eye. The system of reflexes is composed of fixed stimulus-reaction interrelationships which operate even without the assistance of the cerebral cortex, i.e., without an actual perception of danger. The trigger is the physiologically active stimulus and not the cognitive processing of the stimulus as a danger signal. The extent of these self-protecting reflexes is, however, limited. The type of stimuli experienced in the technological world are very different, which means that the working man must learn how to avoid the dangers he encounters.

6.4.1 The Principles of Learning

"Learning is the connection of at least two processes within the organism which will lead to a change in behaviour". In this context a change in behavior denotes not only the creation of new behavioral patterns but also the modification or elimination of existing behavioral patterns.

The connection of processes does not necessarily require conscious association. Nonspecific signals which repeatedly coincide with reflex-triggering stimuli are often sufficient to trigger corresponding reactions through these signals, even if the specific stimulus is absent. Thus one speaks of learned or "conditioned" reflexes; this learning process is termed "classical conditioning." Many human self-protecting reactions are acquired on the basis of classical conditioning, i.e., without any conscious effort. Even the behavioral patterns of physical work, which are not composed of reflexes, are acquired through unconscious imitation or through trial-and-error, and rarely through conscious observation of the motion processes of other or through the anticipation of one's own processes in imagination (mental training). Only in very rare instances are behavioral patterns prematurely executed in a rational manner and then systematically practiced. These three types of learning processes determine the subsequent repertoire of safe and unsafe behavior patterns. As in classical conditioning, these patterns are maintained by the interconnection of processes within the organism, whereby success is associated with a particular behavioral pattern and acts as an intensifier. Successful individual motion processes are thus selected and composed into action. This type of learning is referred to as "operant conditioning." The concept of intensifiers is used within the psychology of learning, both in terms of success associated with certain behavioral patterns and also in terms of failures resulting from them. In order to distinguish them, one speaks of positive intensifiers which lead to a consolidation of behavior, and of negative intensifiers which lead to the elimination of the behavioral pattern.

For all types of intensifier the space-time association rather than logical or causal association is of importance. The conscious processing of such associations in the observation of one's own behavior frequently leads to the reinterpretation of a space-time association as a causal relationship, i.e., events are interpreted as they are perceived. The regularity of such cognitive processes also leads to typical misconceptions whereby a causal relationship may be assumed between two coincidentally occurring events which in reality does not exist. This is one of the many reasons why the intellectual processing of behavior has little effect on the acquisition of safety-related behavior patterns.

Within the man–machine–environment–system certain types of behavior designed to prevent accidents are demanded from the man, which are described as reactions. These are frequently reactions to specific events occurring during the course of working activity which represents a significant danger to the person. Today, these reactions may relate not so much to events endangering the operator himself but to processes which could represent a danger to others.

The type of behavior used for averting danger must be learned. Behavior which still requires planning in the face of danger will always tend to overtax the individual and will then result in failure. The best solution for providing protection against new dangers is still the "crisis management team," in which any inappropriate reaction on the part of an individual can be avoided by means of group processes.

The special problem of planning accompanying the prevention of danger will not be discussed further here. Of greater interest is the question of how behavior patterns for the prevention of danger can be learned and readily put into operation the moment that danger occurs. Therefore, the acquisition of such behavior patterns which are known to be associated with safety will be discussed here.

There is an understandable tendency to put information concerning a known relationship between a certain type of behavior and the avoidance of predictable dangers right at the beginning of training. This would result in a "safety course" which would supplement the training for specific work-related behavior. Typical of this is the introduction of the study of traffic safety and the study of danger, which are added to the driver training curriculum as something extra that is important to learn. A prerequisite for this approach is the ability to learn on the basis of reason, where in this case learning is understood as behavioral change. For most people, however, this ability is so limited that it can be disregarded as a factor contributing to the development of safe behavior within the man–machine–environment–system.

The limited extent to which reason will bring about a change in behavior is demonstrated in a number of harmful and unsafe habits, whereby virtually all of those having such habits are aware of their unsafe nature. When attempting to eliminate unsafe behavioral patterns in the man–machine–environment–system a problem is encountered similar to that found with certain consumer habits detrimental to health in that the undesirable behavior is acquired early, for example through imitation, and is reinforced by excellent intensifiers. If one attempts to prevent such behavior by arguing that it is dangerous and that another safer behavior pattern should be adopted, one must also realize that as a rule safety lacks the quality of an intensifier. Safety is subjectively, at best, the absence of an accident, which only operates as a negative intensifier if the victim actually suffers pain. An important element of behavior therefore is the elimination of undesirable behavior merely through the imagination of pain or injury.

6.4.2 Reward and Punishment as Learning Aids

"Once bitten, twice shy" is a valid maxim for everyday life. For learning desirable work patterns in the actual man–machine–environment–system, such a maxim has little practical value. One cannot use injury systematically or meaningfully as a negative intensifier. Since, furthermore, the reward of safety is ineffective as a positive intensifier, only intensifiers outside the causal relationship can be used as influencing factors. These are rewards and punishments which are introduced to reinforce safe behavior and to suppress unsafe behavior. At the moment there are not dangerous workplaces where no sanctions exist against unsafe behavior, such as fines for not wearing safety helmets or protective clothing where these are mandatory.

As negative intensifiers, only those forms of punishment which exist in a space-time relationship with the undesirable behavior are usually effective, i.e., swift and prompt punishment. Thus, if punishment is to have an improving effect, it must be used in accordance with the principles of learning. This, however, is largely neglected in the field of preventive protection against danger. The effective use of positive intensifiers is similarly connected with the principles of learning, which again is seldom considered in existing training schemes. On the contrary, factory-sponsored reward schemes for many years of accident-free work are often introduced with high expectations of success, which if it does occur is more likely to be due to the influence of other factors.

The central problem with any bonus or sanction system, or even all safety campaigns aimed at work processes in the man–machine–environment–system is that, irrespective of whether positive or negative intensifiers

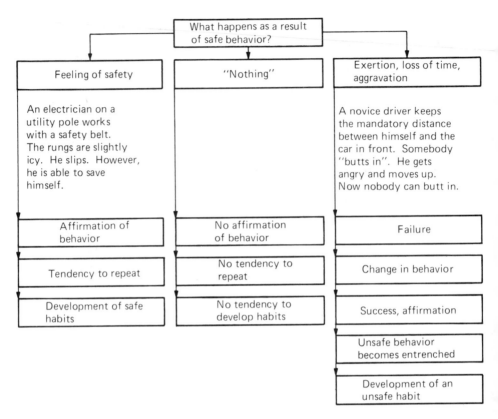

Figure 6.81 Pattern of intensification [6-58].

are employed, deeply entrenched unsafe habits will have to be dislodged. It is very difficult to change or modify such behavioral patterns. Thus the success of safety campaigns is usually doubtful because of the lack of an intensifier to follow-up safe behavior. Behavior which is not acknowledged is not likely to be repeated. Burkhardt, Schubert, and Schubert [6-58] illustrate these learning processes on the basis of two graphic models (see Figure 6.81). On the basis of this figure it can be seen that learning safe behavior from personal experience is dependent on the occurrence of infrequent events.

There remains the question of the intensification of unsafe behavior. The above-mentioned authors [6-58] note (see Figure 6.82) that generally nothing occurs, because the ratio of unsafe behavior patterns to accidents is usually greater than 100:1. The normal occurrence, not the extreme occurrence, is "stored." In many cases unsafe behavior is associated with rewards, be they gain in time, added convenience, or a successful outcome.

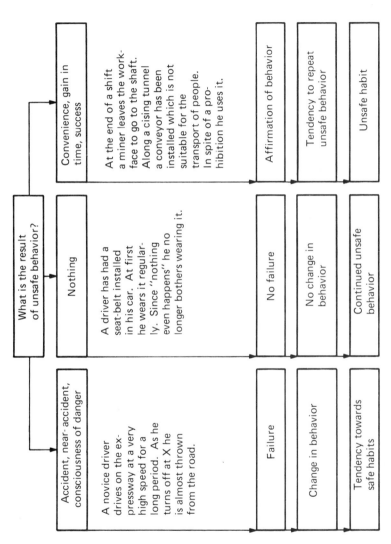

Figure 6.82 Pattern of intensification for unsafe behavior [6-57].

Any behavior resulting in success tends to be repeated. Unsafe behavior is very rarely followed by an accident, a near-accident, or even the feeling that "something might have gone wrong." One must realize that again and again additional unsafe behavior patterns are formed according to the principles of success, thus making it imperative in safety-oriented education and training to eliminate these unsafe behavioral patterns. The possibilities for this are, however, limited.

6.4.3 The Element of Safe Behavior in Training

A better approach to education and training is to teach safe behavior directly as the work behavior, which is then practiced within the man–machine–environment–system. This means that safety is not taught as an additional component of work behavior, but as an integral part of the work procedures being taught. The success in learning the work processes will then also be associated with the integrated safe behavior. Safe behavior will result from this success without the necessity for conscious effort. The learning of certain work processes cannot therefore be left to chance or to a learning process involving the trial-and-error method. Due to the widely spread ability to learn through imitation, all work processes should be examined to see whether this type of learning method can be applied.

Since in this context only the safety-related goals of work, traffic, and dangerous leisure activity are of interest, the following limiting statement for imitation learning should be made: imitation should be used for safety training only when the models whose behavior is to be imitated always themselves act in a responsible and safe manner, and where endangerment occurs with a frequency sufficient for the demonstration of the benefits of safe behavior in a wide variety of situations. These models must possess all the characteristics desirable in the trainee and must be able to praise safe behavior convincingly. Such praise will only be effective, however, when it comes from a model accepted by the trainee. Thus in terms of teaching safety, no teaching system can substitute for a good teacher. Thereby— and of decisive importance because safety is not required for the learner as an independent motive—the teacher does not need to convince the learner of the safety-related advantages of the behavior patterns taught, as long as he acts in accordance with accepted safety goals and turns this type of behavior into a successful and positive experience for the learner. Personal acknowledgement from a superior is an excellent intensifier. Training in safe behavior through imitation does not exclude the possibility that the learner will form some individual behavior patterns. However, when these are inappropriate no major effect will need to be made to overcome unsafe habits. Thus if safety training is to be successful it must form an integral part of basic training.

6.4.4 Program Controlled Instruction

In many cases safety-oriented learning is not possible because the dangers involved in the behavior to be imitated are not adequately represented. This applies, for example, in those behavioral fields where self-teaching can lead to inappropriate behavior. As can be seen in the case of driving behavior, models can be taken whose behavior may seem to indicate proficiency, but which in many cases is unsafe. Such models are likely to undermine the efforts of a teacher to instill safe behavior in his pupils.

In such cases it is often sensible to establish behavioral patterns before the learner has any contact with the actual situation. This pretraining involves mental training using particular media and operates on the principle of observing or acting within a simulated situation. A classical form of such pretraining is programmed instruction. The type of behavior to be learned is represented by pictures and text. The principles to be taught can best be understood if the learner is presented with illustrations. A section of a tried and tested learning program is shown on pages 256–258 [6-59, 6-60]. The design of the teaching material and the control of the learning process in the example follow the principle of small and certain learning steps and the automatic confirmation of the acquired knowledge through intermediate tests. These are designed so that the learner must fill in gaps in a piece of text. The sequence of pictures realize the principle of mental training in so far that imagined movements can be consolidated in the learning situation. Teaching success is associated with the degree to which the written text corresponds to the way in which a teacher would speak in class. Because they follow different models and have differing functional habits, scientifically trained writers are not always able to write in this way.

In many groups of learners no selection is made on the basis of high intelligence. Lack of reasoning power must be expected and basic knowledge should be assessed very low if the instruction is to be successful. This means that with suitably designed learning increments the possibility of confirming the progress made in learning must follow as a matter of course. The designing of learning programs is a question of the will and the emotional penetration on the part of the instructor to write in such a way that he will not put undue stress even on the least gifted pupil.

Intelligent people looking at such learning programs, without wishing to learn from them, often have the feeling that they are long-winded and trivial. Although not false, these feelings are unsuitable for the evaluation of a learning program. Such feelings only produce an effect if an already informed student is forced into the learning process. If, however, the learning process is organized in such a way that the learner can determine his own pace, any effect of triviality in the learning aid will be minimized. Subjectively easy passages can then be passed over quickly and unknown material can be absorbed intensively, using as much time as necessary.

Learning programs without individualization of the learning progress are unprofessional. Working in groups and programming are mutually exclusive. The unqualified advantage of programmed learning lies in the goal achievement. Given sufficient time, even the slowest learner will achieve the goal. This is the precise aim of all safety instruction. As many as possible should learn the desired safety-related behavior patterns. However, because of this emphasis on all learners reaching a goal, programmed learning is of limited value in general education. This is due to the fact that a tendency will occur towards incomplete knowledge or mediocrity in that there is little opportunity for the expression of full understanding of a topic and no reward for making rapid progress. Where, however, all learners must attain complete knowledge, the time differentiation is decisive. This can in practice only be achieved with the use of such learning media. Time differentiation alone makes a fundamental contribution to the achievement of a generally high learning standard.

6.4.5 The Use of Simulators

In many cases the simple visual reproduction of the actual situation is not sufficient because complicated procedures must be exercised or information from senses other than vision must be processed. In such cases training with simulators is often successful. However, in contrast to programmed learning, some basic knowledge is generally a prerequisite for training in simulators and complex behavior patterns are learned.

In early simulators, particularly driving simulators, actual situations were filmed and used in the simulator. Mistakes made by the learning, however, were not distributed realistically over time; neither were their effects displayed optically, nor were the consequences of these mistakes realistically demonstrated. Thus the use of such films in safety training is practically excluded. Whenever a pupil in a simulator "driving" on a winding road steers the car into a ditch, all the film can do is stop. This takes the critical element out of the reality-oriented simulation.

A more favorable method of demonstrating the behavioral situation is by means of synthetic pictures. In this context electronically generated pictures transmitted via a cathode ray tube are usually effective. A street scene, for example, is simulated in four sections (see Figure 6.83): generation and storage of the layout of the road, computation of the position of the car, computation of the visible part of the road, perspective transformation and display of the visible road section.

For this type of simulator and the task to be carried out in it, the scenery around the road remains in the dark. However, the essential signals to be processed, if the task of following the car in front is to be executed faultlessly, are present in the picture. The reduction to fundamentals corresponds precisely to the human behavioral requirements. In general it can be said that in the animal world behavioral triggers are relatively sim-

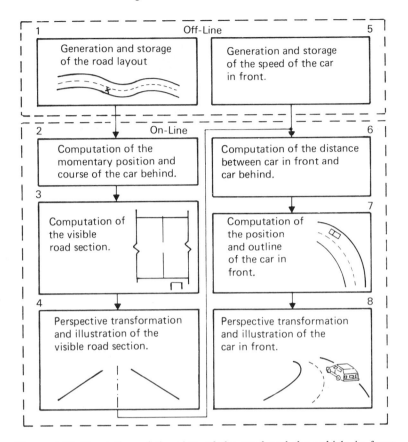

Figure 6.83 Simulation of the view of the road and the vehicle in front.

ply structured and can be easily depicted. Provided that learning situations rather than innate behavioral patterns are involved, it is educationally advantageous to simulate only those signals which are directly relevant to the behavior patterns to be learned, although it is perfectly possible to build distracting elements into the signal system.

Other simulators employ models of the work situation from which cameras relay the relevant view to the trainee. The movement of the camera is determined by the controlling actions of the learner. Driving simulators of this type have, for example, been used for the training of tank drivers. The economical superiority of military training with the aid of simulators is particularly high. Figure 6.84 shows clearly that with optimal use, simulators usually pay for themselves before the end of the first year (intersection of the lines A and B). After this point the training of tank drivers in simulators, as compared with conventional methods, becomes cheaper. In addition to the economic advantage, this technique allows critical situations to be simulated at will and as often as desired.

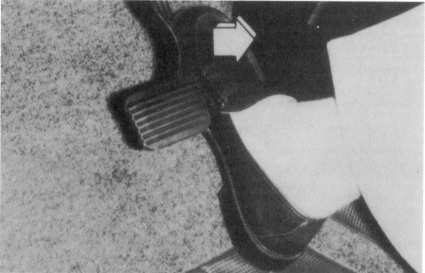

1. Braking and driving on

You are approaching an unmarked intersection. A car may
be approaching from the right, which has right of way.

You remove your foot from the gas pedal.

The car does not simply roll on, but is braked by the
motor, i.e. the motor does not only propel, it can also act as a brake.

When you slow down, the automatic transmission selects a lower gear.

The braking power of the motor is greater in the lower
gear. The lever remains in the D position.

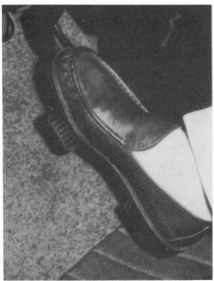

A car approaches from the right, which has the right of way.

You must drive even more slowly. The braking power of
the motor is not sufficient.

Therefore you depress the brake pedal. You now slow down
much more.

The gear lever remains in the D position.

Once the road is clear again, you again depress the gas pedal.
The automatic transmission selects a higher gear.

What does the automatic transmission do if you brake and then drive on?

If you remove your foot from the gas pedal, the car will _____

The car is braked with the _____

If the car is slowed down, the automatic transmission selects _____

The motor's braking power in the lower gear is _____

The gear lever _____ in the

_____ position _____

Solution:

If you remove your foot from the gas pedal the car will slow down.

The car is braked with the motor.

If the car is slowed down, the automatic transmission selects a lower gear.

The motor's braking power in the lower gear is greater.

The gear lever remains in the D position.

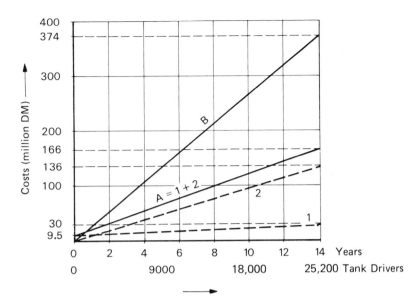

Figure 6.84 Cost comparison diagram for a training center with four simulators, excluding personnel costs and with fully utilized training capacity.

The possibilities of improving safety training as outlined here have only been realized to any great extent in military, civil aviation, and merchant marine settings, and less for operators in the control rooms of large industrial plants or for drivers. The fact that there is always a higher risk associated with learning a new activity should provide an incentive towards the increased use of the described methods in a great number of professional and private fields of activity. An exclusively theoretical training cannot be expected to overcome the element of personal disregard for safety.

6.4.6 Motivation for Safety in Education and Training

If one wishes to limit the negative effects of risk compensation—as described earlier—neither safety education nor training can dispense with the need to create a channel for the safety-oriented use of the knowledge acquired. The value of a higher qualification level can only be applied to one set of goals. If, for example, a higher qualification level for workers is invested in increasing production by accelerating the work cycle, a simultaneous increase in safety can hardly be expected. The same applies for behavioral fields with self-determined work cycles; for example, driving in traffic. If a higher degree of mastering the vehicle is used only to drive at a higher speed, the resulting degree of safety will remain constant, as suggested by the risk compensation theory.

The risk compensation theory is not without importance in safety science, because it allows motivational factors to be considered. If, for example, the driver within the man–machine–environment–system acts with a tendency towards risk homeostasis, past risks will determine future risk. Thus it can be concluded that the number of accidents happening within a specific country depends on the number of accidents the population of that country is prepared to accept. It does not depend on the effects of other elements of the system, at least not in the long term. This argument, described by Wilde as "Trojan Horse" [6-41], makes it clear that a reduction in risk brought about through the machine or the environment will be of little value if this brings about an increased exploitation of the total system for purposes other than safety. In general it can be said that any man–machine–environment–system will provide the degree of safety determined by its users, provided that they can exert a controlling influence. Here the importance of developing collective safety objectives becomes clear. A reduced risk brought about through technology can only be converted to increased safety if the user of the system has the desire to actually use the benefits of the new technology for safety purposes. In other words, he must refrain from using the safety advantage for other purposes such as developing experience or increasing productivity. The practical importance of the risk compensation theory thus lies in demonstrating that an increase in safety achieved within a technical subsystem will only lead to an actual gain in safety when the users of the system are willing to use this technical advantage for safety alone.

Chapter 7

The Environment as a Safety Factor and an Asset Worthy of Protection

Man and machine are embedded in their environment. The latter is the third important cornerstone within the basic cybernetic model. Accidents and loss on the part of the machine, caused by the operation of the machine by man, may have a deleterious effect on the environment. On the other hand, there are a multitude of natural processes within the environment, e.g., earthquakes and disastrous storms, etc., as well as disasters of a technical origin, like fires or explosions, which have a damaging effect on the machine.

"Environment" is understood to comprise all life and matter, man, animals, and plants, air, water, and soil surrounding the technical object in question within the predetermined observation space. Depending on prevailing conditions, the environment can be limited around a vantage point, either nearer or farther. Of decisive importance here is how far the actions effect the operation of the machine in question or, conversely, are caused by it. Below we will first deal with the natural and civilizatory environmental factors affecting the machine from outside. Thereupon we shall look at the influence of the machine on the environment. In Section 7.3 we will examine—in a reverse direction, so to speak—the effect of machines within the environment on technical systems in question.

Given the extraordinary complexity of the environment and its multitude of effects and ways of influencing the environment, only the most important aspects can be dealt with in this connection. In each and every case, one must first determine whether and how the environment is affected by the machine or whether the machine is endangered by the environment. Only through a thorough analysis of the manifold interrelations between man and machine, on the one hand, and environment, on the other hand, can mistakes in the representation of the cybernetic system of man–machine–environment be prevented. Underestimating the importance of the environment may lead to grave consequences for the operational safety of technical systems.

7.1 Natural and Civilizatory Factors

Soil, air, and water in our environment form the basis of life for man, animals, and plants. Due to the natural capability of these important environmental media to regenerate, their composition may fluctuate between certain limiting values without endangering life in any way. These limits, in most areas, have not been reached yet and it is man's duty to see to it that this will never happen.

The three environmental media mentioned above cannot be looked at separately as far as environmental strain is concerned; there are widespread interrelations between them. The following are some convincing examples. In the hydrological cycle (Figure 7.1) water is present in air as well as in the soil. It evaporates, forms clouds, comes back to earth in the form of precipitation, flows back into the oceans, and evaporates again. In the organic carbon cycle (Figure 7.2) carbon and its organic compounds, together with oxygen, burn and form carbon monoxide and carbon dioxide, which are thus removed from the soil and enter the atmosphere. By assimilation in plants, carbon dioxide is again converted into oxygen and carbon compounds which are again introduced into the plant and later into the soil and the air. During assimilation, oxygen is introduced into water and the atmosphere. In the nitrogen cycle (Figure 7.3) combustion— and thunderstorms as well—generates nitrogen oxides (NO and NO_2) which again oxydate into nitric acid (azotic acid) or enter the soil with precipitation. Microorganisms (breathers of nitrate) again degrade these nitrates and pass the nitrate on in the form of pure nitrate (N_2) or laughing gas (nitrous oxide, N_2O) into the atmosphere.

As a side effect caused by man's activities in the manufacture and use of products, the natural composition of the media air, water, and soil is being affected.

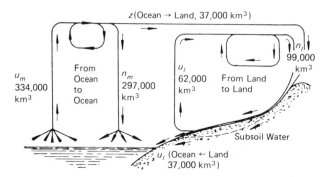

Figure 7.1 The hydrological cycle in its simplest form showing the absolute quantities in km^3.

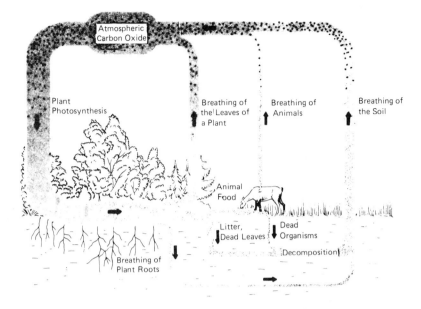

Figure 7.2 The organic carbon cycle represented schematically. The widths of the bands approximate the quantities which move along the various paths [7-2].

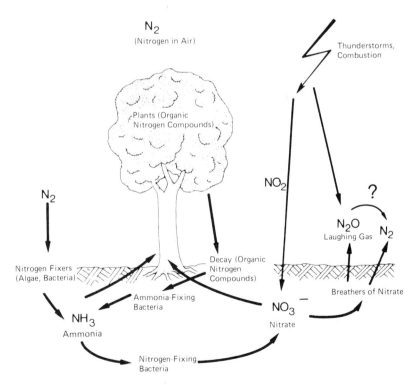

Figure 7.3 The organic carbon cycle. A schematic representation [7-3].

7.1.1 Air

All living creatures are surrounded by the atmosphere, i.e., by the air they breathe. One-fifth of the atmosphere consists of oxygen and four-fifths of nitrogen. A strongly varying share in the natural composition of air, depending on air temperature, is water in its gaseous phase with approximately 0.04–5.4 vol.%. Other gaseous substances, mostly rare gases (Table 7.1), constitute a share of less than 1 vol.%. Several gaseous or dusty substances which enter the atmosphere as a consequence of man's activities are present in small quantities only; however, they are able to influence living organisms adversely as a nuisance or a threat to health. Widely dispersed alien substances in air are sulfur dioxide, carbon monoxide, nitrogen oxides, hydrocarbons, and dust. Locally one will find a number of gaseous or dusty substances, e.g., hydrogen fluoride, hydrogen chloride, vinylchloride, hydrogen sulfide, phenols, soot, lead compounds, and heavy metals. The background level of these substances, as a rule, is very low and, in pure-air areas, shows the following values:

nitrogen dioxides (in the shape of NO_2) less than 10 g/m^3 (4 ppb);
sulfur dioxide less than 10 μg/m^3 (2 ppb);
hydrogen sulfide less than 0.3 μg/m^3 (0.2 ppb);
carbon monoxide less than 1 mg/m^3 (1 ppm);
hydrocarbons (with the exception of methane) approximately 1 ppb.

In contrast to the above, considerably higher concentrations of these substances are found—yearly average and for short periods—in polluted industrial and urban areas (Table 7.2).

Alien substances in air, as far as they are found in large quantities in locally limited areas, are usually the consequence of human activities. In densely populated areas, a constant rise in nitrogen dioxide emissions over

Table 7.1 Natural composition of air (in its dry state) [7-4]

Component	vol.%
Nitrogen	78.084
Oxygen	20.948
Water	—
Argon	0.934
Carbon dioxide	0.0314
Neon	0.0018
Helium	0.0005
Methane	0.0002
Krypton	0.0001
Sulfur dioxide	up to 0.0001
⋮	
Hydrogen	5×10^{-5}
Nitrogen dioxide	up to 2×10^{-6}

Table 7.2 Maximum concentrations of substances alien to air in densely populated areas [7-5]

Substances (in g/m^3)	Yearly average	for a short time, temporarily*
Nitrogen dioxide	70	150
Nitrogen monoxide	130	450
Sulfur dioxide	150	850
Hydrogen sulfide	5	45
Carbon monoxide**	1000	4000
Hydrocarbons (without methane)	700	6000
Hydrogen fluoride	1.3	20
Suspended substances	120	280
Lead	1.3	
Cadmium	0.018	
Zinc	3.2	

*Shown as a 95-centile.
**In busy thoroughfares in downtown areas, carbon monoxide concentrations of up to 20 mg/m³ can be found.

the past 30 years has been quite obvious (Figure 7.4). Worldwide, however, a rise in nitrogen dioxide load cannot be proven (Figure 7.5). The mere fact that a global increase cannot be proven may only be explained by the capability of the environment for self-regeneration.

In contrast to the above, carbon dioxide in the atmosphere shows a distinct and gradual increase worldwide. Measurements of the CO_2 content in the air over Hawaii, the South Pole, and the Rocky Mountains over a

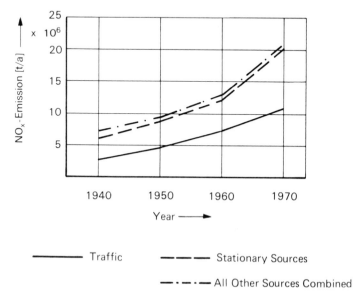

Figure 7.4 Development of nitrogen oxide emissions [7-7].

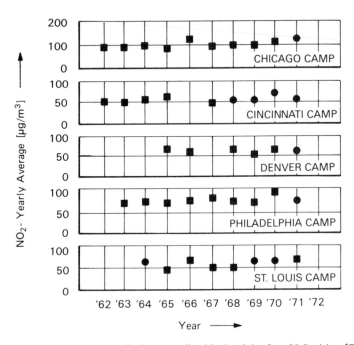

Figure 7.5 Development of nitrogen dioxide load in five U.S cities [7-8].

period of almost 20 years shows an increase of magnitude of 1 ppm per year (Figure 7.6). This increase is likely to originate from increasing CO_2 production through combustion processes (Figure 7.7). A comparison of natural and manmade (anthropogenous) sources for carbon dioxide in the atmosphere (Table 7.3), however, demonstrates that the anthropogenous part of the CO_2 production constitutes only 3% of the natural increase.

A change in climate, i.e., primarily of the mean temperature of the upper atmosphere because of a change on CO_2 content, can at this time only be conjectured; since the last Ice Age, approximately 16,000 years ago, the CO_2 content of the Earth's atmosphere had increased from 135 ppm to 290 ppm by 1880 and to 330 ppm by 1977. This historical review is possible via the existence of relative contents of carbon isotopes (C14). Simultaneously, since the last Ice Age, the mean yearly temperature has risen by approximately 12°C. If the CO_2 content continues to

Table 7.3 Worldwide sources of CO_2 (according to [7-9])

	Source ppm/a	Sink ppm/a
Oceans	40–45	40–45
Biosphere	25	25
Combustion processes	2	0

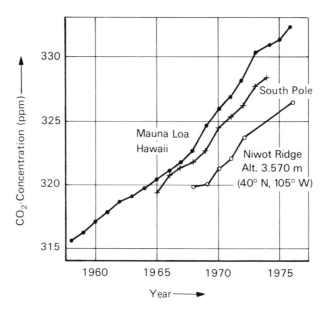

Figure 7.6 Progression of the yearly mean values of carbon dioxide content in the atmosphere at various measuring points (according to [7-9]).

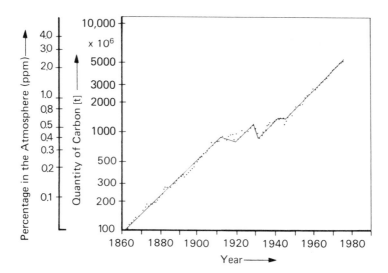

Figure 7.7 Development of global CO_2 production [7-10].

increase at the same rate, the start of the 21st century would witness a CO_2 content of between 400 and 500 ppm and a global temperature rise by 1°C. The results at our disposal, however—because of the existing doubts as to the feedback process—do not permit any prognosis as to a temperature rise, although they cannot be definitely excluded.

Besides air pollution through chemical components like nitrogen, sulfuric and carbon oxides, or dusts, man is exposed to radiation of natural and also artificial origin. The natural exposure, at least within the range of cosmic radiation, has not changed during the past few decades. However, a change is feasible through changes in the atmosphere enveloping the Earth. Thus, e.g., the absorption properties of the layers of air can be changed by fluorocarbons from spray cans.

The genetically significant radiation exposure of the population (Table 7.4) by natural sources (approximately 110 mrem/a) is roughly twice as much as the mean radiation exposure by artificial sources. The radiation load through X-ray diagnosis constitutes the major part of the artificial radiation the population is exposed to. The contribution from fallout from nuclear experiments is less than 1% of the natural radiation exposure. Exposure through nuclear facilities is assumed to be of the same magnitude.

7.1.2 Water

Water is an extremely important component of our environment; it covers approximately 3/4 of the Earth's surface and has a determining influence on vegetation and life. Water is also an essential component of the human or animal body (approximately 60–70%) as well as of plants (up to 90%).

Water found in natural surroundings, i.e., technically uncontaminated water, contains varying amounts of dissolved gases (oxygen, nitrogen, carbon dioxide) and salts (oxides of calcium and magnesia), part of which are useful and vital. Thus, e.g., minerals and trace elements serve as nutrients for plants and animals: oxygen is indispensable for fish and carbon dioxide for green water plants.

A rough classification of the purity of rivers is shown in the following summary [7-12]:

Degree of purity in a river	BSB_5* mg/l	KMnO$_4$ Consumption mg/l	Ammonia-nitrogen mg/l	Content of Substances suspended in air mg/l
Very pure	1	2.0	0.04	4
Pure	2	2.5	0.24	15
Rather pure	3	3.0	0.67	15
Doubtful	5	5.0	2.50	21
Polluted	10	7.0	6.70	35 and more

* BSB = biological oxygen requirements.

In its capacity as food, water is expected to be of a quality essentially described by the chemical properties in Table 7.5. Quality Class A means that surface waters of that type, upon simple treatment, are potable; water of Quality Class B is contaminated to a certain degree and there are certain

Table 7.4 Genetically significant radiation exposure of the population in Germany [7-11]

1. Exposure to natural radiation	approximately 110 mrem/a
1.1 by cosmic radiation at sea level	approximately 30 mrem/a
1.2 by terrestric radiation from outside	approximately 50 mrem/a
while remaining outside	approximately 43 mrem/a
while staying indoors	approximately 57 mrem/a
1.3 by incorporated radioactive substances	approximately 30 mrem/a
2. Exposure to artificial radiation	approximately 60 mrem/a
2.1 through nuclear facilities	<1 mrem/a
2.2 through the use of radioactive materials and ionizing radiation in research and technology	<2 mrem/a
2.2.1 through technical sources of radiation	<1 mrem/a
2.2.2 through industrial products	<1 mrem/a
2.2.3 through jamming devices	<1 mrem/a
2.3 professionally radiation-exposed persons (contribution to mean radiation exposure of man)	<1 mrem/a
2.4 through the use of ionizing radiation and radioactive materials in medicine	approximately 50 mrem/a
2.4.1 X-ray diagnostics	approximately 48 mrem/a
2.4.2 radiation therapy	<1 mrem/a
2.4.3 nuclear medicine	approximately 2 mrem/a
2.5 through nuclear accidents and special incidents	0
2.6 through fallout from nuclear weapons tests	<1 mrem/a
2.6.1 from outside sources while outdoors and unprotected	<1 mrem/a
2.6.2 through incorporated radioactive substances	< mrem/a

Table 7.5 Quality standards for water [7-13] (excerpts)

Component (in mg/l)	Limit values for quality class	
	A	B
Inorganic components		
Total content of dissolved substances	400	800
Suspended inorganic substances	150	200
Calcium (Ca^-)	100	—
Chloride (Cl^-)	100	200
Sulfate (SO_4^{--})	100	150
Magnesia (Mg)	30	—
Nitrate (NO_3^-)	25	50
Barium (Ba)	1.0	1.0
Boron (B)	1.0	1.0
Fluoride (F^-)	1.0	1.0
Lead (Pb)	0.03	0.05
Mercury (Hg)	0.0005	0.001
Organic components		
Suspended organic substances in		
running waters	5	25
still bodies of water	0.5	1
Dissolved organic carbon, total	4	8
Surfactants	0.1	0.3
Synthetic chelate-forming agents	0.1	0.3
Hydrocarbons	0.05	0.1
Organically fixed chlorine in the shape of Cl	0.05	0.1
Polycyclic aromatic substances	0.001	0.003
Chemical oxygen requirements (CSB)	10	20
Biochemical oxygen requirement (BSB_5)	3	5

Table 7.6 Water quality in the lower course of the Rhine (1976) and comparison with limiting values for surface water of quality classes A and B [7-14]

Component (in mg/l)	Measured value 1976		Quality class according to [7-13]	
	Yearly mean	Maximum	A	B
Chlorides	227	354	100	200
Sulfate	94	114	100	150
Nitrate	17	23	25	50
Ammonia	1.2	3.2	0.2	1.5
Iron (dissolved)	1.5	4.2	0.1	1.0
Lead	0.024	0.17	0.03	0.05
Chromium	0.034	0.085	0.03	0.05
Cadmium	0.004	0.007	0.005	0.01
Copper	0.021	0.054	0.03	0.05
Mercury	0.0005	0.0026	0.0005	0.001
Zinc	0.20	0.36	0.5	1.0
Chem. oxygen requirements	20	29	10	20
Dissolved organic carbon				
(total)	7.0	9.3	4	8
Hydrocarbons, dissolved	0.24	0.99	0.05	0.2

Table 7.7 Oxygen content (mg/l) of sweet water in a saturated state (at sea level)

Temperature	0°C	10°C	20°C	30°C
O_2 content	14.6	11.3	9.2	7.6

limits imposed on its use. Sewage from cities and industrial plants tend to pollute water of Quality Classes A and B. Samples taken from the Rhine, at the Lower Rhine, from 1976 (Table 7.6) demonstrate that the content of several substances conformed to the requirements set forth in Quality Class A, other substances, however, did not even meet the Class B requirements.

While water pollution by substances difficult to degrade (lead, arsenic, manganese, cadmium) almost exclusively comes from industrial sources, more and more products of the chemical industry used in households and in agriculture are contributing to water pollution. Among them are detergents, soaps and dishwashing liquids, cosmetics, insecticides, and plant protectives.

Another factor influencing the quality of water and thus life in the water is temperature. By increasing the water temperature, the capability of water to absorb oxygen will be reduced (Table 7.7). Simultaneously, biological/chemical processes are accelerated and algae growth increases. Due to the resulting lack of oxygen, fish are endangered, and by the growth of algae, the treatment of water for drinking is rendered more difficult.

The oxygen economy is dependent on oxygen-providing and oxygen-consuming processes. Through physical and biogeneous aeration, oxygen is supplied. Through breathing processes of plants and animals, degradation processes in microorganisms, and physical degasification, oxygen is used up. Breathing and degrading processes due to organic loads in the water may cause an O_2 deficit. The consequence of this is a lack of solution equilibrium between water and atmosphere.

The speed of oxygen supply into the water is proportional to the magnitude of the CO_2 deficit.

Like the oxygen economy, the carbon dioxide and nitrogen economies in water are temperature dependent. At higher temperatures, the contents of these will decrease and the biotic conditions will deteriorate.

7.1.3 Soil

The third essential medium in the environment, the soil, forms the border area between hydrosphere, atmosphere, and biosphere. The growth and prosperity of plants is essentially influenced by the quality of the soil. Organic substances (plants) are formed from water (rain and subsoil water) and the nutrients in the soil. As a consequence of microbial decomposition of animals and plants after they die, fixed nutrients are again released and added to the soil. On one hand, the soil has a cleansing effect on precipitation by filtering out and buffering contaminants and harmful substances. The addition of minerals, organic or inorganic, is effected by man, somewhat intentionally by fertilization and somewhat unintentionally by dumping by-products from technical installations.

The degradation of harmful substances in the soil can be effected by chemical decomposition, absorption and chemical precipitation, absorption through plants, or by dissolution in the subsoil water.

Important properties of soil are its abundance of bacteria—this makes possible the exchange of gases within the soil (breathing of the soil), small fungi, rainworms, etc., the content in mineral salts (nitrogen, calcium, potassium, and phosphor salts as well as trace elements), and acid, neutral, or alkaline soil reaction.

The contents of several heavy metals in exposed and unexposed soils is demonstrated in Table 7.8. For purposes of comparison, contents in river sediments and sewage sludge are shown. The higher values in exposed soils are generally found in the upper 5–10 cm of soil. The differences in content between exposed and unexposed soil can be 100 times or more.

Due to industrial activities and the resulting emissions, especially of sulfur dioxide and nitrogen dioxides, up to 100 kg of sulfur and 30 kg of nitrogen are deposited by rain into the soil in the vicinity of densely populated areas, mostly in the form of acids and salts. The pollution load capacity of the soil depends on the properties of the soil itself as well as on the physical/chemical properties of the substances introduced. In general, it increases with an increasing content of organic substances. In most cases of soil pollution, the buffering capacity of the soil is sufficient to prevent contamination. In specific cases, however, relatively small amounts of pollutants will be enough to trigger damage, especially to the fauna and flora, yet without rendering the soil unusable. Many of the substances accumulated in the soil thus reach man indirectly via plants and animals.

Table 7.8 Heavy metal content in exposed and unexposed soils, taken from sediments from the Elbe and Rhine rivers and from sewage sludge [7-15]

| Element | Exposure (mg/kg dry mass) in | | | | |
	Unexposed soil	Heavily exposed soils (near metal refinery)	Elbe sediments	Rhine sediments	Sewage sludge (mean value)
Zinc	5–100	2900	4600	2900	2500
Chromium	5–100	—	600	1240	1500
Lead	5–40	2000	600	800	600
Copper	2–30	70	1150	600	1000
Arsenic	2–20	—	275	220	15
Cadmium	0.5	40	42	45	15
Mercury	0.2	—	28	23	10

7.1.4 Food

Besides water, plants and animals constitute man's principal food. Plants and animals need organic carbon compounds in order to develop their body substance and to meet their energy requirements. In contrast to the animals, plants can build these carbon compounds from the basic substances water and carbon dioxide (assimilation). For this purpose, sufficient energy must be made available, e.g., by sunlight and its utilization via chlorophyl (green leaf).

Animals, however, must utilize already available organic substances. Without plant life, no animal life would be possible, since plants constitute the sole connecting link in the carbon cycle able to convert carbon dioxide into carbon compounds. Besides carbon dioxide and oxygen from the air, plants absorb water and mineral substances from the soil.

The dry substance of plants (elementary analysis less the water content) mainly contains carbon (40–50%), oxygen (30–45%), hydrogen (5–6%), and nitrogen (2–5%), plus a large number of additional chemical substances like sulfur, phosphorus, potassium, calcium, sodium, magnesium, chlorine, and silicon [7-16]. The distribution of these substances varies considerably from one plant to another. A different breakdown of these substances, other than according to the aforementioned elementary analysis, reveals a high content of albumens, fatty substances, carbohydrates, and inorganic substances. The presence of all these groups of substances is decisively important for the prevention of disturbances in these dissolution processes (disturbances of metabolism). For the health of these plants, environmental conditions such as sunlight, temperature, and humidity are of significance equal to that of the supply of nutrients from soil and air.

Nutrients for man and animals are solid foodstuffs, water, and air. Solid food especially, contains the required albumens, carbohydrates, fats, mineral salts, and vitamins. Together with the foodstuff "air" (oxygen), the energy necessary to survive is generated within the body through combustion. Like plants, animals (elementary analysis less the water content) consist mostly of carbon, oxygen, hydrogen, and nitrogen as well as a vast number of other chemical elements. Depending on the animal, the meat contains the following chief components, the relative quantities of which fluctuate considerably [7-17]:

	Lean meat	Fatty meat
Water	50–75%	35–70%
Albumen	14–22%	10–20%
Fat	2–30%	30–55%
Mineral substances	0.8–1.8%	0.5–1.0%

Among the mineral substances in meat are potassium phosphate, sodium chloride, calcium phosphate, phosphate of magnesia, as well as traces of iron, zinc compounds, sulfates, and silicic acid.

Thus the foodstuffs meat and plants contain mineral substances which are necessary for man and animals to survive. However, they can also contain substances of a type which, as alien or noxious substances, may either directly or indirectly cause toxic symptoms. By way of the food chain, these pollutants can be introduced into the human body. The food chain is either very simple or it contains a number of different transmission links. The simplest form of the food chain is the absorption of air through breathing or of water by drinking. More complicated processes are the absorption of harmful substances from the soil via the plant, which in turn is eaten by the animal, whereupon animal products are ingested by man.

7.1.5 Noises—Sound

An essential environmental factor which may affect safety is sound. Vibrations of the air of frequencies from between 16 to approximately 20,000 Hz are perceived by the human ear in the form of sound. The hearing threshold lies approximately at 20 μPa (0.2×10^{-9} bar) of sound pressure, corresponding to a sound intensity level of 0 dB. At a sound pressure (intensity) level of approximately 130 dB, man's pain threshold will be reached. Some actual sound levels, measured at various distances, can be taken from the following table:

Origin	Sound level (mean frequency)	Distance
Ticking of an alarm clock	30 dB (A)	1 m
Rustling of poplar leaves in a wind of 5 m/s	40–50 dB (A)	10 m
Normal conversation	50–60 dB (A)	1 m
Thunder	65–70 dB (A)	2000–5000 m
Motor vehicle	80–85 dB (A)	7 m
Air hammer	90–100 dB (A)	7 m
Jet plane	120–130 dB (A)	50–100 m

The range of conventional noise levels is demonstrated in the noise level scale (Figure 7.8).

Sound volume is a subjective perception which has no relationship to the actual noise intensity levels. Reference data between subjectively perceived sound volume and measured sound-level difference are as follows: 1–2 dB difference is not yet registered, 8–10 dB difference in level is perceived as a doubling or halving of the sound volume. The acoustical environment perceived by man as agreeable is a sound level, devoid of any information, of from 15 to 20 dB (A). Lesser sound levels as well as essentially higher ones are perceived as a form of molestation. Any sound perceived by man as a subjective molestation or disturbance is called noise.

As an evaluation scale for sound, certain emission values have been determined which vary—depending on the time of day or night and on the

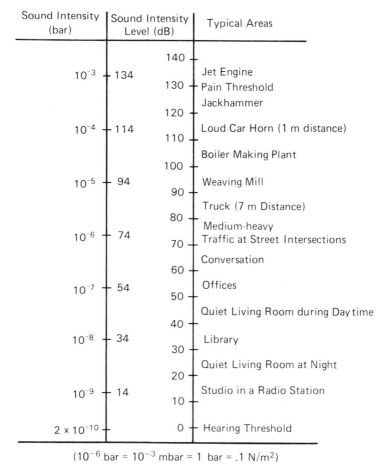

Sound Intensity (bar)	Sound Intensity Level (dB)	Typical Areas
	140	
10^{-3} — 134		Jet Engine
	130	Pain Threshold
		Jackhammer
	120	
10^{-4} — 114		Loud Car Horn (1 m distance)
	110	
		Boiler Making Plant
	100	
10^{-5} — 94		Weaving Mill
	90	
		Truck (7 m Distance)
	80	
10^{-6} — 74		Medium-heavy Traffic at Street Intersections
	70	
		Conversation
	60	
10^{-7} — 54		Offices
	50	
		Quiet Living Room during Daytime
	40	
10^{-8} — 34		Library
	30	
		Quiet Living Room at Night
	20	
10^{-9} — 14		Studio in a Radio Station
	10	
2×10^{-10} —	0	Hearing Threshold

$(10^{-6}$ bar $= 10^{-3}$ mbar $= 1$ bar $= .1$ N/m$^2)$

Figure 7.8 Sound intensity (pressure) and sound level of various noises [7-18].

characteristics of the area to be protected—between 30 and 70 dB (A). Noxious sound from outside noises may override useful sound as well as the sound of speech and may thus inhibit the "ability to comprehend entire sentences," i.e., voice communication. Within the living area, complete voice communicability must be guaranteed; this is the case at noise levels of less than 40 dB (A). Due to the noise abatement achieved through closing windows (10 and 25 dB, respectively), the inside sound level of 40 dB (A) corresponds to an outside noise level of 50 and 65 dB (A), respectively. Voice communicability decreases with increasing noxious noise level within a room (Figure 7.9). At 65 dB (A) it only amounts to approximately 95%, at 70 dB (A) only to 60%, and at 75 dB (A) it is completely lost.

During communication outdoors (outside of rooms) the voice sound level decreases at twice the distance by 6 dB (A). At 50 dB (A) and

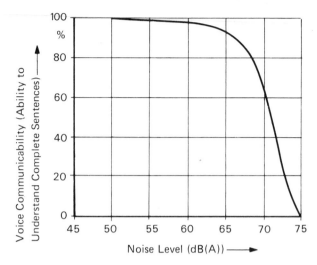

Figure 7.9 Voice communicability as a function of the noise level in normal conversation [7-19].

relaxed conversation, a voice communicability of 100% is found at 0.6 m, and of 99% at 2 m. At 55 dB (A), the voice communicability at 0.35 m is 100%, at 1 m, 98%, and at 3.5 m, 95%. At a very low noxious noise level of 40 dB (A) and normal conversational tone, there is still 90% voice communicability at 20 m. In the urban environment, many areas have communication functions (shopping malls, squares, plazas, meeting points, street cafés, etc.). At 60 dB (A) and normal conversation, there is 99% voice communicability at 0.7 m and 95% at 2 m.

Sleep disturbances usually present themselves in the form of an inability to fall asleep, falling asleep very late, or inhibited sleep behavior. We may safely accept the finding that mean levels at the sleeper's ear of up to 35 dB (A) still remain within the range of not affecting sleep. Any increase in these levels should not exceed these values by more than 10 dB (A). This means that the sleeper will still remain below the threshold for waking reactions and changes in the various stages of sleep. With outside noise levels of 35–45 dB (A), it is thus still possible to sleep with the windows open. The daytime course of noises affecting man is directly coupled to human and technical activities. Night-time decreases of the mean levels by 15 dB (A) will only be found at a great distance from areas and streets where there are night-time noises. The higher the traffic-related noise level, the lower, as a rule, the night-time level decrease; for Autobahns (expressways) and busy highways, it is usually less than 5 dB (A) (Figure 7.10). Within the range of constantly emitting technical or natural sources of noise (e.g., power or chemical plants or the thunder of waterfalls or rapids), a probable increase in the emission levels can be determined at night. In this case, the usually more favorable noise propagation conditions at night will tend to raise the noise level.

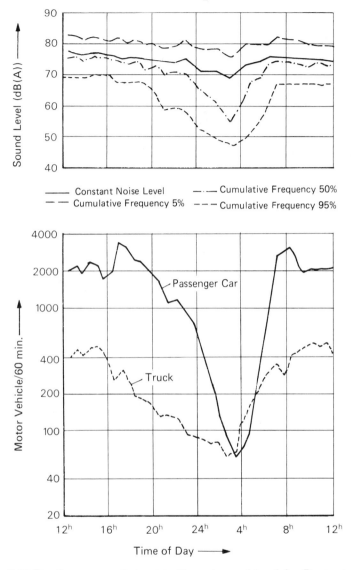

Figure 7.10 Daytime curve of street traffic and sound level for Reuterstrasse in Bonn, Germany [7-20].

Differing from one location to the other, weekly changes in the noise situation may occur. The cause of more noise is often increased traffic on Monday mornings and Friday afternoons. Counteracting factors are factories closing down for the weekend and decreasing density of traffic during the weekend, e.g., in the downtown area. Yearly changes may be caused by the use of leisure facilities or by the work schedules of commercial enterprises.

7.1.6 Vibration and Shock

Periodic shocklike and irregular mechanical movements are described as vibrations or shock. In technical systems they may produce deformations or stress and in man, molestation, unconscious effects (e.g., on the vegetative nervous system), or even damage to health. This may entail changes in a person's well being or decreasing performance levels.

Interrelations can be found between the perception of vibrations and the measurable vibration intensity (Figure 7.11). The intensities of perception develop along steps A through F and are characterized, in accordance with the effect of these vibrations along the x, y, and z axes, by the letters KX, KY, and KZ respectively; KB stands for the structure-related perception intensity. Characteristic control parameters for vibration are vibration path, oscillation frequency or acceleration, frequency, duration of exposure, and frequency of occurrence. So-called KB-values have been suggested as an evaluative scale for vibrations in apartments and comparable rooms, whereby these values vary, depending on the character of the area in question (strictly residential area to industrial areas), between 0.1 and 12 [7-22].

Vibration and shock are largely caused by traffic and industrial plants. The frequencies range from 3 to 25 Hz and the vibration frequencies between 0.2 and 1.0 mm/s.

Stage	Intensity of Perception KX, KY, KZ, KB	Description of Perception
A	< 0.1	not perceptible
	—— 0.1 ——	perception threshold.
B_1		
	—— 0.2 ——	barely perceptible
B_2		
	—— 0.4 ——	
C_1		
	—— 0.8 ——	good perception
C_2		
	—— 1.6 ——	
D_1		
	—— 3.15 ——	strongly perceptible
D_2		
	—— 6.3 ——	
E_1		
	—— 12.5 ——	
E_2		
	—— 25 ——	very strongly perceptible
E_3		
	—— 50 ——	
E_4		
	—— 100 ——	
F	> 100	

Figure 7.11 Scale of vibration and shock [7-21].

7.1.7 Climate and Weather

Climate and the weather are factors which influence our living conditions to a large degree and which may change human activity within a narrower or a wider range. Essential elements of weather are sun and celestial radiation, temperature, humidity, wind, clouds, and precipitation. These constitute the climate in the shape of mean values dispersed over prolonged periods of time. The various effects of the individual elements on organisms demonstrates the importance of probable changes in climate, e.g., by the operation of technical installations. Thus the rays of the sun within the visible range of light are of the greatest importance for the assimilation of plants (generation of oxygen) and thus for life on Earth. The radiation within the infrared and visible range plays an important part in the Earth's thermal economy.

Central and Western Europe are part of the humid-temperate zone, i.e., precipitation is evenly distributed over all seasons and the monthly temperature mean in the warmest month is below 22°C and in the coldest month above −3°C. The climatic description of the Central/Western European area according to Table 7.9 is only intended for a general characterization; it is necessary for the detection and evaluation of the influence exerted by the technical activities of man. The magnitude of long-term air temperature fluctuations can be seen in Figure 7.12.

A similar development, qualitatively speaking, is demonstrated by mean values over several years (five-year mean) of the surface temperature in the Northern and Southern Hemispheres (Figure 7.13). The absolute change of the yearly mean values during the past 100 years has amounted to

Table 7.9 Climatological data from several German cities [7-23]

	Hamburg	Essen	Kassel	Stuttgart
Temperature (°C) (mean values)				
1931–1960	8.5	9.6	8.7	9.9
1962–1969	8.3	9.3	9.0	9.0
1970–1978	8.7	9.6	9.2	10.0
Hours of sunshine (yearly mean)				
1951–1960	1628	1496	1594	1734
1962–1969	1474	1387	1430	1679
1970–1978	1611	1453	1440	1630
Precipitation (mm/year)				
1891–1930	738	892	594	660
1931–1960	713	931	644	700
1962–1969	765	939	744	733
1970–1978	722	829	616	672

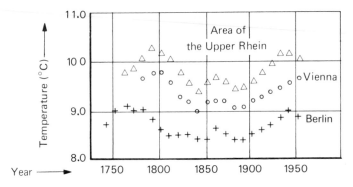

Figure 7.12 Changes in time of the thirty-year mean temperature [7-24].

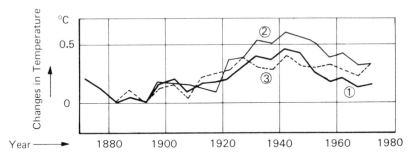

Figure 7.13 Five-year mean of the Earth's surface temperatures (changes in relation to 1880–1884): 1 = the entire Earth; 2 = 0–80°N; 3 = 0–60°S [7-25].

approximately 0.5–1°C. In comparison, the seasonal temperature fluctuations of nearly ±10°C are distinctly higher (Table 7.10). Year-to-year changes in one location are higher in winter by ±6°C than in summer with −3°C (Figure 7.14)

Increasing technicalization and mechanization and the resulting higher energy consumption may interfere with the local climate by means of increasing emission of alien substances into the air and of waste heat. Globally as well as regionally, a change in the composition and heat content of air—and thus a climatic change caused by anthropogenous activities—has yet to be proven.

Local changes in climate (city climate), however, have been determined. Thus, the yearly mean temperature in cities is 1–2°C higher than that of its immediate environment; global radiation is lower by approximately 10% and the wind velocity is approximately 40% lower. The cause for these changes are buildings erected in the area as well as the vast amounts of heat released during the generation of energy and the emission of alien substances into the air. In energy generation—whether in a power plant or in a household—only 40–60% of the primary energy is actually used up; the rest is released in the form of waste heat into the atmosphere.

Table 7.10 Monthly and yearly mean air temperatures (observation period 1880–1930) [7-26]

Period	Cologne	Hamburg	Munich
January	2.4	0.3	−1.3
February	3.4	1.0	0.2
March	5.9	3.5	3.8
April	9.4	7.5	7.8
May	14.1	12.3	12.9
June	16.8	15.4	15.9
July	18.4	17.1	17.8
August	17.7	16.2	17.0
September	14.9	13.6	13.5
October	10.4	8.8	8.3
November	6.0	4.2	3.1
December	3.4	1.6	0.0
Year	10.2	8.5	8.2

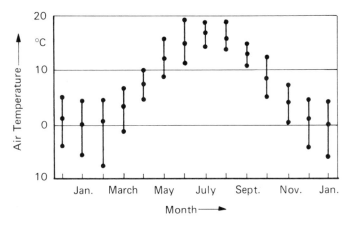

Figure 7.14 Air temperature—long-term monthly means as well as highest and lowest monthly means (Hamburg 1881–1930) [7-26].

Released substances which are alien to air, mainly carbon dioxide and water vapor, are capable of effectively absorbing long-wave radiation and thus influencing the radiation economy.

As in cities, an insular heat effect can also be observed in the vicinity of large industrial complexes. As a result of waste heat, the air above an industrial complex is transported upward, while on the ground a simultaneous "air intake" takes place. However, this does not produce a climatological interference of a considerable magnitude; changes in the climate's parameters, especially those of temperature, are restricted to strata maximally 100 m above ground.

Heat emission in power plants up to 100 MW of output, with power flux densities of approximately 0.1 MW/m^2 compared to the power flux density of sunlight of 160 W/m^2, can be called high; however, they are essentially restricted as to space. Thus, because of their often isolated location, they will not have any local, not to mention climatic, effects. In large power plant parks with over 10,000 MW, however, influencing the climate is certainly possible unless the distances between locations of high-energy transfer are sufficiently great [7-27].

Temporarily higher loads of substances normally not carried in the air can be expected mainly in densely populated areas (cities); here, orography and the local structure of the area, as well as the resulting microclimatic conditions constitute an essential influence factor. Locations in valleys demonstrate preferred air currents (winds along the valley); locations in natural bowls show a strong tendency towards meteorological situations with restricted exchange of air. Unfavorable atmospheric conditions like this will occur in situations where there is very little wind (wind velocities below 1 m/s) or inversion-type meteorological situations. Inversions are characterized by the normal absence of temperature decrease in the atmosphere at higher altitudes. On the contrary, at specific heights, a temperature increase will be noted. Inversions like this are especially noted during the morning hours, because the cold ground is heating up as a result of the sun's rays. During the course of the morning, these inversions will dissolve and lift off the ground (lifted-off inversions). This impedes vertical air exchange. Accumulations of air pollution can occur, together with inversion-type or low-exchange meteorological situations, and lead to smog which is an artificial word created from the two concepts of smoke and fog. Smog will form the moment large amounts of sulfur dioxide and soot accumulate in the atmosphere (London or winter smog) or whenever nitric oxides and oxygen-enriched carbohydrates are subjected to photochemical reactions under solar irradiation (Los Angeles or summer smog). Man's well-being—and thus, in a wider sense, safety in his occupational environment—depends on the climatic factors of air temperature, humidity, and wind velocity. Higher temperatures with high relative humidity bring forth a feeling of mugginess and thus discomfort. Higher temperatures in conjunction with low relative humidity which are locally limited and act on man abruptly and without warning are found during Föhn situations and also influence man's well being.

7.1.8 Catastrophic Natural Phenomena

Elementary forces such as floods, excessively high tides, earthquakes, electrical and other storms, may influence the environment and its safety to a considerable degree. Floods, storms, and extreme high tides which have occurred during this century are shown in Table 7.11. As a supplement, the frequency of catastrophic storms around the world during this century is represented in Table 7.12. In Europe, these events are far less

Table 7.11 Storms and floods in Western Europe since 1900 [7-28]

Event	Location	Date	No. of fatalities	Damage in millions of dollars (U.S.)
Flood	Danube area	Aug.1920		20
Flood	Rhine area	1925/1926		
Flood	Rhine area	June/July 1926		
Flood	Po delta (Italy)	Nov. 1951		
Extreme tide	Holland	Jan./Feb. 1953	1787	3000
Flood	Danube area	July 1954		50
Extreme tide	North Sea	Feb. 1962	347	600
Flood	Barcelona (Spain)	Sep. 1962	474	80
Flood (after mudslide	Langarone (Italy)	Oct. 1963	1800	
Flood	Arno (Italy)	Nov. 1966	113	1300
Hurricane	North Sea	Feb. 1967		300
Tornado	Northern France	June 1967		20
Flood	Lisbon (Portugal)	Nov. 1967	457	
Flood	Po Valley (Italy)	Oct. 1970		200
Flood	Alicante (Spain)	Nov. 1972		20
Hurricane	Lower Saxony (Germany)	Nov. 1972	52	420
Hurricane	Palermo (Sicily)	Oct. 1973		120
Flood	Granada (Spain)	Oct. 1973	350	135
Hurricane	Germany	Jan. 1976		1300
Flood	Canton of Uri (Switzerland)	Jul./Aug. 1977		50
Flood	Po Valley (Italy)	Oct. 1977		150
Flood	Athens (Greece)	Oct./Nov. 1977	25	30

Table 7.12 Catastrophic storms worldwide since 1900 [7-28]

Continent	Frequency	Type of storm
Africa	1	Cyclone
Asia	29	Typhoons
	16	Cyclones
	1	Tornado
Europe	4	Hurricanes
	2	Extreme tides
	1	Tornado
North America	17	Hurricanes
	11	Tornadoes
Central America	8	Hurricanes
Australia	1	Cyclone

frequent than in Asia and America, but they cannot be altogether excluded.

High-water marks along rivers have been registered for decades. Thus, in 1970, the Rhine showed the highest water level ever recorded with 40.6 m above normal level [7-29]. Statistics show that a high water level of 41.6 m above normal can be expected every 100 years and a level of 42.6 m every 1000 years.

The distribution and the frequency of *earthquakes* worldwide is shown in Figure 7.15. Earthquakes in Europe (Table 7.13) are less frequent than in Asia, as can be seen from Table 7.14. Until now, Europe has not experienced any larger-scale earthquakes.

The effects of earthquakes are classified according to various scales (Table 7.15). The Richter scale is allocated to varying amounts of energy in

Table 7.13 Severe earthquakes in Europe since 1900 [7-28]

Location	Date	Number of dead	Damage in millions of dollars (U.S.)
Messina/Italy	Dec. 1908	83,000	
Avezzano/Italy	Jan. 1915	30,000	
Naples/Italy	July 1930	1883	
Bucharest/Rumania	Nov. 1940	980	10
Skopje/Yugoslavia	July 1963	1070	300
Sicily/Italy	Jan. 1968	281	320
Friuli/Italy	May 1976	978	2000
Bucharest/Rumania	March 1977	1581	800
Montenegro/Yugoslavia	April 1979	130	
Southern Italy	Nov. 1980	3200	

Table 7.14 Severe earthquakes worldwide since 1900 [7-28]

Continent	Date	Frequency	Fatalities
Africa		6	21,000
Morocco	1960		13,000
Asia		38	1.3 Million
China	1976		665,000
	1920		100,000
	1927		80,000
	1932		77,000
	1970		20,000
Japan	1923		143,000
Turkey	1939		33,000
Pakistan	1935		30,000
Iran	1928		25,000
USSR	1948		20,000
Europe		10	120,000
Italy	1908		83,000
	1915		30,000
North America		5	1100
California	1906		750
Central America		11	40,000
Guatemala	1976		23,000
South America		8	100,000
Peru	1970		52,000
Chile	1939		28,000

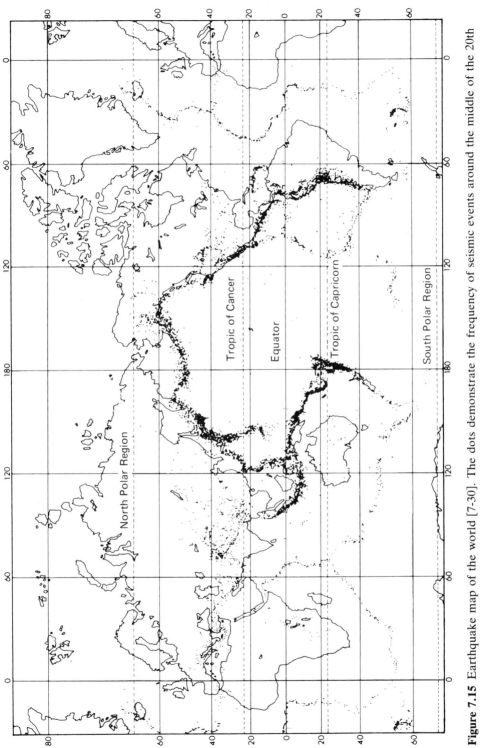

Figure 7.15 Earthquake map of the world [7-30]. The dots demonstrate the frequency of seismic events around the middle of the 20th century.

Table 7.15 Scales for measuring earthquake intensity [7-31]

Modified Mercalli intensity scale	Characteristics	Richter scale	Earthquake zone according to [7-32]
I	Registered by seismological equipment only	2	
II	Only sporadically registered by persons at rest	3	
III	Felt by a few persons only		
IV	Felt by many, windows and china rattle	4	0
V	Hanging objects start to swing, many sleepers wake up		
VI	Slight damage to buildings, fine cracks in the plastering	5	
VII	Cracks in the plastering, gaps in walls and chimneys	6	1
VIII	Large gaps in the masonry, gables and roofs cave in		2 3
IX	In some buildings, walls and roofs collapse, mudslides	7	4
X	Many buildings collapse, cracks in the ground up to 1 m wide		
XI	Many cracks in the ground, avalanches in the mountains	8	
XII	Marked changes in the Earth's surface	9	

Table 7.16 Other natural catastrophes in Western Europe since 1900 [7-28]

Occurrence	Location	Date	Damage in millions of dollars (U.S.)
Volcanic eruption	Vesuvius (Italy)	Apr. 1906	
Volcanic eruptions	Etna (Italy)	1928	
Hailstorm	Bavaria/Württemberg (Germany)	July 1929	55
Volcanic eruption	Etna (Italy)	1950	
Hail	Bavaria	July 1953	25
Cold wave	Spain	Jan. 1971	400
Hailstorm	León (Spain)	Aug. 1971	55
Hail	Lower Saxony (Germany)	Aug. 1974	22
Forest fire	Lower Saxony (Germany)	Aug. 1975	15

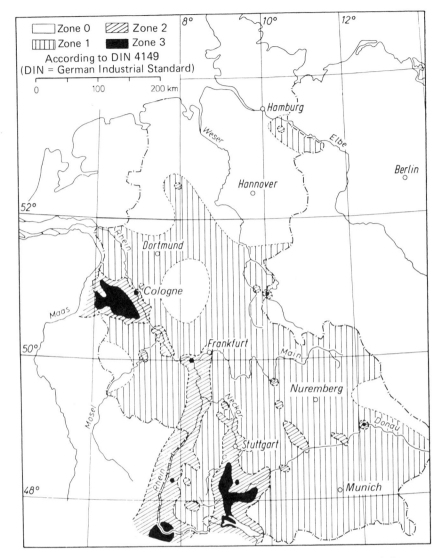

Figure 7.16 Map showing earthquake areas in the Federal Republic of Germany [7-33].

the epicenter; it corresponds to 2×10^{15} erg of intensity III, 10^{19} erg of intensity V, and 10^{22} erg of intensity VII. There is abundant information available on the danger of earthquakes in the Federal Republic of Germany. The earthquake zones shown in Figure 7.16 are meant for easier orientation. According to that map, the critical zones are the areas East of Aachen, around Lörrach and the Schwäbische Alb.

Other *natural catastrophes* such as volcanic eruptions and hailstorms have caused damage in Western Europe during this century; however, they have not resulted in a loss of human lives (Table 7.16).

7.1.9 Structure of the Area and Density of the Population

The importance and scope of environmental pollution, i.e., of molestation and endangerment as well as the safety factor, depend on the density of the population, the industrial infrastructure, and their interrelationship. The density of population varies greatly within any given country; thus, in Germany, only approximately 100 people/km² live in the Stade or Lüneburg areas while in Cologne the corresponding figure is 15,000 inhabitants/km². In the Federal Republic of West Germany, the average population density is 244 inhabitants/km². The State of North Rhine-Westphalia (Nordrhein-Westfalen) serves as an example for the density of population and its changes over the past 30 years (see Table 7.17).

In general, industrial complexes are located in the vicinity of man's residential areas. However, great differences can be detected from the comparison of two selected largely industrialized areas: while in the Cologne area, heavy industry is situated peripherally and approximately 10 km from the center of the inner city (Figure 7.17), industrial installations practically dissect the cities of Duisburg, Mülheim, and Oberhausen. In Cologne, this has resulted in a concentration of the population in the inner city (of up to 170 inhabitants/hectare) (1 ha = 2 1/2 acres), while in Duisburg/Mülheim/Oberhausen, the population is distributed over larger areas, with 40–80 inhabitants/hectare. The density of the population in the Duisburg area is shown in relationship with the district boundaries in Figure 7.18.

As a consequence of differing population density and the structure of the built-up areas, the traffic density will vary as well. While in the Cologne

Table 7.17 Density of population [7-36]

State (land)/district	Area (1978) km²	Population density (Inhabitants/km²)			
		1950	1961	1970	1978
Nordrhein-Westfalen	34,100	388	467	497	499
Nordrhein	12,700	535	668	716	719
Westfalen	21,400	301	348	367	369
Ruhr Area	3900	1071	1322	1312	1257
Bochum	200	2823	3361	3193	2997
Dortmund	1050	891	1095	1121	1112
Duisburg	2500	387	486	502	491
Essen	380	2587	3151	3051	2836

Nievenheim
Dormagen
Monheim
Opladen
Worringen
Lever-kusen
Fühlingen
Sinners- dorf
Merkenich
Wies-dorf
Longerich
Stammheim
Höhen-haus
Bickendorf
Riehl
Mühl-heim
Nippes
Ehrenfeld
Kalk
Ostheim
Köln
Deutz
Rath
Sülz
Bayenthal
Kletten-berg
Heumar
Frechen
Zollstock
Marienbg.
Eil Porz
Rodenkirchen
Hürth
Rondorf
Sürth
Knapsack
Godorf
Lülsdorf
Berzdorf
Wesseling
Niederkassel

■	Industrial Areas
▨	Commercial Areas
░	Residential Areas
⫶	Recreational Areas (Forest, etc.)

0 1 2 3 4 5 km

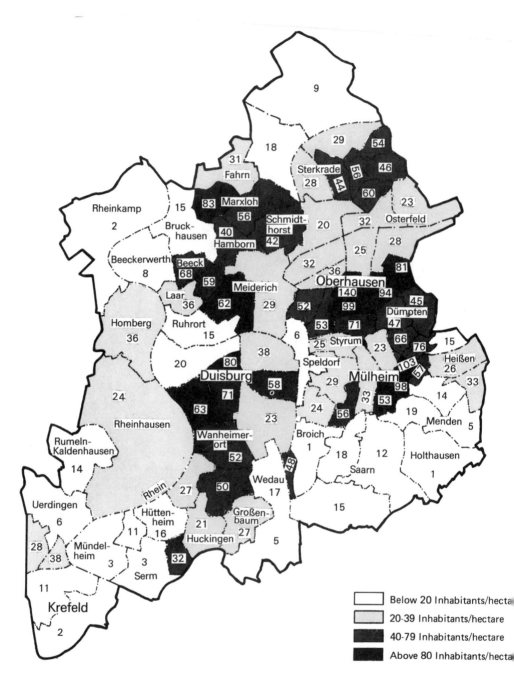

Figure 7.18 Density of the population in the Duisburg/Mülheim/Oberhausen region [7-35].

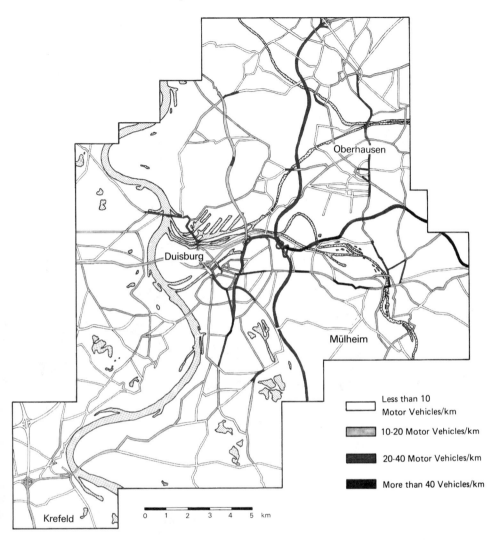

Figure 7.19 Traffic density in the streets of the densely populated area Duisburg/Mülheim/Oberhausen [7-6].

area, the interstate highways (Autobahns) touch the city only peripherally, the Autobahns in the Duisburg/Mülheim/Oberhausen region practically dissect the settled areas (Figure 7.19). The mean vehicle traffic densities on the main traffic arteries are highlighted in the illustration. Traffic at various hours of the day makes it obvious that the streets, at certain hours, are heavily congested, and at other hours are used very little (Figure 7.20).

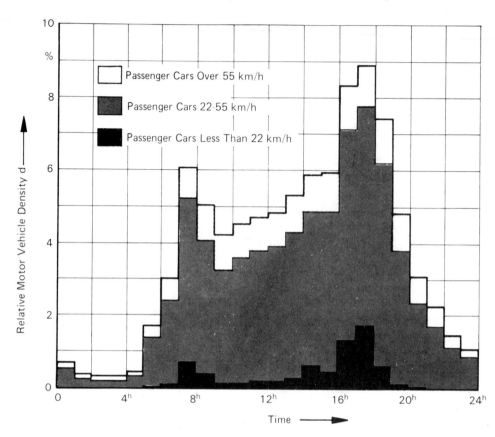

Figure 7.20 Example of the traffic density for passenger car traffic as a function of the time of day [7-35].

7.2 Environmental Strain Through Accidents in Technical Installations

Experience and accident statistics alike demonstrate that, apart from accidents within technical installations, losses affecting human beings and the environment may occur outside the technical installation itself. There may be interrelationships between accident and environmental strain. The effect of an accident in a technical installation becomes obvious the moment toxic or flammable substances spread in the air, the water, or on the ground, or whenever pressure waves propagate in the air or in the ground and lead to environmental damage. The causes, e.g., may be mechanical damage to machinery, explosions, or fires. The influence of environmental stress on the evolution of the accident, on the other hand, is not equally obvious. There is, however, the possibility of human error while handling a technical installation and as a consequence of impaired reaction in a situation which can be traced back to technical processes. Man may become hypersensitive because of high sound, vibration, or olfactory levels and produce a negative type of behavior within the man–machine–environment–system.

7.2.1 Environmental Stress Under Normal Operation

Environmental stresses through technical installations are usually transmitted through the carrier media of air and water which are both part of the environment. The noxious substances are introduced in gaseous or dusty form into the air or, as a solution, into the water. Through the air we breathe and the water we drink, these noxious substances bring about a direct endangerment or, via the food chain, stress on man and animals.

For the normal operation of technical installations, the emission of noxious substances can be reduced to a certain degree by appropriate devices; however, in most cases, they cannot be excluded completely. Systematic checks on the emission of substances alien to the environment permit the establishment of emission files for air and water pollution. On the basis of these recordings, and via analyses of the origin of these substances, an improved environment can be created by purposeful measures. The effects of pollution on air under normal operating conditions in densely populated areas is shown in Table 7.18.

The establishment of and adherence to emission limits at the source is designed to ensure that environmental stress is kept within certain predefined limits. Beyond these minimum requirements in the form of emission limits (Table 7.19), another concept is of great importance, namely that the limiting of emissions must be tuned to the possibilities open to us within the scope of the state of the art (technology).

As a form of protection from environmental stress on man, animals, and plants, emission limits have been established which are designed to prevent health hazards and molestation to man from the operation of technical installations and to guarantee the protection of animals, plants, and property (Table 7.20). The interrelationship between environmental stress and impairment to health is a rather complex one and can be represented statistically in a generalized form only (Figure 7.21). The values established by law in Table 7.20 can only be designed on a long-term basis. While the long-term value should not be exceeded as a yearly average, the short-term value may be exceeded during 5% of the (operational) hours per year which, however, are not required to form a single period of time.

In order to be able to evaluate short-term environmental stresses, the values of the maximum emission concentration (MIK) offer a better way. The state of knowledge is still limited, a fact which is expressed in a

Table 7.18 Examples of emissions of pollutants in densely populated areas [7-6], [7-34], [7-37], [7-38]

| | Densely populated areas | | | | | | | |
| | Rhine Route South | | Ruhr Area West | | Ruhr Area East | | Ludwigshafen | |
	1000 t/a	t/a km²*	1000 t/a	t/a km²*	1000 t/a	t/a km²*	1000 t/a	t/a km²*
Inorganic gases								
Industry	298	460	1674	2350	455	567	96	857
Household coal	107	165	163	231	159	198	13	116
Motor vehicle traffic	134	206	111	156	125	156	24	214
Organic gases								
Industry	84	129	151	211	4.4	5	11	98
Household coal	6.3	10	12	17	4.9	6	0.96	9
Motor vehicle traffic	4.8	7	9.1	13	10	12	2.0	18
Dust								
Industry	25.2	39	192	271	19	24	17	152
Household coal	5.0	8	9.1	13	3.1	4	0.24	2
Motor vehicle traffic	0.47	1	0.29	1	0.32	1	0.04	1

* The emissions are related to the total area of the densely populated region.

Table 7.19 Examples of emission limits for dusty substances as well as vapors and gaseous organic compounds in waste gases [7-40].

The following mass concentrations in exhaust gas and at a mass flow of less than or equal to Q must not be exceeded:

Class	Dust	Organic compounds
I	20 mg/m^3 (Q = 0.1 kg/h)	30 mg/m^3 (Q = 1.0 kg/h)
II	50 mg/m^3 (Q = 1 kg/h)	150 mg/m^3 (Q = 3 kg/h)
III	75 mg/m^3 (Q = 3 kg/h)	300 mg/m^3 (Q = 6 kg/h)

Dusty substances	Class	Organic compounds	Class
Bitumen	III	Acetone	III
Lead	I	Acrolein	I
Cadmium	I	Acetic Acid	II
Cobalt	II	Formaldehyde	I
Quartz ($<5 \ \mu$m)	II	Mercaptan	I
Tar	II	Methanol	III
Uranium	I	Phenol	I
Bismuth	III	Xylol	II

Table 7.20 Emission limits in μg/m^3 of air [7-39]

Foreign matter in air	Long-term value	Short-term value
Sulfur dioxide	140	400
Nitrogen dioxide	100(80*)	300
Nitrogen monoxide	200	600
Carbon monoxide	10,000	30,000
Hydrogen sulfide	5	10
Chlorine	100	300
Hydrogen fluoride	2(1*)	4(3*)
Lead and its inorganic compounds	0.2	—
Cadmium*	0.04	—
Dust suspended in air	100	200

* In accordance with the draft for the Amendment TA Luft (Air), Sept. 1978 [7-40].

Table 7.21 Maximum emission concentrations [7-42]

Substance	Mean value in μg/m^3 over		
	1/2 h	24 h	1 year
Sulfur dioxide	1000	300	100
Nitrogen dioxide	200	100	—
Nitrogen monoxide	1000	500	—
Carbon monoxide	50,000	10,000	10,000
Hydrogen fluoride	200	100	50
Lead and its inorganic compounds	—	3	1.6
Cadmium compounds	—	0.05	—
Fine dust	300	200	100

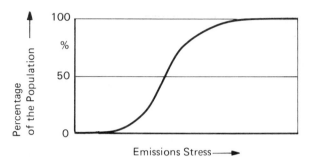

Figure 7.21 Percentage of the population exposed to health hazards as a function of the emission load (e.g., concentration of foreign matter in air in mg/m^3).

relatively small number of published MIK values, especially for organic substances [7-41], [7-42]. As for the air, there are concentration limit values for noxious substances in water as well. Table 7.22 contains several critical concentration limits for substances contained in waste water, i.e., the minimum concentrations for which a toxic effect has been established. Here it should also be stated that besides degree of concentration, the toxic effect of substances on man also depends on a number of other factors connected with water, e.g., oxygen content, temperature, and pH-value. Just as in air, there are synergistic effects in toxic substances in water. Thus the ammonia ion, at lower oxygen tensions, is more toxic than at higher tensions, and its toxic effect is strongly dependent on the pH-value of the water.

The "critical concentration limits" in waste water (Table 7.22) must thus not be mistaken for "highest permissible concentrations" in drinking

Table 7.22 Concentration limits for the toxic effects of various components in waste water (sewage) [7-43]

Substance	mg/1
Ammonia	0.2
Aniline	20.0
Benzene	5.0
Prussic acid	0.03
Cadmium	0.05
Chlorine	10.00
Formaldehyde	0.01
Copper	0.02
Phenol	1.0
Toluol	10.0
Xylol	10.0

water. Thus, e.g., the critical concentration limit of cadmium is fivefold higher than its highest permissible concentration [7-44].

7.2.2 Environmental Stress During Incidents (in Nuclear Facilities)

Environmental stress and endangerment of exceptional magnitude may occur whenever, during incidents, larger amounts of energy or considerable amounts of substances dangerous to man's health are released. The harmful effects as a consequence of incident-related release of health-endangering substances may be extremely violent, as has been demonstrated in the past by a variety of accidents.

For a number of installations where hazardous substances must be present due to the nature of the operation, or which may originate during an incident, safety analyses are required by law [7-45]. The purpose of analysis is to institute preventive measures in order to forestall accidents or to limit their consequences as much as possible. If suitable measures are taken at an early point in time, incident-related releases of toxic or flammable substances usually result in short-term environmental stresses only.

The degree of knowledge of the harmful effects of toxic substances as a function of the duration and degree of stress is still unsatisfactory. As far as the toxic effects on human beings inside technical installations are concerned, the nature of the substance and the amount ingested or absorbed is of decisive significance. Therefore, it is important to know the relationship of toxic effects and absorbed dosages (Table 7.23). In many cases, animal research is conducted for that particular purpose (Table 7.24).

In the past, the public's interest has been centered almost exclusively on acute poisoning as a consequence of the operation of technical installations. Today we distinguish between acute, subchronic, and chronic

Table 7.23 Toxic effect of harmful substances at varying concentration levels, measured in $\mu g/m^3$ as well as exposure times [7-46]

| Toxic substance | Toxic effect | | | | |
| | None | Dangerous | | Fatal | |
	(1)	(2)	(3)	(4)	(5)
Carbon monoxide	110	440–550	1600–2200	4000	
Chlorine (Cl_2)	1–3	12	120–180		3500
Hydrogen fluoride (HF)	1–2	7	35–170		
Prussic acid	2.2	5.5–6.5	100–250	220–500	33,000
Nitrogen dioxide (NO^2)	20–70		180–900		360–1200

(1) After several hours of exposure; (2) after one hour of exposure; (3) after $\frac{1}{2}$ to 1 hour of exposure; (4) within $1\frac{1}{2}$ hours; (5) immediately.

Table 7.24 Results of animal research on the toxicity of fluoride in rats, mice, and guinea pigs [7-47]

Concentration (ppm)	Exposure	Effect
30–90	1 h	No symptoms of affected health
93–150	1 h	
115–140	30 min	Stronger irritation (urge to vomit)
190–200	15 min	
150–180	1 hr	
225–276	30 min	lethal concentration, LC_{50} within 14 days
375–395	15 min	
600–820	5 min	

toxicity. Acute toxicity is the toxic effect of a substance within a short time after absorption or ingestion. The usual unit for acute toxicity is the mean lethal dose (LD_{50}). That is the dose which causes half of the treated research animals to die within a predetermined period of time. For subchronic toxicity, harmful effects as a consequence of repeated exposure are registered over a period of time starting within a few days to three months. Chronic toxicity describes the effect of a toxic substance over a prolonged period of time during which acutely effective toxic doses are applied repeatedly. When testing subchronic and chronic toxicity, determination of cumulative effects is of the utmost importance. Figure 7.22 shows the process of substance accumulation. As long as the ingested substance, during repeated ingestion, manages to degrade the preceding ingested substance in each case, the level of concentration in the organism will always remain below that of acute toxicity (Figure 7.22a). If the preceding dose is not degraded every time, the substance will slowly

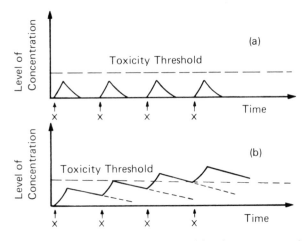

Figure 7.22 Repeated ingestion of substances x with subsequent complete degradation (a) and with accumulation of the substance (b).

accumulate in the organism and will finally cross the threshold of toxic effects (Figure 7.22b).

The concentration limits contained in the pertinent guidelines and regulations [7-39], [7-41], [7-42], [7-44] are not suited for the evaluation of such short-term effects. For toxic substances, the following examinations or stresses can be used for the evaluation of acute or subchronic damage:

1. Experience values in human beings (Table 7.25).
2. The dose which causes 50% of the research animals to die; the research animals (rats), by oral ingestion, receive a specified amount of the toxic substance in mg/body weight (LD_{50} orl-rat).
3. The dose which causes 50% of the test animals (rabbits) to die after resorption (absorption through the skin) (LD_{50} skn-rbt).
4. The concentration which, in 50% of the test animals (rats) and after an exposure over 1 h; ingestion is effected through inhalation of the offered toxic concentrations in mg/m$_3$ (LC_{50} ihl-rat).

Table 7.25 Experience gained from human beings [7-47]

Chlorine

Concentration (ppm)	(mg/m^3)	Symptoms
0.02–0.5	0.06–1.5	Smell threshold for young and healthy persons
0.27	0.81	Lower threshold of olfactory perception
0.5	1.5	No smell
0.5–5.0	1.5–15.0	Smell threshold
1.0	3	Nontoxic, somewhat irritating and slightly noxious
1.0–2.0	3–6	No interference with work
2.0–3.0	6–9	Work still possible; however, stay in vicinity is becoming noxious
4.0	12	Impossible to continue working
17.0	51	Tolerance limit

Fluorine

Number of exposed persons	Concentration level (ppm)	Duration of exposure	Symptoms
5	10	3–15 min	No irritation
9	10	Repeatedly Briefly	No irritation
9	15–25	2–3 breaths	Slight irritation of nose and eyes
5	25	5 min	Slight eye irritation
2	25	1 × briefly	Strong irritation
5	50	3 min	Irritation of eyes and nose
2	50	1 × briefly	"Intolerable without respirator"
2	67–78	1 min	Increased irritation
5	100	$\frac{1}{2}$–2 min	Considerable irritation "Intolerable without respirator"

Table 7.26 Classification of substances hazardous to health as to their effects on rats and rabbits

Form of application	Limiting value (unit)	Very toxic	Effect Toxic	Less toxic
Oral ingestion	LD$_{50}$ (mg/kg)	25	25–200	200–2000
Application to skin	LD$_{50}$ (mg/kg)	50	50–400	400–2000
Inhalation per 4 h	LC$_{50}$ (mg/l)	0.1	0.1–0.5	0.5–5

5. The lowest determined concentration or dose which has, in the past, caused test animals to die (LCLo or LDLo).
6. The substance including a carcinogenous, teratogenous, or mutagenous risk.

The law on the composition of chemicals (Chemikaliengesetz) [7-48] discusses limiting values as a function of the forms of application listed in Table 7.25.

7.2.3 Dispersion of Toxic Substances in Air

Environmental stress caused by the normal operation of technical installations can be determined by measurements, i.e., subsequently, or prophylactically through computational models. Simulated computations of the dispersion of toxic substances in the air are usually based on the assumption that releases of alien substances over a prolonged period of time are constant and emanate from fixed sources—mostly smokestacks [7-49]. The exhaust air is carried away by the wind and spreads gradually throughout the atmosphere. The exhaust air funnel (Figure 7.23), with the most widely used dispersion model, is described by a Gaussian distribution of the concentration either in a horizontal or vertical direction, or vertically to the direction of flow. The dispersion parameters (dispersions of the Gaussian distributions) represent the segregation of the exhaust plume on a time average. In addition, if the temperature of the exhaust air is higher than its surroundings, it will experience a thermal lift. The basic equation for the concentration of toxic substances C in the atmosphere at a specific location, depending on the structure of the model, is as follows:

$$C(x, y, z) = \frac{Q}{2\pi\sigma_y\sigma_z\bar{u}} \exp\left(-\frac{y^2}{2\sigma_{y^2}}\right) \cdot \left\{\exp\left(-\frac{(z-h)^2}{2\sigma_{z^2}}\right)\right.$$

$$\left. + \exp\left(\frac{(z+h)^2}{2\sigma_{z^2}}\right)\right\}. \tag{7-1}$$

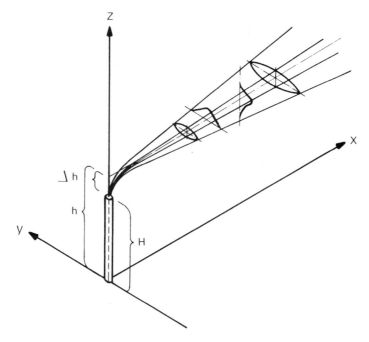

Figure 7.23 Idealized exhaust air plume (?).

Here:

$$Q = \text{intensity of emission;}$$
$$x,\ y,\ z = \text{Cartesian coordinates of the emission point;}$$
$$x = \text{horizontal direction of drift;}$$
$$y = \text{horizontal-perpendicular to direction of drift;}$$
$$z = \text{vertical-perpendicular to direction of drift;}$$
$$\sigma_y = \sigma_y(x),\ \sigma_z = \sigma_z(x) = \text{dispersion parameter (dependent on the state of turbulence in the atmosphere);}$$
$$h = H + \Delta h = \text{effective emission level;}$$
$$H = \text{actual emission level;}$$
$$\Delta h = \text{plume rise because of thermal lift;}$$
$$\bar{u} = \text{representative wind velocity.}$$

By applying Equation (7-1) to the various meteorological conditions (direction and velocity of wind as well as turbulence in the atmosphere), and including the statistical frequencies of these conditions, the expected stresses can be computed in the form of frequency distributions of the concentrations of pollutants in air; for the evaluation of technical installations under normal operating conditions, yearly mean values as well as short-term values are determined. Short-term values are concentration values which are not exceeded during 95% of h/year. Due to the use of meteorological conditions which remain the same with regard to space and time, this dispersion model can only be utilized for distances from the emission source not exceeding 15 km. A lower limit of 100 m distance from the source is justified because of the uncertainties during the determination of dispersion parameters in the immediate vicinity of the source. Large-scale experiments require different dispersion models (e.g., box model, trajectory model).

The results of an application of the dispersion model for a cluster of emission sources consisting of 30 sources with emission levels of between 0 and approximately 50 m is demonstrated in Figure 7.24. Shown here are the lines of equal excess frequencies of a predetermined concentration of a pollutant.

Incident-related releases of air pollutants often occur within seconds or minutes. In order to be able to estimate these stresses, the model of an exhaust air plume, which is the basis for the basic equation (7-1), is replaced by a model which, in addition to a turbulent diffusion at right angles to the wind direction, also includes a turbulent diffusion in the direction of the wind. In this manner, short-term emissions or emissions

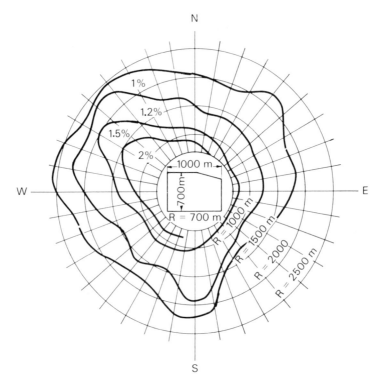

Figure 7.24 Example for the computation of environmental stress in the vicinity of a technical installation [7-50].

which change over time $Q(t)$ may be taken into account. The concentration on the ground ($z = 0$) as a consequence of the released toxic substances can be described as follows:

$$C_{x,y,0,t} = \int_0^t dt' \frac{2 \cdot Q(t')}{\sqrt{(4\pi(t-t'))^3 K_x K_y K_z}} \exp\left\{-\frac{(x - \bar{u}(t - t'))^2}{4K_x(t - t')}\right\}$$
$$\times \exp\left\{-\frac{Y^2}{4K_y(t - t')}\right\} \exp\left\{-\frac{h^2}{4K_z(t - t')}\right\},$$

(7-2)

where t is the point in time starting with the incipience of the emission period; t' is the integration period, maximum emission period; and K_x, K_y, K_z, are diffusion coefficients.

If time-dependent functions $Q(t')$ are inserted into Equation (7-2), the integral defies explicit solution. In that case, a numerical solution will be necessary. The equation will become simpler the moment we are dealing with a short-term emission only (e.g., a flash evaporation) or whenever the emission period T is small in comparison to the period of time necessary to transport the pollutants to the receptor. Assuming an emission level of h as well as a short-term emission of the entire toxic mass of Q', the integral in Equation (7-2) can be solved, and the result is a concentration on the ground of

$$C(x, y, 0, t) = \frac{2 \cdot Q'}{\sqrt{(4\pi t)^3 K_x K_y K_z}} \exp\left\{-\frac{(x - \bar{u}t)^2}{4K_x t} - \frac{Y^2}{4K_y t} - \frac{h^2}{4K_z t}\right\}.$$

(7-3)

The diffusion coefficient corresponds to a representation of the dispersion parameters as follows:

$$\sigma_x^2 = 2K_x \cdot \frac{x}{\bar{u}} = 2K_x \cdot t;$$

$$\sigma_y^2 = 2K_y \cdot \frac{x}{\bar{u}} = 2K_y \cdot t;$$

$$\sigma_z^2 = 2K_z \cdot \frac{x}{\bar{u}} = 2K_z \cdot t.$$

The following example demonstrates an estimate of a predictable air pollution level as a consequence of an incident [7-51]. With a sudden release of approximately 500 tons of ethylene, we may assume that approximately half of the free liquid ethylene will evaporate spontaneously (flash evaporation) and be injected into the atmosphere. The residual amount of ethylene will evaporate gradually (Figure 7.25) and will contribute only insignificantly to acute short-term pollution. With a dispersion over flat ground, the concentrations near and above ground for stable turbulence conditions and a wind velocity of 1 m/s were computed

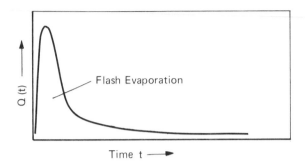

Figure 7.25 Qualitative emission behavior of hazardous substances as a function of time t.

with the help of this model (Figure 7.26). These computations demonstrate that, in the case of such an incident and over an area of approximately $1/3$ km^2, the result would be an ethylene–air mixture which would be ignitable up to a height of approximately 50 m.

The same data permit the computation of the progress of these concentrations as a consequence of a flash evaporation at distances of 100 and 500 m from the emission source (Figure 7.27). Due to the rate of dispersion of the very rapidly released amount of ethylene, the ethylene/air mixture would remain ignitable for approximately 350 s at a distance of 500 m from the emission source, i.e., the lower ignition threshold would have been exceeded and the upper level would not yet have been reached.

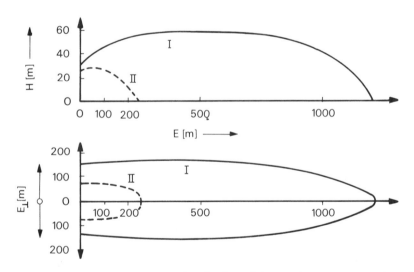

Figure 7.26 Vertical and horizontal distribution of the maximum concentration of an ignitable gas (ethylene) downwind of an emission source at a wind velocity of 1 m/s and stable atmospheric stratification.

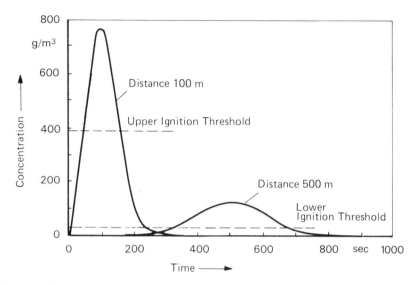

Figure 7.27 Progress of the concentration over time of 100 and 500 m distance from the emission source, at a wind velocity of 1 m/s and stable atmospheric stratification as a consequence of the rapid release of a flammable gas.

At a point approximately 100 m from the source, a cloud of ethylene/air mixture would pass by for 200 s, where it would be ignitable for only 100 s.

Under neutral turbulence conditions in the atmosphere and a wind velocity of 3 m/s, the area in which a flammable ethylene/air mixture must be expected will be reduced to approximately 1/6 km^2 with a maximum height of 100 m (Figure 7.28). At 100 m from the emission source, the passing mixture would still be flammable during approximately 40 s, and at 500 m during approximately 60 s (Figure 7.29).

In the same manner, stresses caused by short-term incident-related releases of toxic substances can also be computed. In that case, curves II would correspond to a concentration limit for the lowest determined toxic concentration, and curves I would correspond to a different concentration limit (e.g., LC_{50}), which are exceeded in the areas between the source and the curve in question.

In most cases, such a release of dangerous substances will not occur spontaneously but in the form of a time-dependent emission. In the simplest of cases, the rate of release equals the rate of emission. During most incidents, however, liquid substances will be released, and will evaporate more or less rapidly, depending on their physical properties. At the outset, as a rule, there will be a flash evaporation, usually if substances stored under pressure above boiling temperature are involved (Figure 7.24). During cooling of the boiling temperature under normal pressure, the released enthalpy becomes available in the form of evaporation heat.

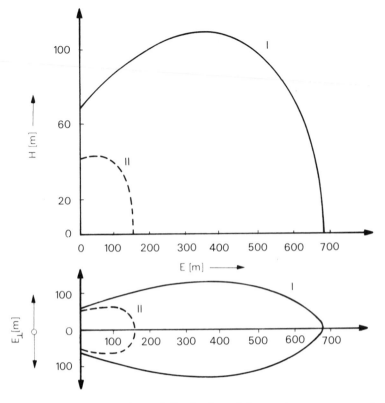

Figure 7.28 Vertical and horizontal distribution of the maximum concentration of an ignitable (flammable) gas (ethylene) downwind of an emission source, at 3 m/s wind velocity and stable atmospheric stratification.

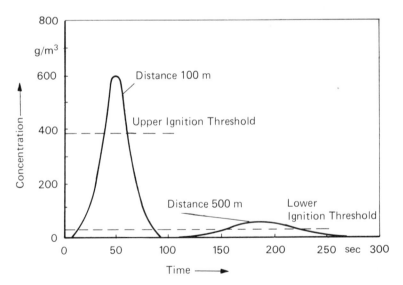

Figure 7.29 Progress of the concentration over time at 100 and 500 m distance from the emission source at 3 m/s wind velocity and neutral atmospheric stratification as a consequence of a sudden release of flammable gas.

The energy required for the evaporation or vaporization must be withdrawn from the environment (soil, air, water). This means that, depending on the individual case, vastly different heat exchange processes should be taken into account. If the emission behavior of the released toxic substance is known as a function of time, then the dispersion equations (Equation (7-2)) must be solved numerically.

The model concepts for the determination of the dispersion of harmful substances during accidents (incidents) explained in this connection can do no more than provide an estimate of the impact (concentration levels and dose load) within certain limits. Extensive assumptions and presuppositions which usually do not correspond to actual conditions, severely limit credibility. At the present time, complicated simulation models are in the development stage. However, the necessary input parameters are missing for their successful implementation. If we use the identical input parameters as employed in the above-described dispersion models, we cannot expect the results to improve significantly.

During the release of highly concentrated, e.g., supercooled heavy gases within seconds, one may expect that the heavy gas will first spread on the ground as a consequence of gravitation; here the volume is assumed constant, i.e., an initial volume (e.g., a cylindrical shape with a radius r_0 and a height h_0) will change its shape in such a way that the radius will increase with time t while its height will decrease.

$$
\begin{array}{ll}
r = r(t) & \text{and } h = h(t) \\
r = r_0 & \text{and } h = h_0 \text{ for } t = 0
\end{array}
\qquad (7\text{-}5)
$$

with

and the auxiliary condition

$$\pi \cdot r^2(t) \cdot h(t) = \text{constant.}$$

During a later phase, this cylindrical volume will deform into a torus, and only thereafter will the dispersion become effective in accordance with the turbulent diffusion (Equation 7-2).

7.2.4 Dispersion of Toxic Substances in Water

The dispersion of harmful substances in water can likewise be represented mathematically via the geometry of flow and the ensuing chemical and physical processes [7-52], [7-53], [7-54]. Thereby the description of the oxygen content, biochemical oxygen requirement (BSB), and water temperature which are essentially characteristic for the state of the water, are of the utmost importance. On the basis of these model computations, the state of the water emerges in the shape of longitudinal profiles and time profiles.

The simulation model described below, for the time being, assumes homogeneous conditions for the cross section of the river—i.e., no streak formation. All infusions are assumed as emanating from distinct points with subsequent complete dissolution at the point of discharge, and it is further assumed that all harmful substances are transported by water with uniform speed. Weirs are considered oxygen suppliers or artificial aerators. Whenever a certain river section is fully known, local and time fluctuations can be simulated in this manner.

Even in undisturbed stretches of flow a number of partial processes exert a certain influence of the five magnitudes determining the state of the water (Figure 7.30), namely water temperature, sedimentable and non-sedimentable biochemical oxygen requirement, oxygen content, and oxidizable nitrogen. The model is represented by five equations:

$$\frac{dT}{dt} = -k_T \cdot (T - E); \tag{7-6a}$$

$$\frac{dL_{CNS}}{dT} = -k_{1C} \cdot L_{CNS}; \tag{7-6b}$$

$$\frac{dL_{CS}}{dt} = -(k_{1C} + k_{1CS}) \cdot L_{CS}; \tag{7-6c}$$

$$\frac{dL_N}{dt} = -k_{1N} \cdot L_N; \tag{7-6d}$$

$$\frac{dD}{dt} = -k_2 \cdot D + Z + D_B/H - P + k_{1C} \cdot (L_{CNS} + L_{CS}) + k_{1N}L_N. \tag{7-6e}$$

Equation (7-6a) describes the temperature behavior of the drainage ditch during thermic discharges. Here, T is the water temperature; t is the time of flow (days); k_t is the velocity constant for the temperature decrease (1/day); E is the natural water temperature (°C).

While the chemically unstressed drainage ditch shows the natural water temperature E, a higher water temperature will prevail if there is a constant discharge of substances, and over the time period of flow, this temperature will eventually revert to the natural water temperature E.

Equation (7-6b) shows the degradation of the nonsedimentable carbon-SBS. However, the latter will not appear according to the formula, but instead the attendant full carbon-BSB, where L_{CNS} is the nonsedimentable full carbon-BSB (mg/l) and k_{1C} is the velocity constant for the degradation of the full carbon-BSB (mg/l). The magnitude of L_{CNS} is connected with the nonsedimentable BSB$_5$ (BSB$_5$) in the following way:

$$BSB_{5NS} = L_{CNS} \cdot (1 - e^{-5 \cdot k_{12}}).$$

The degradation (decomposition) constant k_{1L} relates to the degradation curve determined in the lab, while the degradation constant k_{1C} characterizes the degradation curve in the drainage ditch under natural conditions. The model uses $k_{1L} = 0.1$ (1/day) throughout.

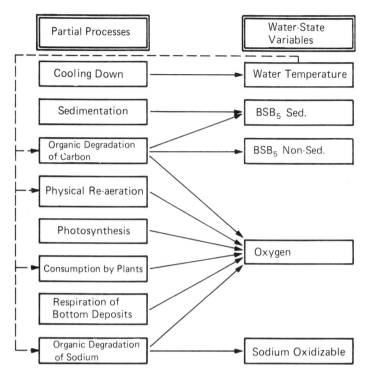

Figure 7.30 Partial processes within the river and their influence on the water characteristic (model) [7-56].

Equation (7-6c) describes the progress of the sedimentable BSB (BSB_{5S}) which, in a manner similar to the nonsedimentable BSB, is replaced by the full sedimentable carbon-BSB, L_{CS}, where the two magnitudes again interrelate through the equation

$$BSB_{5S} + L_{CS} - (1 - e^{-5 \cdot k_{12}}).$$

Furthermore, k_{1CS} is the sedimentation rate constant (1/day).

Equation (7-6d) describes the progress of the oxidizable nitrogen N by the oxygen consumption rate L_N where

$$L_N = 4.3 \cdot N \text{ (mg/l)}.$$

Furthermore, k_{1N} is the velocity constant for the oxidizable nitrogen (1/day).

Equation (7-6e) shows the oxygen graph, where D is the oxygen deficit (mg/l); k_2 is the aeration constant (1/day); Z is the consumption by plants and algae (mg/1 · day); D_b is the consumption by the bottom sludge; H is the water depth; and P is photosynthesis.

There is an interrelationship between oxygen deficit D and oxygen concentration via the saturation concentration C_S:

$$D = C_S - C.$$

The saturation concentration C_S is computed from the water temperature T:

$$C_S = \frac{475}{33.5 + T} \; (mg/l).$$

A solution of the equation system (Equation (7-6)) is possible for an undisturbed stretch of flow with a uniform intensity of the partial processes, i.e., as long as the parameters k_T, k_{1C}, k_{1CS}, etc., characteristic for these partial processes, remain constant.

If necessary, the river section can be subdivided into sufficiently small subsections. Thereby—and without any loss of computational accuracy—a constant water temperature within one such subsection may be assumed. This means decoupling the equation system. Equations (7-6a) through (7-6d) can be solved explicitly, and for (7-6e), a numerical process is employed. The examples in Figures 7.31 and 7.32 show the local course of the BSB$_5$ and oxygen contents determined with the help of a model, as well as values obtained by actual measurements along the Alzette River beginning at the France–Luxembourg border and ending at the point where it empties into the Sauer River. Here, actual emission conditions have been considered.

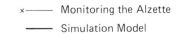

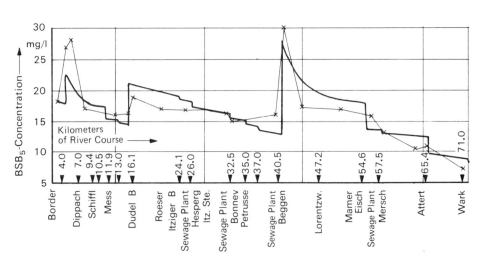

Figure 7.31 Example for the BSB$_5$-levels in the Alzette River, Luxembourg [7-55].

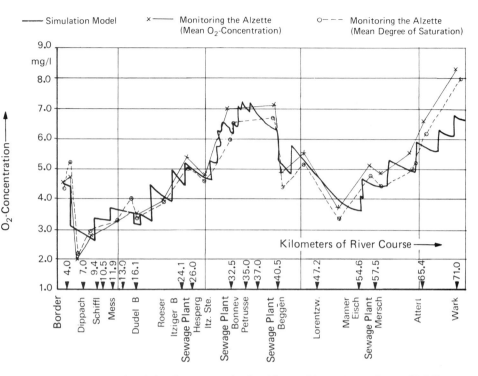

Figure 7.32 Example of the O_2-content in the Alzette River, Luxembourg [7-55].

7.2.5 Dispersion of Toxic Substances in Soil

Harmful substances introduced into the soil often remain stationary because soil itself rarely moves. However, with water as a vehicle, they may penetrate to deeper strata of the soil. Water usually moves in a vertical direction as long as the ground has not yet been saturated. Transport is effected by gravitational forces and through capillary effects. The removal of harmful substances from the water and their deposition into soil is effected by physical, chemical, and biological processes, e.g., filtration, absorption, ion exchange, precipitation, hydrolysis, and biotic accumulation. Once these harmful substances have reached the subsoil water level, they may be carried along with the subsoil water. The path of these substances in a subsoil water carrier (water-saturated soil) usually runs along a horizontal plane. The transport mechanism in water can be described with the help of a model similar to that used to compute the dispersion of harmful substances in air [7-56]. Here a distinction must be made between convective transport with the flow velocity of the water and the transport of substances through dispersion. These transport processes are influenced by overriding reactions which may lead to temporary changes in the concentration of water-borne substances.

7.2.6 Propagation of Sound in the Atmosphere

Sound varies locally and over time as a function of the fluctuations of sound emission as well as of changes in the conditions favoring or inhibiting sound propagation. Besides the measurement of noise levels by technical means, the computation of sound intensity is possible with the help of propagation models. This type of computation is especially useful for emission prognoses, for large-area determinations of sound level distribution patterns, and for the preservation of the reference source for propagation computation in causal analyses.

The sound intensity level generated by a source of sound within its effective range depends on the properties of the source of sound (sound output, directional characteristics, sound spectrum), the location of the source and the receiver, the terrain (and on any obstacles in the path of sound), the local propagation conditions determined by local topography, growth and cultivation, and from meteorological conditions.

In these sound propagation models, the propagation of sound is first related to several basic interrelations, so that additional effects may be considered. For a spherical type of propagation, the sound intensity amplitude p, located at a distance s from the source, can be described by the following equation:

$$\frac{P}{P_0} = \frac{1}{s} \cdot e - iks \tag{7-7}$$

where P_0 is the sound intensity (pressure) at a distance of 1 m; k is the number of sound waves; i is the imaginary number ($i^2 = -1$).

Due to the quadratic relationship between sound intensity (pressure) and the intensity of a sound wave, the result is a $1/s^2$ formula or:

$$L_s = L_w - \Delta L_s \tag{7-8}$$

where L_s is the sound intensity level at a calculated point; L_w is the sound output level; $\Delta L_s = 10 \log 4s^2/s_0^2$ is the measurement of the distance; S is the distance between source and receiver in meters; and $d_0 = 1$ m.

For punctiform coherent sound sources this will result in a decrease of the sound level by 6 dB per doubled distance. A corresponding approach will result in a sound level decrease by 3 dB per doubled distance for liniform sources with cylindrical propagation patterns, and 0 dB for a level wave of an area source. If linear or area sources are of a limited extent, the short-range propagation will assume the shape of a spherical wave.

During sound propagation in air, sound energy is converted into heat, either through absorption or dissipation. The resulting decrease in sound level depends on the temperature, degree of (relative) humidity, the air and, very strongly so, the frequency, whereby it will remain proportional to the length of the acoustic path:

$$\Delta L_L = \alpha_L \cdot s \qquad (7\text{-}9)$$

where ΔL_L is the unit of air absorption and α_L is the damping coefficient for the absorption in air.

The linear relationship between absorption and distance renders this effect negligible at an insignificant distance from a source of sound (approximately 200 m), while at greater distances it will become a determining factor and will limit to several thousand meters the effective range even of sound sources of a considerable acoustic output (see Figure 7.33).

The decrease in sound level shown in Figure 7.33 applies to uninhibited sound propagation high above ground, e.g., airplane noise. If the sound is propagated at a low level above ground, growth, and cultivated or built-up areas, additional level decreases must be taken into account. Therefore shielded areas, deflection as well as reflection of the sound waves, play an essential part, where the sound emission at a calculated point is the result of the superimposition of the direct, deflected, and reflected sound waves. Any computation of sound emissions requires detailed knowledge of the location, size, and surface structure of obstacles like buildings, elevations, and growths.

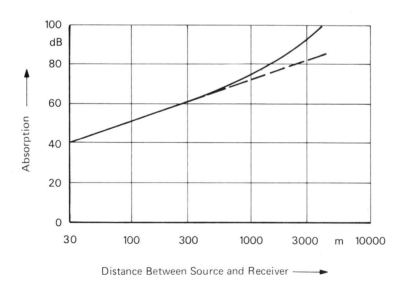

Figure 7.33 Interdependence of sound absorption and acoustic path between source and receiver.

The models developed here [7-20], [7-57], [7-58] permit a highly precise computation for existing or planned construction. The example of a computed noise chart (mean level during daytime) is demonstrated in Figure 7.34 [7-20]. With increasing distance between sound level and calculated impact point (approximately $S > 200$ m), however, meteorological conditions (direction of wind, wind velocity, and wind temperature gradient) will influence the propagation rate of sound. The temperature-constant speed of sound and the vectorial superimposition of the speed of air and sound will produce curved sound paths in the atmosphere (Figure 7.35).

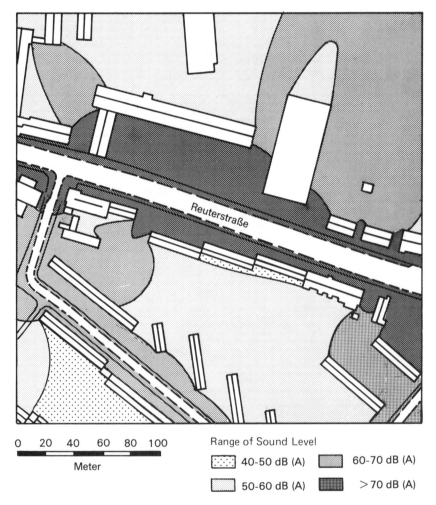

0 20 40 60 80 100	Range of Sound Level	
Meter	40-50 dB (A)	60-70 dB (A)
	50-60 dB (A)	>70 dB (A)

Figure 7.34 Section of the noise pollution map of the City of Bonn [7-20] around Reuterstrasse (Federal Highway No. 9) during daytime (7:00 A.M.–6:00 P.M.).

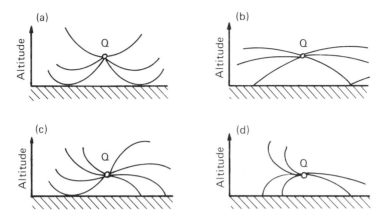

Figure 7.35 Influence of air temperature and wind on the paths of sound (Q = source).

When the wind is in the opposite direction and negative temperature gradients prevail, receivers are reached via sound paths extending close to the ground which, of course, are influenced by level-reducing factors of topography, growth, and buildings. With positive temperature gradients (temperature inversions near ground level) and through wind direction, obstacles are "jumped" via sound paths high above the ground, thus resulting in higher emission intensities.

At a distance of approximately 1000 m from high-output sound sources, and despite constant noise emission, emission level fluctuations of up to 20 dB were determined under varying metereological conditions. For a selected meteorological situation (increasing wind velocity with increasing altitude and positive temperature gradient), the result will be the curved acoustic paths shown in Figure 7.36.

For a safe prognosis and a reproducible measure of sound emissions, meteorological fringe conditions are usually selected which afford stable and ascertainable propagation conditions along the acoustic path. One of these is "the mean downwind situation" (wind low over ground < 3 m/s from source of sound to receiver ($\pm 45°$), quiet, a clear night with a possibility of temperature inversions). It is suggested [7-57] that acoustic paths are assumed which are approximated by an arc of a circle with a radius of 5 km. The sound level computed in such a manner is rarely attained or exceeded, even over longer periods of observation. Through deduction of a distance-dependent adjustment value for specific topographic influence factors under given meteorological conditions, it will be possible to compute the averaging level in wind direction. Differing "prevailing meteorological conditions" at a specific location can then be adjusted by additional level reductions within the scope of the evaluation.

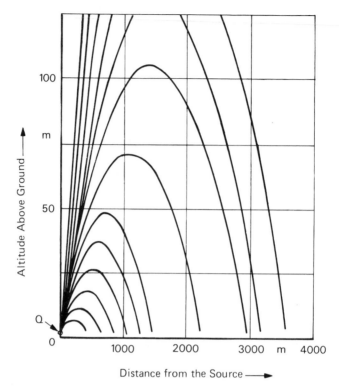

Figure 7.36 Sound propagation in the atmosphere.

The results of application of the sound propagation model for a technical installation situated in a level environment without any structures are shown in Figures 7.37 and 7.38. Therein, the lines of equal sound intensity level are depicted in increments of 5 dB.

7.2.7 Preventive Measures in Case of Incidents

Incidents cannot be completely excluded in a technical installation through preventive measures, therefore secondary measures designed to minimize the impact of an incident become imperative. As far as possible impacts of incidents are concerned, a distinction should be made between the consequences of explosions or conflagrations within a technical installation and those resulting from the release of substances hazardous to health or flammable substances involving the possibility of subsequent explosions or fires outside the technical installation. Flooding of an installation because of high water may also be of significance.

Once the hazard potential inherent in a technical installation exceeds specific limits—which may be defined by the use of specific amounts of substances—systematic planning of accident prevention measures will become mandatory.

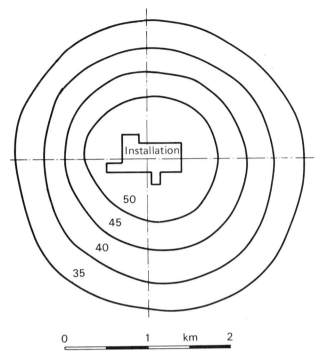

Figure 7.37 Sound propagation emanating from a technical installation defined by lines of equal sound intensity level at a light SE breeze (.2 m/s).

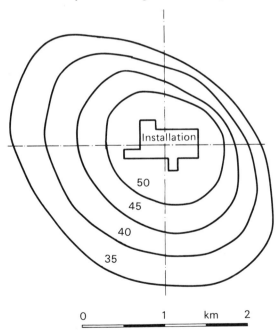

Figure 7.38 Sound propagation emanating from a technical installation defined by lines of equal sound intensity level with the wind from NW (wind velocity 1–2 m/s).

Depending on the planned structure of the measures designed to control an accident, the most important problems are: early recognition of the incident, the alarm, the warning signal, and the resulting active counter-measures. For the recognition of an incident, it is of the utmost importance that verified information about the extent of the incident be available without unnecessary delay. This, in turn, requires continuous monitoring of technical installations or those parts of the installation which are known to possess inherent critical hazard potential (e.g., by means of fire or smoke detectors). Furthermore, for monitoring the environment, an alarm cen-ter, as well as the determination of the sensitive areas of impact, either through actual measurements or through estimates, is necessary.

The alarm is designed to inform responsible personnel and security forces of an incident or accident, simultaneously appraising them of the protective measures to be taken. If outsiders are involved, they should be alerted and advised of any protective measures they are required to undertake.

The protective measures indicated depend on the nature of a specific danger situation as well as on the possible manner in which the danger may develop further. Typical countermeasures in the vicinity of a technical system are: seeking shelter in protected space, evacuation of personnel, decontamination of individuals and property, destruction of contaminated food and drinking water, or other countermeasures of a medical nature. For the planning of extensive and difficult measures, e.g., the evacuation of larger areas, specific system engineering methods are required.

The progress of an evacuation of large population groups from the vicinity of technical installations with a high hazard potential can be determined with the help of evacuation models [7-59]. These models are based on descriptions of the area in question, whereby distribution of the population, the available road network, and available means of transporta-tion are included. The simulation model observes the time-space move-ment of the population, i.e., their progress aboard conveyances across the network of railroad lines and roads. An essential prerequisite for models of this type is the local allocation of specific population groups to predeter-mined starting points for the evacuation. The model determines evacuation routes for specific population groups as well as the required means of transportation. Furthermore, it regulates the flow of evacuation traffic.

7.3 Environmental Factors Influencing a Technical Installation

External effects endangering a technical installation may originate from natural phenomena or from "the influence of civilization". In general, technical installations with a high hazard potential are protected against such effects.

Special hazards from outside may include pressure waves generated by explosions, conflagrations, or the ingress of noxious material. Moreover, in specific cases, technical installations must be protected against airplane crashes, earthquakes, and floods. A complete survey of all external effects should also include possible violence and sabotage designed to damage the technical installations in question. The protection concept usually includes a combination of structural, installation-related, and organizational measures. The buildings and safety-relevant parts of the installation must be designed accordingly. Important parts of the installation having identical functions should be locally separated in order to minimize the consequences of effects from outside and to assure maintenance of essential operational functions.

For evaluation of the potential hazard to the technical installation or to the staff, it will be necessary to conduct a comprehensive site analysis. This should include the listing of all potential sources of danger within the installation itself, on the plant premises as well as in the installation's immediate environment. Sources of danger in the vicinity may include depots for flammable liquids, depots for chemicals, pipelines above or below ground level carrying explosive or harmful substances, and busy air lanes (e.g., lanes of approach to an airport).

7.3.1 Pressure and Explosion Shock Waves

The release of energy during explosions or the sudden release of gas will create continuous pressure waves originating at the source and propagating in the atmosphere. Where buildings must be protected against the effects of these waves or where a sufficiently great distance must be ensured between the installation and locations where explosives are stored, handled, or transported, a detailed knowledge of the possible development of pressure waves is indispensable.

The dynamic process of pressure propagation can be described by the laws of the conservation of mass, momentum, and energy. From these laws, a theoretical representation may be derived for simple cases, as described below for pressure waves with small pressure amplitudes. Such a pressure wave can be described approximately by the following equation:

$$\frac{\partial^2 P_s}{\partial t} - c^2 \Delta P_s = 0, \qquad (7\text{-}10)$$

where

$$\Delta P_s = \frac{\partial^2 P_s}{\partial x^2} + \frac{\partial^2 P_s}{\partial y^2} + \frac{\partial^2 P_s}{\partial z^2}.$$

P_s is the pressure difference $P-P_\infty$ against the environmental pressure; c is the speed of sound in the atmosphere; and x, y, z are the space coordinates for the propagation of the pressure waves. For a wave which spreads

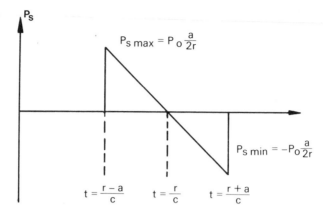

Figure 7.39 Pressure progression with small pressure amplitudes at a calculated impact point at a distance of $r \cdot a$ from the center of a bursting vessel.

symmetrically in the form of a sphere, using spherical coordinates will produce the following differential equation (instead of Equation 7-10):

$$\frac{\partial^2 (r \cdot P_s(r, t))}{\partial t^2} = c^2 \cdot \frac{\partial^2 (r \cdot P_s(r, t))}{\partial r^2}, \qquad (7\text{-}11)$$

where r is the distance from the starting point of the pressure wave. If the pressure wave emanates from a bursting vessel with radius a, then at time $t = 0$ the following equations hold for the differential pressure: inside the vessel $P_s(r, 0) = P_0$; outside the vessel $P_s(r, 0) = 0$, $\partial P_s(r, 0)/\partial t = 0$.

Given these initial conditions plus the condition that, at location $r = 0$, $r \cdot P_s(r, t)$ always equals zero, then the result will be as shown in Figure 7.39. Although Figure 7.39 merely represents an idealized graph of the

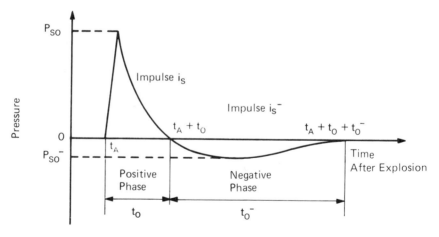

Figure 7.40 Typical pressure-time progression of an explosion-induced pressure wave at an impact point near the origin of the explosion.

pressure/time relationship of a pressure wave of small amplitude, it still demonstrates several characteristic attributes of pressure waves. A positive pressure phase is followed by a negative pressure phase, and the amplitudes of the pressure wave at greater distances from the point of origin are approximately proportional to $1/r$.

Actual pressure waves caused by the bursting of a pressure vessel or by an explosion demonstrate the characteristic pressure/time relationship shown in Figure 7.40. First, the pressure rises steeply to a maximum value, then drops and falls below the environmental pressure. The duration of the positive pressure phase is shorter than the negative pressure phase. The maximum value of the pressure front decreases with increasing distance from the source, whereas the pulse lengths of the positive and the negative phases increase. If the pressure wave strikes a larger obstacle, the wave is reflected at the surface. The superposition of incoming and outgoing waves leads to positive and negative pressure amplitudes greater than those of the incoming wave. This is demonstrated in Figure 7.41. The level of the pressure amplitudes is dependent on the angle of impact of the pressure wave onto the surface of the obstacle. It is greatest at an angle of 90°. With a sufficiently small amplitude, incoming and outgoing waves will superimpose directly. The positive and negative pressure amplitudes in front of the obstacle, with an impact at right angles, will become twice as large as those of the incoming wave. Pressure waves with large amplitudes, when reflected, may generate amplitudes of more than twice their original magnitude.

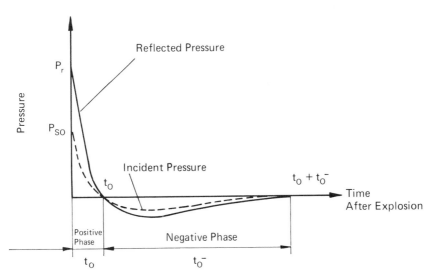

Figure 7.41 Pressure/time relationship of an explosion-generated pressure wave after being reflected by an obstacle.

For trinitrotoluene (TNT) used as an explosive, extensive measuring results are available for the various pressure wave effects. These measurements demonstrate that the pressure waves for various amounts of explosives follow a law of similitude. The pressure amplitudes of explosion waves and other parameters, for a distance related to the amount of energy, the so-called reduced distance $Z_R = R/W^{1/3}$, have one and the same value; R = the distance from the location of explosion, and W = the mass of TNT. The pressure wave parameters as functions of the reduced ground level distance for explosions at ground level are demonstrated in Figure 7.42.

The relationship shown in Figure 7.42 is the basis for assessing the behavior of pressure waves using the TNT equivalent method. The effect of an explosion of any explosive substance or the bursting of a gas-filled pressure vessel is estimated on the basis of the effect of an explosion of an equivalent TNT mass. For flammable gases and liquids the equivalent TNT mass is determined from Table 7.27. For gas-filled pressure vessels, the stored energy is first computed from the vessel's pressure and volume:

$$E = \frac{P_0 \cdot V}{\kappa - 1}\left[1 - \left(\frac{P_\infty}{P_0}\right) \exp\frac{\kappa - 1}{\kappa}\right], \qquad (7\text{-}12)$$

where P_0 is the pressure in the vessel; P_∞ is the environmental pressure; V is the volume of the vessel; and κ is the isentropic exponent.

The equivalent mass of TNT is determined from this energy. Since 1 kg of TNT corresponds to an energy of 4.25×10^6 J, the TNT equivalent is obtained at

$$W = E/(4.25 \times 10^6 \text{ J/kg}). \qquad (7\text{-}13)$$

The reduced distance $Z_R = R/W^{1/3}$ is computed from the equivalent TNT mass W and the distance from the point of explosion measured on the ground. The parameters of the pressure wave at a distance R may then be taken from Figure 7.42. The pressure data in Figure 7.42 are the differences between the pressures generated by the explosion and the environmental pressure, i.e., either positive or negative. A comparison between computed pressure waves according to the TNT equivalent method and measured pressure waves is demonstrated in Figure 7.43.

The procedure used for determining the load from an explosion pressure wave is demonstrated here in the example of a liquefied propane vessel. In a spherical vessel of 21,000 mm diameter, propane gas is stored under a pressure of 18 bar (Figure 7.44). Because of a manufacturing defect, the vessel is ripped below the liquid level.

The liquid or a liquid/gas mixture escapes from the crack and the liquid evaporates due to its temperature above the boiling point at atmospheric pressure. Because of the rapid evaporation of the liquefied gas, an explosive cloud of gas is formed. The evaporated mass M_v can be determined from the stored mass M_0, the specific heat c_{pm}, the heat of

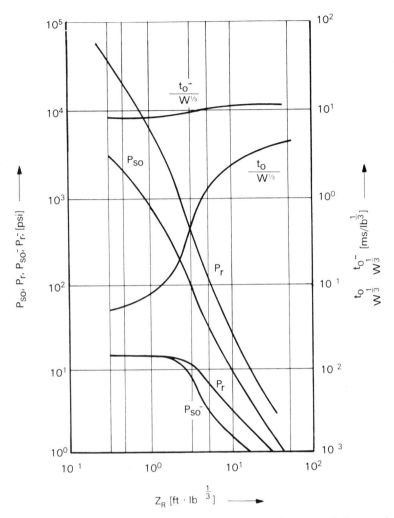

Figure 7.42 Pressure wave parameters according to [7-60]. Reduced distance from ground $Z_R = R/W^{1/3}$.

Table 7.27 TNT equivalents for flammable gases and liquids

	TNT equivalent
Gas under pressure (gaseous)	4 kg of TNT/kg
Gas under pressure (liquefied)	1000 kg of TNT/m³
Deep-cooled liquefied gas, stored at ambient pressure	200 kg of TNT/m³
Flammable liquid with a flame point below 21 °C	10 kg of TNT/m³

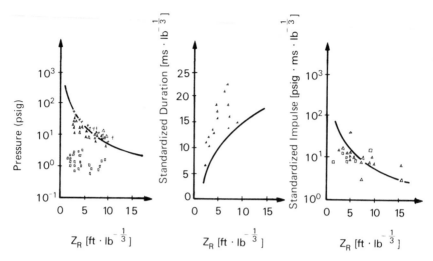

Figure 7.43 Comparison between measured pressure waves from a burst pressure vessel and advance computations according to the TNT equivalent method [7-61].

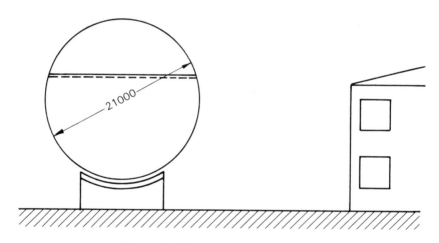

Figure 7.44 Liquefied propane vessel.

evaporation k_v, the boiling temperature t_s of the liquid, and the liquid's temperature t_0 prior to the accident:

$$M_v \cdot k_v = M_0 \cdot c_{pm}(t_0 - t_s). \tag{7-14}$$

In the case of propane, with

$$c_{pm} = 2.26 \text{ kJ/kg } K, \ t_0 = 50 \text{ °C}, \ t_s = -42.6 \text{ °C}, \ k_v = 447 \text{ kJ/kg},$$

approximately 50% of the liquid originally stored in the vessel will be evaporated. The liquefied gas vessel shown in Figure 7.44 holds 5000 m³.

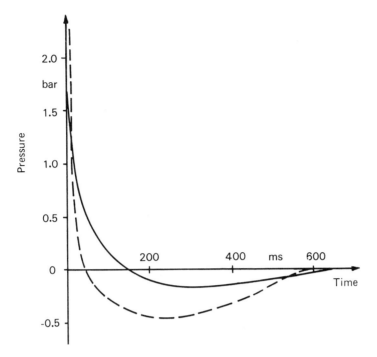

Figure 7.45 Pressure-time progression of the reflected pressure wave at a distance of 165 m(——) and 65 m(– – –).

At a specific weight of $\gamma = 0.5$ kg/dm^3, 1,250,000 kg of propane are thus being released. It can be assumed that only approximately 0.75 – 10% of the released amount will actually participate in the chemical reaction which leads to the explosion [7-62]. In the present example a value of 10% was used. The explosion pressures as well as the duration of the various pressure phases can be read from Figure 4.42. Figure 7.45 shows the pressure/time relationships for distances of 65 and 165 m, respectively.

On the basis of the pressures computed for the respective distances from the pressure vessel, a determination of whether exposed buildings are sufficiently protected against pressure waves can now be made. An idea of the damage to be expected as a result of the action of the pressure waves is provided by Table 7.28.

7.3.2 Fire

Published statistics on the incidences of fires reveal that 11% of all fires occur in industrial or commercial establishments. The event "fire" constitutes an essential type of accident within the technical environment, even if not all fires are of a spectacular magnitude.

Table 7.28 Damage caused by explosion-generated pressure waves according to [7-63]

Object	Type of damage	Pressure above atmospheric (bar)
Window panes, large and small	Usually cracking, and sometimes destruction of the frames	0.035–0.07
Corrugated asbestos walls	Cracking	0.07–0.15
Corrugated sheet metal or aluminum wall sections	Joints separate; after that, buckling	0.07–0.15
Brick wall sections 20–30 cm thickness, not armored	Failure through shearing and bulging	0.5–0.6
Wooden side wall sections, standard housing construction	In general, failure at the main places of abutment; entire wall section can be blown away	0.07–0.15
Concrete or scoria (slag) brick wall sections, 20–30 cm thickness, not armored	Wall will fragment	0.15–0.21

A meaningful protection program against fire demands an analysis of the fire protection-related situation and the possible dangers and losses which may occur as a consequence of a fire, either within a technical installation or in the form of the impact on the installation's environment. When examining possible causes of fire, external impacts like forest fires, fires caused by explosions, or airplane crashes should be considered side by side with fire hazards from inside the installation if the particular type of situation warrants this. It should be remembered that the probability of an external fire affecting a technical installation or a fire erupting inside the installation because of external effects is usually much less than the probability of a fire being caused be internal causes, e.g., through the failure of technical components or through human error. In order to be able to incorporate the possibility of fire into the entire safety concept of a technical installation in a meaningful way, it will help if the process of burning, the type of condition which makes it possible, and its consequences are described in detail. From this, fire parameters and the necessary detection, prevention, and firefighting measures must be derived.

Burning is defined as an independent exothermic reaction, involving a flame and/or glowing embers, between a flammable substance and the oxygen in air. In order to induce combustion, the requirements are a source of ignition and the correct ratio of flammable substance to its reaction partner oxygen. Consequential phenomena like heat, smoke, flying sparks, and decomposition products, must be expected to ignite other materials, damage components, cause component failure, or create the danger of smoke poisoning. These consequential phenomena are often helpful for the recognition of the outbreak of a fire so that proper countermeasures can be taken.

Starting a fire requires a specific ignition energy which makes possible a reaction between the flammable substance and the oxygen contained in the air. A simple evaluation formula for this process does not exist because too many different influence factors are involved. The degree of flammability as well as the amount of energy required in order to ignite a specific substance depend on its chemical composition, the physical condition, the specific surface of the substance, and the "thermic history", i.e., the duration and magnitude of a temperature acting on a substance, thereby causing a drying process, the release of flammable gases, etc., as well as the type of ignition source. Moreover, the presence and effect of heat dissipation, heat capacity, and catalysts should be taken into consideration.

The combustive behavior of a substance upon ignition depends on the ratio of required ignition energy vs. the energy released by the exothermic reaction. If the released energy of an ignited mass particle is less than the required ignition energy of a neighboring mass particle, the reaction will be terminated. This also applies to cases where both amounts of energy are equal. Due to energy loss, for example through heat irradiation, the released energy must at all times be greater than the amount of energy necessary to ignite, so that the substance will be able to continue burning once the ignition source has been removed. Said effect is also demonstrated by the fact that the necessary temperature for continued burning is higher than the original ignition temperature.

The released energy of the exothermic reaction (and thus the temperature of the fire) is directly dependent on the rate of oxidation. This means that, for the purpose of analyzing a fire hazard, a chemically slow-reacting substance can be regarded as relatively harmless, whereas a fast-reacting substance may require additional protective measures. The rate of oxidation is influenced by the essential factors of "degree of flammability" and "fire temperature". Characteristic quantities for flammable substances are the upper and lower ignition limits for gaseous mixtures or dusts, and, for flammable liquids, the flash or ignition point.

Due to the possible side effects of a fire, the following phenomena should be closely observed: high temperatures entailing the danger of burns, the propagation of fires, explosions, blasts, collapsing structures, poisonings, corrosion of material, smoke, developing fumes, and breathing poisons. The temperature of a fire is the direct result of the rate of reaction between the flammable substance and its reactive partner. Smoke and fumes are mixtures of fire-generated gases and particles of a diameter of approximately 1–0.01 μm—soot particles in the case of incomplete combustion of organic matter. A model of the evolution of a fire is demonstrated in Figure 7.46. During phase I we usually speak about a smoldering fire. It is characterized by slow decomposition of the flammable substance and, if the process is not disturbed, a heating-up period until ignition temperature is reached.

Phase II, the incipient fire, begins with the actual ignition. Progress of the fire is still limited to a narrow area around the point of ignition. As a consequence of the incipient fire, and because the room temperature is rising, slowly at first and more rapidly afterwards, as well as the additional

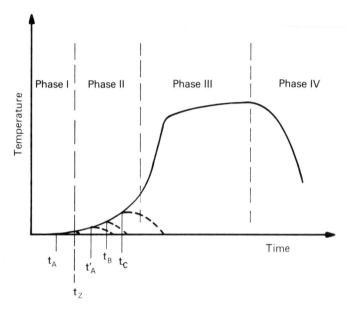

Figure 7.46 Model of the progression of a fire [7-64].

heat irradiation from the burning substances, other materials, first those nearby and later others located farther away, will begin to heat up. This is followed by a drying process due to the evaporation of the water contained in those substances, the release of flammable gases, and similar phenomena. The incipient fire then turns into the fully developed conflagration of phase III. The latter is accompanied by the swift propagation of fire across the entire room and, if there are no fire walls, into adjoining rooms or areas. Under special conditions, e.g., the sudden addition of oxygen, this transitory phase is characterized by the so-called flashover, i.e., nearly all flammable parts will simultaneously erupt in flames. As a function of the fuel-fed fire, the dying fire will then follow as phase IV, where the temperatures and other consequential phenomena will again diminish.

It is quite obvious that an effective firefighting effort must, at the latest, be implemented during phase II of the incipient fire. Only this will afford protection for the various endangered areas. While the fire is at its peak, the only measures available are the protection of adjacent areas and, if possible, an earlier termination of the fire. However, it will hardly be possible to save materials within the fire's sphere of influence.

Depending on the interaction of influence factors, varying progressions of the fire will develop. Thus a thermal overload on cables with low-energy ignition sources and flame-retardant substances will lead to a distinct preponderance of phases I and II within the above conflagration model, while in the presence of high-energy ignition sources in combination with easily flammable materials, phase I will not take place at all and phase II will be shortened considerably. Points t_A through t_C in Figure 7.46 constitute possible interference points in the fire's progression, where— through automatic fire alarms and subsequent action by fire-fighting personnel—the temperature/time curve of the fire's progression can be influenced decisively. Fire detection systems are based on the various fire characteristics. Consequently, these systems consist of thermo-material and thermo-differential heat detectors, optical detectors and ionization alarms for smoke particles, and decomposition products as well as flame detectors.

If explosion-type progressions are part of the development of the conflagration, i.e., the abrupt combustion of dusts or vapor/air mixtures, a further criterion which must be considered is the pressure emanating from the explosion. The pattern of installed fire alarm systems is determined on the basis of the available criteria. Thereby an unequivocal recognition of the area affected must be ensured at a secure location and there must be a minimum of time delay so that effective fire-fighting measures can be instituted quickly. Fire-fighting measures are based on factors determining the nature of the fire. Cooling decreases the rate of the chemical reaction. Below a certain minimum temperature, there will not be any automatic exothermic reaction. By the addition of nonreactive substances (e.g., CO_2), the proportion of ingredients can be reduced to such a degree that the fire will die down. Another possibility is the addition of so-called inhibitors. Constituents of the extinguishing substance react with radicals necessary for the combustion process chain reaction. As a consequence, the fire dies down.

The determination of measures by means of which the danger of a fire and of its propagation can be diminished should be an essential component of any fire analysis. This should include reflection on how—by the separate arrangement of ignition sources and flammable materials, the use of flame-retardant substances as well as through organizational measures— the outbreak of a fire can be prevented in the first place. Structural measures such as the provision of fireproof partitions and adequately dimensioned smoke and heat exhaust ducts are also part of the above. If fire-fighting measures should impede with other protective measures in any way, a basically desirable extraction of fire-related gases may be inadmissible in an area where the release of toxic substances in these gases may endanger the entire environment. In that case, substitute measures must be found.

Because of these complicated interrelationships, the creation of a complete theoretical model of a fire is not possible. Technical fire protection measures are thus instituted based on experience gained during actual events.

The physical separation between parts endangered by fire and other components may be termed the best among all conceivable structural fire prevention measures. If this is impossible or insufficient, prevention can be achieved by sealing systems off against one another. In each case, however, it should be ascertained that, in the case of fire, the extent of damage must be kept at an absolute minimum.

Individual buildings in a technical installation should also constitute individual fire cells which, if fire protection considerations should demand it, could be subdivided into additional subcells and areas with increased fire hazard. Such cells or rooms would be enclosed by fire-resistant walls (fire walls) and have cable and pipe ductwork corresponding to the fire resistance of the confining fire walls. For the protection of persons in case of fire, escape corridors and stairs outside the plant premises (fire escapes) must be provided in the form of independent fire cells with separate fire walls. In specific cases it would be helpful if, in addition to structural measures, for components requiring protection, additional measures were instituted which are designed to minimize the impact as well as the extent of a fire. This includes the use of flame-retardant fluids in control systems, encasing oil lines in special ducts, the placement of oil-transporting components in oil trays, separate routing of cables, of steam, and oil lines or of additional bulkheads, the use of small oil-volume encapsulated switches in electrical switchgear, etc. The concept of "ventilation-related fire prevention" includes measures designed to prevent the spread of fire, heat, and fumes through ventilation devices to areas of the installation not originally endangered and, at the actual scene of a fire, would result in effective heat and smoke removal. This permits swift and accurate fire-fighting measures, with consequential damages kept to an absolute minimum. Ventilation systems connecting two different fire cells or areas should be avoided. If this is not possible, adequate protection in case of fire must be provided by automatically closing fire shutters. In fire stairs and corridors, separate smoke exhaust and ventilation systems must be provided if the building's ventilation system is not able to keep these areas smoke-free at all times.

7.3.3. Noxious Substances

The endangerment of technical installations through the impact of noxious substances acting from outside may directly affect the installation to be protected. Such effects may include decomposition of material through aggressive substances, interruption of contacts by chemical/physical reactions as well as various other effects. Once these reactions manage to disable safety-relevant parts or devices, the safety of the technical installation is impaired. Noxious substances act on a technical installation through the atmosphere, water, or soil.

The various effects on a technical installation through noxious substances acting from outside may be represented according to the schematic illustration in Figure 7.47. Here it is demonstrated that endangerment of

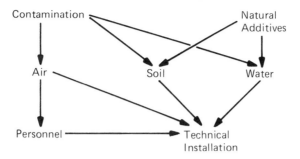

Figure 7.47 Schematic representation on the possible impact of noxious substances.

technical installations is caused not only by civilization-related pollution, but also through natural additives

The most commonly known type of reaction from the atmosphere is the phenomenon of atmospheric corrosion, which, for non-alloyed and low-alloyed steels, are connected with the generally used concept of "rust". It is estimated that approximately 80–90% of all rust formation can be attributed to atmospheric corrosion [7-65]. An essential prerequisite for atmospheric corrosion is the presence of humidity. For steel surfaces, the critical relative humidity—beyond which corrosion will set in—ranges between 60–70% (Figure 7.48).

The most aggressive component is SO_2, 90% of which is a result of the combustion of sulfurous fossil energy carriers (coal, coke, oil), especially from coal fired stoves, conventional power plants, and other furnaces and which, by oxidation into sulfuric acid, essentially hastens the corrosion of metallic materials. SO_2 becomes especially noxious the moment it is

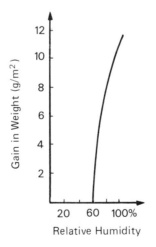

Figure 7.48 Corrosion rate in iron as a function of relative humidity at .01% SO_2 content [7-66].

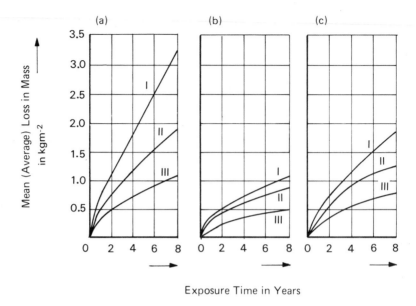

Figure 7.49 Rate of corrosion of non-alloyed (I) and low-alloy (III) construction steel in site-specific atmospheres [7-67].

combined with aerosols, which will remain suspended in air for very long periods of time and thus become condensation nuclei for aggressive gases. Also, the salt contained in ocean water (NaCl) becomes active as an aerosol, especially in coastal areas, thereby aiding corrosion. Depending on the type and quantity of noxious substances, a distinction is made between city, rural, industrial, and ocean climates (Figure 7.49). Moreover, it must be taken into consideration that the influence of SO_2 may sometimes cover wide distances. Thus, e.g., SO_2 is carried from the Bohemian coal fields into the upper strata of the Erzgebirge; corrosion damages in the South of Sweden are partly attributed to SO_2 from the industrial areas of Northeastern Great Britain and from the German Ruhr area. Other aggressive noxious substances in the atmosphere may be hydrohalogens, nitric gases, or ammonia which originate from specialized industrial installations, e.g., the combustion of chloric plastics, from metal finishing plants, or from fertilizer factories, and produce a specific, location-related influence on corrosion.

Of interest in this connection is the fact that atmospheric corrosion mainly occurs in the form of a more or less uniformly distributed scar-shaped erosion. However, other forms of corrosions are also encountered. Thus, for instance, the impact of a humid atmosphere on zinc alloys may produce intercrystalline corrosion, and the impact of a humid atmosphere containing NH_3 may lead to stress cracks in brass components.

The dust content of the atmosphere may have a wide variety of effects. Airborne ashes and soot usually contain water-soluble salts, especially sulfates, often sulfuric acid as well, and thus accelerate corrosion. On the other hand, alkaline dusts from the lime and cement industries may form hard incrustations and may thus sometimes have a protective effect. Mineral dusts like sand, e.g., during sandstorms in desert areas, may produce damage through abrasion or, by the mere formation of deposits, may inhibit the function of mechanical or electrical switching elements.

Water, together with the substances contained in it, may have a variety of effects on technical installations, externally through precipitation and internally through the process of cooling and the creation of waste water. Precipitation usually carries the substances present in the atmosphere, in which those of an acidic character usually produce most of the damage. Subsoil water and surface water which, treated or untreated, is used as cooling water, contains acids, salts solids, gases, organic substances, and microorganisms which may be of a natural or an anthropogenic origin. Of special significance for the safety of a technical installation are acids and gases aggressively attacking metallic materials. Industrial waste water is less dangerous than the natural carbonic acid and oxygen contents of subsoil and surface water. The action of water, depending on the pH value and the content of hardness-inducing agents, CO_2, chlorides, and other noxious substances, may result in a variety of corrosive actions. The induced damage will range from a uniform to a trough-shaped erosion, via crevice corrosion and pitting onto the perforation of pipelines. Alien rust and sand contained in the medium may form deposits and corrosion layers; at high flow rates, damage by abrasive corrosion may occur. In mining, aggressive pit waters may cause corrosion damage on shaft cables and thus impair safety to a considerable degree.

The chloride-based increased corrosivity of sea water demands the use of higher-grade materials, such as CrNiMo steels, Cu–Ni alloys, titanium, or the introduction of special protective measures.

The fact that organic water components should not be neglected is demonstrated by the following example. Braunstein and Hochmüller [7-68] report about a great deal of damage caused in steam generating plants by chlorine-organic compounds. These substances pass the desalinization plant of the installation in the form of contaminants in untreated water, and produce corrosion in the evaporator tubes. The untreated water is taken from the Rhine river which, in turn, had absorbed industrial waste water containing these noxious substances. The effect of precipitation combined with industrial salt—and probably the simultaneous effect of wire rope vibrations—led to extensive corrosion damage on the suspension ropes of the Köhlbrandbrücke in Hamburg, which was discovered in 1978.

The reaction of humidity also plays a decisive role in corrosion within the soil. Of greatest importance here are air and water permeability as a consequence of composition of the soil, as well as the different drying rates of the various types of soil which promote the formation of corrosive aeration elements. The extent of the damage depends on the presence and quantity of natural components like salts and humic (ulmic) acids, and contamination by urban and industrial waste waters as well as by fertilizers. Moreover, in moist soil and in stagnant lakes, canals, and waste waters, damage through aggressive microbiological action may occur. The impact of aerobic and anaerobic hydrogen-oxidizing or sulfate-reducing bacteria can lead to locally increased corrosion and to perforations in subterranean oil or water pipelines. Anaerobic, sulfate-reducing bacteria sometimes cause extensive corrosion damage in gasoline and petroleum tanks in the area where sludge is settled on the bottom. Mothballed ships sustain grave damage by noxious bacterial substances in the form of pitting below the waterline.

Technical installations can be endangered not only by the effects of corrosion but by noxious substances, which would adversely affect the personnel in an installation, inhibiting their efficiency. Accidents may release noxious substances into the atmosphere, strongly inhibiting a person's perception in general, his ability to react, etc. Effects of this kind should be expected mainly in situations where large amounts of hazardous substances are being transported or stored in the technical installation under review.

7.3.4 Earthquakes

Earthquakes are natural shocks originating from the Earth's interior. From the epicenter of the quake, shock waves are propagated which may act on technical installations in the form of excitation. Free-field excitation magnitudes of the uninfluenced construction ground in Germany were established on the basis of earthquakes which occurred in the United States and whose variations over time have been recorded (see Figure 7.50, examples El Centro, Golden Gate, Taft).

Earthquakes cause vibrations or oscillation in buildings or similar structures as a whole or in technical installations housed within them. The excitation of the machine components is transmitted via the machine beds and foundations. Figure 7.51 shows the load determination for technical installations housed inside buildings. Analysis of the sustained stress is carried out step by step.

During the first step, oscillation behavior of the building (or any technical structure) is determined with the help of the time function for acceleration, velocity, or shift (translation) as a consequence of earthquake excitation. As shown in Figure 7.52, the building can be replaced by a bar graph or a multiple-mass vibrator. Because of movement at the base points, structures are caused to vibrate or oscillate. The continuous movement and degree of excitation for these oscillations can be computed,

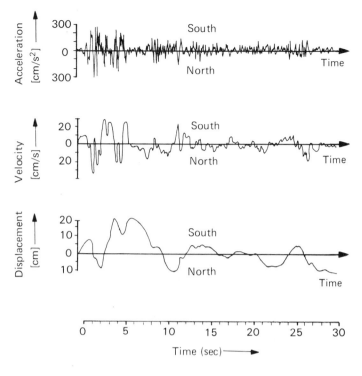

El Centro, Cal, Earthquake of May 18th, 1940

Figure 7.50 Time histories for acceleration, speed, and displacement through earthquake excitation [7-69].

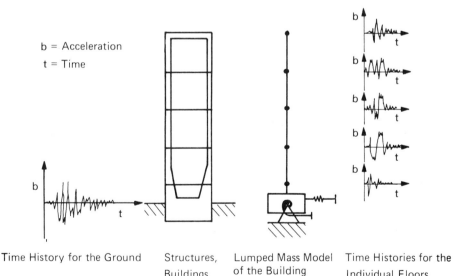

b = Acceleration

t = Time

Time History for the Ground   Structures, Buildings   Lumped Mass Model of the Building   Time Histories for the Individual Floors

Figure 7.51 Determination of the time histories for the individual floors based on a time variation for the ground.

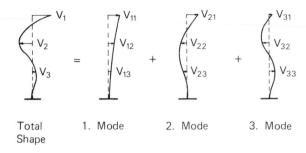

Total   1. Mode   2. Mode   3. Mode
Shape

Figure 7.52 Inherent (natural) forms of clamped bar and total excursion.

as will be demonstrated below. The result is the time function for acceleration, velocity, or shift. The computation factors have been described extensively in literature [7-71], [7-72], [7-77]. Because of their importance for the analysis of dynamic behavior of structures as a consequence of external stress, we intend to describe them briefly. If a stress variable in time is applied to a technical installation, e.g., a building, the equation of motion for dynamic behavior of the installation can be expressed in the form of the following matrix differential equation, provided the technical installation is suitably broken up into a finite number of mass elements.

$$(M)\ddot{v} + (C)v + (K)v = P(t). \tag{7-15}$$

Therein, (M) is the mass matrix of the system; (C) is the matrix of the damping coefficient; (K) is the rigidity matrix of the system; $v(t)$ is the displacement vector of the system; and $P(t)$ is the time-dependent load vector.

In cases where the the allocation in time must be known, or the influence of the phase position (wave angle) is of importance, the time variation method is employed. Displacement-time curves, velocity-time curves, or acceleration-time curves may be established. If Equation (7-15) for the vibration system is solved directly, this is called a time variation method with direct integration. This method permits an allowance for nonlinear material behavior.

Due to the huge amount of data needed and the complexity of the method, the computation requires a considerable amount of work and time. Therefore, in cases such as this, the so-called "modal time variation method" is being used. Here natural frequencies and modes of the system must be known beforehand. The solution of the differential equation (7-15) is determined from the inherent (natural) vibrations (oscillations) of the structure and its degree of excitation. Inherent vibrations (oscillations) are the solutions of the homogeneous equation which—disregarding the damping effect—appear as follows:

$$(M)\{\ddot{v}\} + (K)\{v\} = 0. \tag{7-16}$$

Inherent vibration behavior constitutes a simple harmonious motion which can be expressed by the following equations:

$$\{v\}_i = \{\hat{v}\}_i \sin(\omega_i t) \tag{7-17a}$$

and

$$\{\ddot{v}\}_i = -\{\hat{v}\}_i \omega^2 \sin(\omega_i t) = -\omega_i^2 \{v\}_i, \tag{7-17b}$$

where v is the oscillation amplitude and i the index for the ith intrinsic shape. When Equations (7-17a, b) are inserted into Equation (7-16), the result is an intrinsic value equation for a natural frequency plus the allocated mode:

$$(K)\{\hat{v}\}_i = \omega_i^2 (M)\{\hat{v}\}_i, \tag{7-18a}$$

or for all natural frequencies and the pertinent natural forms, respectively:

$$(K)(\Phi) = (\omega^2)_D (M)(\Phi), \tag{7-18b}$$

where (Φ) is the matrix of the mode forms and $(\omega^2)_D$ the diagonal matrix of the natural frequencies. From this equation, the natural frequencies and forms can be determined.

The motion equation (7-15), taking into account the orthogonality of the mode, may be reduced to a noninteracting system with a degree of freedom for the amplitudes of the inherent oscillations. Whenever there is excitation through earthquake-induced acceleration $b_E(t)$, the result will be the equation

$$\ddot{v}_i^* + 2\xi_i \omega_i \dot{v}_i^* + \omega_i^2 v_i^* = B_i b_E(t). \tag{7-19}$$

In this equation, ξ stands for the damping factor and v_i^* for the generalized displacement for the mode i. It is a measure for the response of the ith mode to the earthquake-induced excitation. B_i is the so-called participation factor which constitutes a measure or unit for the participation of the ith mode in the overall excursion during an external excitation. Usually, the equation is solved by means of the Duhamel integral, and the response of the ith mode in relation to time t can be established from

$$v_i^*(t) = \frac{B_i}{\omega_i} \int_0^t b_E(\tau) e^{-\xi_i \omega_i (t - \tau)} \sin \omega_i(t - \tau) \, dt. \tag{7-20}$$

The total response at a mass point (internal point) j can then be established as the sum of contributions by the individual inherent forms

$$v_j(t) = \sum_{i=1}^{N} \Phi_{ij} v_i^*(t), \tag{7-21}$$

where Φ_{ij} stands for the relative displacement amplitude of the mass point (internal point) j for the intrinsic (natural) form i.

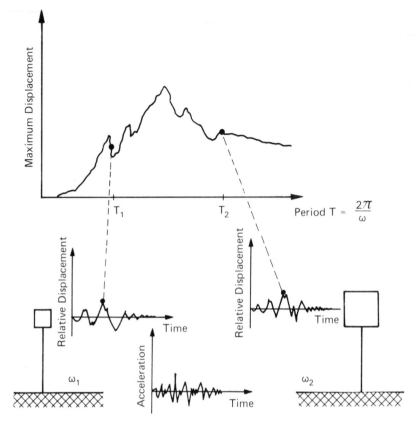

Figure 7.53 Schematic drawing for the determination of modal response spectra (spectrums) [7-31].

In Figure 7.53, the mathematically displayed method is once more explained schematically. An acceleration time variation is used to determine the responses of the components. If the maximal acceleration responses to earthquake excitation by single mass resonators are taken from the established response spectrum—as a function of the natural frequencies and damping effect—these are then called response spectra (spectrums). In an analysis according to the *response spectrum method*, the generalized displacement v_i^* does not have to be established for each point in time t. The response spectrum constitutes an evaluation of the Duhamel integral for maximum values established under specific variations of the natural frequency, in the presence of damping, and of an excitation function. Response spectrums are the most common form of representing earthquake-induced stresses.

For actual buildings, response spectrums are often stated as a function of the height above ground on various floors (floor response spectrum). The

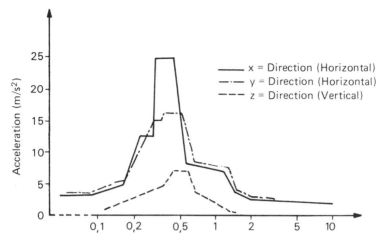

Figure 7.54 Response spectra for earthquakes, damping 1% [7-72].

computation work for determination of the component is reduced to the determination of inherent frequencies (natural frequencies) and natural shapes as far as they are able to contribute essentially to the degree of displacement.

The figure in [7-72] represents the result of the modal response spectrum for an evaporator vessel. The response spectrums for the part of the building wherein the vessels are installed are sketched in Figure 7.54, and the vessel design proper is shown in Figure 7.55. The distribution and magnitude of the inertial, damping, restoring, and stress forces are simulated in a "finite" mechanical model. Figure 7.56 shows that—for the type of construction in question—beam elements with three translatoric and three rotational degrees of freedom are utilized at the nodal points.

The cylinder is represented by beams of equivalent rigidity. The interaction of individual masses is assumed, whereby masses of neighboring areas are concentrated at the nodal points. For the system itself, 69

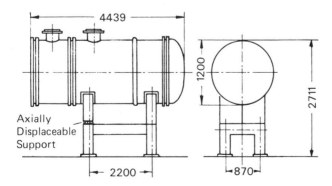

Figure 7.55 Evaporator vessel.

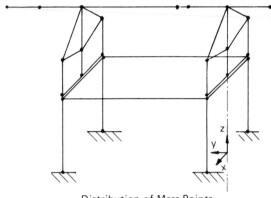

Distribution of Mass Points

Figure 7.56 Lumped mass model of the vessel.

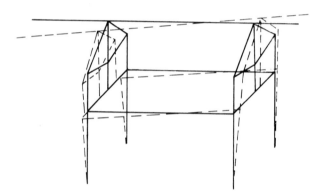

Figure 7.57 First mode $F = 7,6\,\text{HZ}$ $T = 0.13\,\text{s}$.

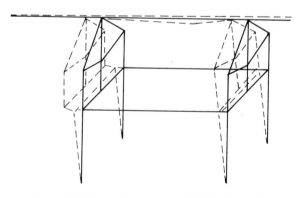

Figure 7.58 Second mode $F = 8.1\,\text{HZ}$ $T = 0.12\,\text{s}$.

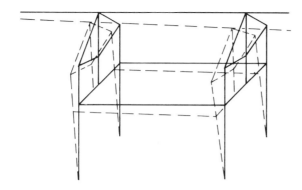

Figure 7.59 Third mode $F = 10.2\,\mathrm{HZ}$ $T = 0.10\,\mathrm{s}$.

modes (see also Figures 7.56–7.59) and frequencies can be determined on the basis of the degrees of freedom. In actual cases, however, the determination of higher frequencies is usually dispensed with, since their contributions towards the total excursion are insignificant and the computation itself extremely time-consuming. For the evaporation vessel, the six lowest natural frequencies and allocated modes were determined.

Table 7.29 shows a summary of the decisive natural frequencies, cycles, and modal participation factors. The maximum average magnitudes of importance for a stress analysis can be seen from Table 7.30. Figures 7.57–7.59 show the plots of the first three natural shapes.

Table 7.29 Natural frequencies, cycles, and modal participation factors

No. of natural frequency	Circular frequency (s^{-1})	Frequency (Hz)	Cycle (s)	Vibration factor		
				x-direction	y-direction	z-direction
1	47.8	7.6	0.131	1.22	$3 \cdot 10^{-15}$	$-6 \cdot 10^{-14}$
2	50.7	8.1	0.124	$7 \cdot 10^{-13}$	0.25	4.05
3	64.1	10.2	0.098	3.99	$5 \cdot 10^{-14}$	$-2 \cdot 10^{-13}$
4	188.4	30.0	0.033	$6 \cdot 10^{-10}$	2.37	-0.08
5	241.4	38.4	0.026	$-1 \cdot 10^{-8}$	2.41	0.50
6	294.7	46.9	0.021	-0.13	$-8 \cdot 10^{-9}$	$-3 \cdot 10^{-9}$

Table 7.30 Cross section loads at the support gables

	Left-hand support gable	Right-hand support gable
Normal force N	$2.85 \cdot 10^4$ N	$3.28 \cdot 10^4$ N
Transverse force Q_x	$1.38 \cdot 10^4$ N	$1.7 \cdot 10^4$ N
Transverse force Q_z	107 N	$1.23 \cdot 10^4$ N
Torsional moment M_y	50 N mm	$1.37 \cdot 10^4$ N mm
Moment of flexion (bending moment) M_x	$5.18 \cdot 10^4$ N mm	$1.12 \cdot 10^7$ N mm
Moment of flexion M_z	$3.96 \cdot 10^6$ N mm	$5.4 \cdot 10^6$ N mm

The stresses obtained from the analyses—one example of this is demonstrated in Table 7.30—must be superimposed by the operational stresses of the component in question. Depending on the function and safety-related importance of the installation, other stress factors, like those resulting from accidents etc., can be superimposed as well. Stress assumptions are used to document the static stability and functionality of the technical installation. In general, this is achieved by the computation of stresses occurring in the supporting cross sections. For this purpose, simple methods are stated in the regulations and standards set up to the fields of construction and mechanical engineering.

7.3.5 Airplane Crash

Protection of a technical installation from the possible consequences of an airplane crashing into the installation may be of importance in cases where a high hazard potential for human beings and structures may be expected. Here it is necessary to start with the assumption that practically any technical installation above ground can be hit by an airplane or airplane debris. Fuel fires caused by the crash may result in extensive damage in technical installations.

The extent of danger to a technical installation through an airplane crash is defined by the extent of damage to be expected as well as by the degree of probability of such an event occurring. The overall effects of an airplane crash are determined by the force of the impact, the mass involved, and the mass distribution of the aerodyne or parts of its wreck. Hazard classifications can be made on the basis of the takeoff weight and air speed of the plane. Therefore, reports found in the literature usually separate data on the frequency of airplane crashes into those for fast-flying military aircraft and civil aviation aircraft with varying takeoff weights.

An estimate for the probability of an airplane crash at a specific location is usually subdivided into three phases or types of air traffic [7-73]:

—takeoffs and landings;
—air-lane traffic with automatic navigation;
—"free" air traffic.

In Germany, there is no need for a distinction between these three types or classes of air traffic because of the relatively high air traffic density and the insignificant distances between takeoff/landing sites. In the case under review, the category "takeoff and landing traffic" will suffice.

The extremely low probability of a crash displayed in Table 7.31 demonstrates (for civil aviation aircraft and fast-flying military aircraft) that designing a technical installation especially for such an event, including provisions for protection against an airplane crash, would only be justified in situations where the consequences would cause damage of a catastrophic extent. If, in these exceptional cases, the protection of a

Table 7.31 Probability of a crash of a civilian or military plane on an area of 10,000 m^2

	Crash probability
Civil aviation aircraft flying on regular air lanes (takeoff weight in excess of 200 kN)	2×10^{-11}/a
Civil aviation aircraft in "free" traffic (takeoff weight of less than 200 kN)	9×10^{-7}/a
Fast-flying military planes	1×10^{-6}/a

technical installation is regarded as absolutely necessary, this can be achieved in a simple way by the erection of protective structures, the walls of which cannot be penetrated by the crashing plane. For the design of such a protective structure, the forces which the crashing plane, debris, etc. may exert on the structure must be fully known. The first examinations in this field were conducted by Riera [7-75] and Lorenz [7-76]. Their work, however, treated the deformation capability of the airplane in a rather general manner. The results were exaggerated dimensions of the protective structures. Improved computation methods have been developed by Rice and Bahar [7-77]. These take into consideration the force-crush interrelation of the plane in a very responsible manner. Drittler and Gruner [7-78], in addition to the behavior of the materials involved, also examined the behavior of the entire airplane structure. All of the authors tried to keep the extent of the computational effort within tolerable limits in that—in their assumed models—they based the geometry of the airplane, the position of the plane during impact, and the reaction of the building wall impacted on the following simplifying premises:

—The longitudinal axis of the airplane is the same as the direction of travel. During impact, it is at a right angle to the tangential plane of the building wall. Under this assumption, the greatest possible total impulse is transmitted onto the structure, so that the most unfavorable case should be used for safety-related reflections.
—The structure wall is assumed to be a rigid body.
—The speed of the aerodyne's impact is constant across the entire structure and is braked down to zero by the rigid wall.

Figure 7.60 shows the idealized model of the plane during impact on the rigid wall. Here, the x-axis extends along the longitudinal direction of the airplane. The origin of the coordinate has been determined at the tail of the airplane at time $t = 0$. The mass distribution along the x-axis is designated $m(x, t)$. $t = 0$ is the time of first contact between the airplane and the wall. Drittler and Gruner have selected Lagrange's (bounded input) presentation for the time-motion sequence. This means that the shift of a specific mass point is being traced. Figure 7.61 is a schematic presentation of the above. For the derivation of the equation of motion,

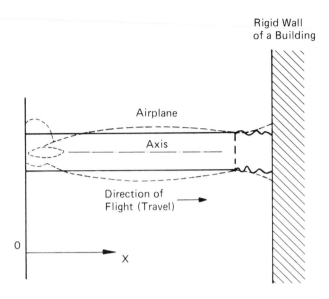

Figure 7.60 Geometric configuration during the impact of a crashing airplane.

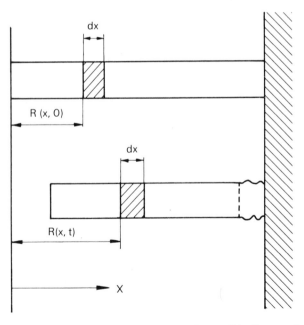

Figure 7.61 Coordinates of a mass element at $t = 0$; $x = R(x,0)$ at $t > 0$; $x = R(x,t)$.

the mass element is considered which, at time $t = 0$, is located between x and $x + dx$. According to Figure 7.60, this mass, at time $t > 0$ is located between $R(x, t) + dR(x, t)$, so that—based on the principle of conservation of mass—the following equation will emerge:

$$m(x, t) \, dR \, (x, t) = m(x, 0) \, dx. \tag{7-22}$$

Taking into consideration the velocity of one mass element,

$$u(x, t) = \frac{\partial R(x, t)}{\partial t}, \tag{7-23}$$

the total force acting on the mass element can be determined with the help of Newton's law of motion:

$$m(x, t) \, dR \, (x, t) \frac{\partial u(x, t)}{\partial t} = -dK(x, \, t). \tag{7-24}$$

From this, a system of soluble partial differential equations can be derived where the accompanying conditions are the following:

—The aerodyne strikes a rigid wall, where the velocity of the aerodyne components is braked down to zero velocity.
—The total air travel beginning at time $t = 0$ of contact is not greater than the length L of the aerodyne, since the wall of the structure is assumed as rigid.
—The aerodyne, at time $t = 0$, is traveling at the same speed over the entire length of the structure.

Drittler and Gruner, like other authors, have used numerical methods to solve these differential equations. The result in the shape of a braking force can be represented in the simplest manner in the form of a time-load-sequence. Figure 7.62 shows the time-load curves have been entered as simplified approximation curves for different aircraft types. Here, the stresses which result from the impacting fuselage and simultaneously impacting debris are superimposed. It is clearly demonstrated that fast-flying military aircraft present the greatest danger. Detached parts of the aircraft are considered in the computational model in such cases where the stress at failure has been exceeded for one element of the entire aerodyne structure. Figure 7.63 demonstrates the braking force created by the aerodyne proper in combination with the total force resulting from the superimposition of the impact forces of the aerodyne debris.

Buildings, by using the proper design as well as suitable materials, can be constructed in such a way that the impacting aerodyne or debris cannot penetrate their walls. At any rate, however, the impact causes the building to vibrate (oscillate). The forces of acceleration generated are especially great in the area of impact; they decrease with increasing distance. The forces of acceleration which are transmitted onto the technical installations

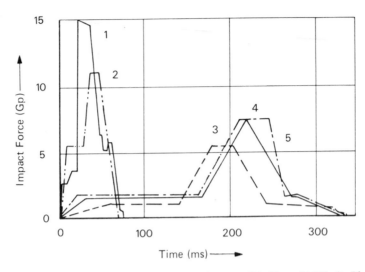

Figure 7.62 Idealized load-time graphs. *1, Phantom F4, Riera [7-75]; 2, Phantom F4, Drittler [7-78]; 3, Boeing 707, Fuzier [7-79]; 4, Boeing 707, Riera [7-75]; 5, Boeing 707, Rebora [7-80).*

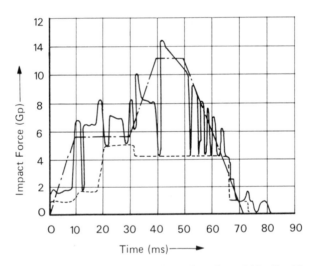

Figure 7.63 Impact force (——), braking force (– – –), and idealized load (—.—.) for the crash of a Phantom onto a rigid wall at $v_0 = 215$ m/s; $M = 20,000$ kg; $\int_F dt = 4.475 \cdot 10^5$ kp·s.

housed inside the building via ceilings and lofts, are progressively weakened.

In order to determine these forces, the floor response spectrums for the building under review are determined in accordance with the method of modal response spectrums described in Section 7.3.4. Figure 7.64, as an example, demonstrates the horizontal response spectrums of a building at

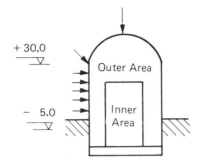

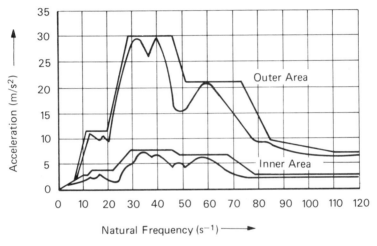

Figure 7.64 Horizontal floor response spectrums for airplane crash horizontal acceleration at 2% damping.

various floors which were determined by the load-time-pulse shown in Figure 7.63. Whenever response spectrums for buildings are employed in this manner, it signifies that components housed inside those buildings can also be suitably dimensioned of specific dynamic loads.

7.3.6 Floods

Technical plants often require huge amounts of industrial water for technological processes. Thus they are preferably located near rivers; in special cases coastal sites are selected. Among the natural phenomena endangering technical plants, flooding due to high-water levels should be considered. In river systems with well-balanced flow characteristics, there is very little fluctuation of the water level, thus either requiring little or no control actions if technical plants are located along their sides. In the case of a site analysis, sufficient information can be derived from the evaluation of statistical data on the respective water levels. Whenever considerable level fluctuation is expected, water level control of rivers is of decisive importance.

In the past, the most common and simple way of regulating a river was by damming the water with lateral dykes. Another convincing technical method is the creation of a landfill well above the highest water level ever observed and building the technical plant thereon.

Modern flood control systems are barrage weirs, clusters of barrage weirs, or high-water reservoirs. Besides protection from floods, these often present advantages of an ecological or economic nature. Difficulties in planning and dimensioning flood control systems are often caused by the fact that a number of random variables must be taken into account as input data.

Control of water flow by means of dams and clusters of dams aim at adapting the actual water level $W(t)$ as closely as possible to an ideal level $W_s(t)$ during periods of high or low water. This objective is formulated by the mathematical function:

$$Z = \int |W(t) - W_s(t)| \, dt = \text{minimum.} \qquad (7\text{-}25)$$

The solution of the equation is a fluid mechanics and process control-related problem, which is determined by the performance function (7-25) itself.

Flood control devices, under safety aspects, serve to lower the crest of high water by releasing the amount of wavelike arriving water in the most equally distributed manner possible. This can be expressed by the performance function

$$Z = \int_0^{t_H} Q_A^2(t) \, dt = \text{minimum.} \qquad (7\text{-}26)$$

Q_A stands for the amount of water released from the high-water retaining reservoir and t_H denotes the duration of the high-water situation including the full time interval of control action.

The way a high-water storing reservoir works is explained by the following example. A flood has to be controlled by damming it up in two reservoirs located at rivers A and B in such a way that in river C, formed by the junction of rivers A and B, the crest of the high water is kept at a minimum. The two headstreams A and B have a damming capacity of $k_1 = 0.81 \times 10^6 \, \text{m}^3$ and $k_2 = 0.18 \times 10^6 \, \text{m}^3$, respectively. The course of the rivers, the locations of the dams, and the assumed site of the technical plant are shown in Figure 7.65. This example has been computed by the Institut für Wasserbau at Karlsruhe with the help of a computer program developed by the Institute [7-81].

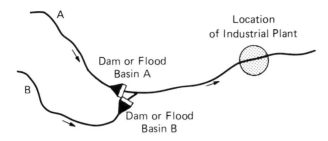

Figure 7.65 System drawing of an example for flood control dams.

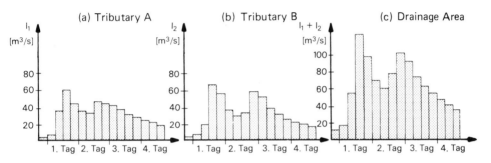

Figure 7.66 Tributaries and drainage areas without flood control.

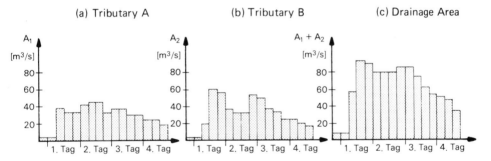

Figure 7.67 Tributaries and drainage areas with flood control.

The amounts of incoming and outgoing water without flood control are shown in Figure 7.66, while those with flood control can be found in Figure 7.67. Figure 7.67c shows a significant reduction of the crest, a fact which has been achieved by filling the reservoirs in rivers A and B.

7.3.7 The Impact of Organized Violence

By intentional acts of violence and sabotage, a technical installation may sustain considerable damage. Installations with a high hazard potential must be protected against these eventualities. Acts of sabotage and

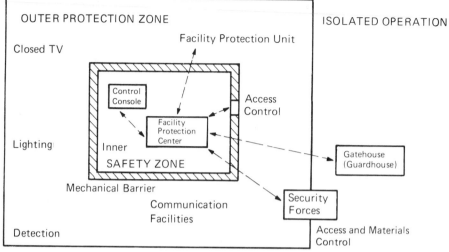

Figure 7.68 Inner and outer safety zones.

organized violence may be encountered in different ways:

—They may be initiated by persons normally not having access to the installation and thus forcing their way in. During demonstrations, spontaneous actions of large groups of people may cause grave damage to technical installations.

—Organized manipulation of a technical installation, however, may also be initiated by employees of the installation who have access to the plant and are familiar with it.

For the protection of facilities and installations against sabotage, safety plans have to be devised. The protection of installations does not constitute an individual or separate measure within the installation; it is an integral part of the overall planning. Redundancies, the sealing off of certain areas, or the physical separation of endangered parts* of the installation will help to protect facilities and installations effectively. The most important measures for the protection of facilities against sabotage, in the narrower meaning of the word, are essentially supplementary measures. Within the framework of installation planning, the watchwords for the protection of facilities are: "Recognition, delaying action, and control measures". "Recognition" denotes the installation of a detection threshold around the entire facility. Access to the protected area encircled by the detection threshold should be fully controlled, e.g., by a guardhouse (Figure 7.68).

---

* As is necessary due to technical safety aspects.

There, any persons or materials about to enter the installation would be investigated.

The systems and components of special importance for the safety of the facility should be provided with mechanical protection in the form of a barrier. This zone is called the "inner safety zone." It is not necessary to limit the inner safety zone to one single area as is the case in Figure 7.68. The most decisive component of the inner safety zone is the mechanical barrier designed to delay or even prevent access by third parties. Within the safety zone, a facility protection center should be established which monitors and releases all alarms, reports, access control from the guard-house, and access control to the inner safety zone. The facility protection center must provide several communication facilities

—two-way communication with the guardhouse;
—communication with access control to the inner safety zone;
—communication with the monitoring facility or the control room of the entire technical installation;
—communication with the facility protection service;
—communication with security forces (e.g., police precincts).

The facility protection service controls the outer safety zone and, if necessary, goes into action in order to avert damage to the installation. During the planning of a technical plant, specialists must develop the facility protection concept and become thoroughly familiar with every aspect. Of greatest importance is a clear and concise definition and delimitation of individual safety zones. Only then can the specific technical measures for the protection of the facility be initiated and the organization established.

Chapter 8

Safety Legislation

Foreword*

In this chapter the role of safety legislation will be discussed. Although some international issues and foreign legal systems will be touched upon, it will primarily be the interdependence between safety science and the legal system of the Federal Republic of Germany, the authors's home country, which will be reviewed. The legal system of the Federal Republic of Germany is commonly described as a civil law system and manifests numerous distinctions to the common law legal system prevalent in most of the Anglo-Saxon countries. However, a closer scrutiny of the reactions of common law legal systems to the dangers inherent in modern technology will reveal striking parallels, as well as distinctions, to the ways in which the German legal system tries to adapt to scientific and technological developments. The legal system of the United States of America may serve as an example. Penalties, fines, and liability for damages are common and well-recognized means in the U.S. legal system for ensuring that products are designed safely and manufactured carefully. The sometimes exorbitant awards granted by American courts in product liability cases are widely known. Product liability, when combined with the possibility of having to also pay "punitive damages" has become a major deterrent in the U.S. against unsafe products and has even turned out to be fatal to some businesses. In addition to putting pressure on manufacturers and other users of technology through the courts, in the U.S. there is also the vast field of administrative law on the federal as well as the state level. The U.S. Congress, for example, found it necessary to establish motor vehicle safety standards for motor vehicles and equipment in interstate commerce. It promulgated the "National Traffic and Motor Vehicle Safety Act" which requires the U.S. Secretary of Transportation

---

* By Dr. Wilfried Witthuhn, Attorney at Law (New York), Rechtsanwalt (Hamburg).

to establish, by appropriate order, federal motor vehicle safety standards. Numerous similar examples of administrative involvement could be added (e.g., Food and Drug Administration, Aviation Administration, etc.). All of them show the understanding in the U.S. that it is the proper function of the legislative branch of government, in the exercise of police power, to consider the problems and risks that arise from the use of technology and to attempt to adjust private rights and harmonize conflicting interests by comprehensive statutes for the public welfare. As shown by this brief comparison to the U.S. legal system, safety legislation follows similar patterns throughout the industrialized world. Therefore, the following chapter, though focusing on the German legal system, does have a meaning which goes beyond the national safety legislation of the Federal Republic of Germany: it will provide the reader with a general idea about the possibilities and limitations of safety legislation.

Introduction

In the first few chapters, the "man–machine–environment–system" (MMES) was described in a system-theory-related approach and later described in detail. The role of safety legislation in this connection was not considered. We will now proceed with an examination of the importance of the legal system for safety legislation and shall begin with an attempt to determine where and how it can be integrated into the MMES.

Safety legislation, to a vast extent, is engaged in an examination of the actual state of affairs within the MMES as well as the impact of changes occurring therein. The legal system, however, as the totality of the rules and regulations in force within a community, is a system based on targets yet to be achieved. The rules of law determine what should be. This antithesis, which has been carried to extremes for the sole purpose of clarity, could give rise to the impression of a largely unrelated side-by-side coexistence of the science of safety and the legal system. Nothing could be further from the truth. The legal system can never abandon the life and health of the citizens to the dangers inherent in modern technology. On the contrary, it must determine—as a reflection of the basic political decisions as well—whether and to what extent modern technology can be accepted by law. Thus the legal system is placed outside the MMES proper inasmuch as it does not constitute a variable of one of its system components. However, it is a force acting upon each and every part of the MMES in a modifying way, thus developing the capability of being instrumental in effectively changing the above-mentioned variables.

On the other hand, the legal system is also affected by the findings of the science of safety. These findings are intended to ensure the increased safety of technical products and the protection of man and his environment from the hazards and dangers inherent in modern technology. The latter can

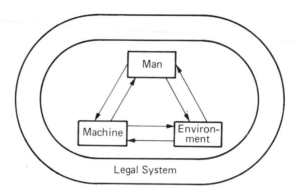

Figure 8.1

only be achieved in an effective and comprehensive manner if the findings of the science of safety are incorporated into the existing body of laws. The science of safety, in effect, intends to—and must!—help shape the legal system. Thus the role of the legal system within the framework of the MMES is a dual one. It constitutes the means by which the political intent of the voter or citizen can shape the components of the MMES. However, it is much more than merely a means to achieve a certain purpose. It is also the object of the examination of processes evolving within the MMES, since the knowledge gained from these findings is meant to be incorporated into the body of laws, rules, and regulations designed to benefit the public. Figure 8.1 demonstrates how the legal system and the science of safety influence one another.

8.1 The Concept of Safety Legislation

8.1.1 Developments in Safety Legislation as a Form of Governmental Accident Prevention

Safety legislation, in the narrow sense of the term, is usually understood to constitute the totality of all legislative provisions by means of which the planning, construction or manufacture, and operation of technical products are regulated and thus subjected to governmental control. Legislative provisions as well as governmental control are intended to avert accidents or dangers connected with technical products and to prevent possible losses. Regarded in this light, safety legislation is a form of governmental loss prevention.

Safety legislation by means of governmental loss prevention originated in the 19th century. Numerous boiler explosions in the recently invented steam engines necessitated the establishment of rules and regulations governing the building and operation of steam engines. Their main purpose was to protect the public from the dangers of future explosions. Within the framework of loss prevention-oriented safety legislation during the first half of the 19th century, social considerations were hardly an issue and environmental protection played a minor role. Furthermore, protection of the individual in this connection was purely incidental and not the aim or goal of governmental activities solely intended to serve the interest of the public. Thus the neighbor of the owner of a steam engine had no way of obtaining an interlocutory decree against the licensing and/or the operation of the installation. Liability legislation entailed only responsibility on the part of the immediate causer or occasioner of the loss or damage. Any entrepreneur operating a steam engine could refer the victims of a boiler explosion to the stoker or to the foreman of the engine crew, thus declining any liability on his part. With the progress of industrialization, the growing number of people employed in the industry and the resulting social problems, as well as the increasing differentiation as to the employment of labor rendered these conditions increasingly untenable. Thus, for the first time in history, Art. 107 Industrial Code of 1869 made it an obligation of the entrepreneur to provide the necessary devices and means for the protection of the employee's life and health in his factory. With that, the idea of "safety in the place of work" was introduced into safety legislation for the very first time.

The "Reichshaftpflichtgesetz" (Reich Liability Law) of 1871, among other things, required the operator of a steam boiler plant to assume liability for any losses or damage caused by vapors, it being immaterial whether he was personally at fault or not. Victims of accidents were now in a position to turn directly to the owner or operator of the plant for redress. The socially indefensible and factually unjustified liability of the stoker or guard was now relegated to a secondary position. At the same time, this cause-unrelated liability took into consideration the increased hazard potential of the plant itself. The following decades, especially the period following World War II, saw a hitherto unknown expansion in governmental responsibilities. The classic type of administration by fiat (decree) was joined by the activities of a modern service-providing administration, which frequently are summarized under the heading of "existential precaution" (precautionary measures to ensure a person's existence) as well as a planning administration as a means of shaping the individual's future in a meaningful manner.

Additionally, in the area of safety legislation, the government no longer felt responsible for the prevention of losses to the public alone, but also for an active and comprehensive safeguarding of the life, health, property, and assets of each individual citizen in practically every domain of life. Environmental protection and consumer protection are nothing but two catch phrases under which numerous laws and legislative projects may be summarized as an expression of the government's changed attitude with regard to the scope of its own responsibilities.

8.1.2 Safety Legislation as the Totality of all Safety-Promoting Rules and Regulations

The concept of "safety regulation" must adapt to the changed and expanded comprehension of the responsibilities of government. Thus, we should understand all standards which are directly or indirectly able to increase the degree of safety when handling technology.

Penal statutes prohibiting especially dangerous forms of handling technical products will fall just as much under this definition as would liability standards which enable the victim to recover damages.

Not only are laws in the strict sense of the word included under this heading, but also all forms and shapes of semilegal rules and regulations, from a statutory order down to guidelines and directives issued by the various authorities. Included, finally, are not only all laws and regulations in force but also all technical standards and rules above plant level, irrespective of their judicial value.

Safety legislation attempts to promote technical safety in numerous fields, in a variety of forms, and through various means. Thus the representation of its essential sectors and legal bases in the form of a separate summary suggests itself. Table 8.1 offers a rough survey in which completeness is not claimed.

8.1.3 International Influencing Factors

In all industrialized societies, the legal system must react to the dangers inherent in modern technology in a similar manner. A glance at the national safety legislation of other countries will therefore reveal a structure similar to that of the Federal Republic of Germany. A detailed representation of this structure within the framework of this survey may be dispensed with. Conversely, a look at the national safety legislation of the Federal Republic of Germany may likewise afford the viewer an idea of the importance, basic features, possibilities, and limitations of safety legislation in general.

Apart from natural and technical necessities, safety legislation is also formed by the various intergovernmental relations existing between these countries. Through numerous agreements and treaties, these countries enter into obligations which are important from the standpoint of safety engineering and which must subsequently be incorporated into their own national safety legislation. The Federal Republic of Germany, for instance, as a member of numerous international organizations, is limited in its freedom to act independently to the same degree as the organizations are endowed with the prerogative of setting their own rules and regulations. The most important one among these international organizations is the European Economic Community (EEC). Essential portions of Germany's national safety legislation, especially in the areas of atomic energy and traffic legislation, have come from initiatives and resolutions of the EEC or as a consequence of guidelines, binding the Federal Republic of Germany. Agreements and guidelines of the EEC are binding for member countries only. They are not directly binding for the average citizen. The member countries, through their own legislative bodies, must make the regulations a part or portion of their national safety legislation before it becomes binding for the average citizen. At this point it will be more meaningful to briefly present the main goals of international influencing factors, e.g., on German safety legislation, without reference to any individual regulations.

A. Freedom of Trade—Avoidance of Competitive Imbalance. As a consequence of the "General Program of the EEC for the Removal of Technical Obstacles to Free Trade" of May, 1969, and its consequential programs, a number of guidelines have been published which are intended to simplify the varying national laws governing the registration of motor vehicles. In addition, there are regulations passed by the European Economic Commission of the United Nations (EEC rules).

Frequently these regulations which are extremely important from a safety-related point of view, transcend the present scope of national safety legislation. Once these rules and regulations have been incorporated into the body of national laws, progress in the field of safety legislation has been made. However, it would be erroneous to regard international rules and regulations primarily as an instrument designed for the improvement of national safety legislation. Adapting national rules and regulations, on a European as well as a worldwide level, is primarily designed to remove obstacles in the form of different registration provisions detrimental to free trade, in order to avoid imbalances in the process through varying production conditions. Here, any gain in safety is only of secondary importance.

B. Protection of the Public—Environmental Protection. Naturally, the thesis advocating the priority of economic considerations does not apply uniformly to all international rules and regulations. In the case of several European rules and regulations, one may certainly assume a balance in importance as far as environmental and safety-related motives vs. economic motives are concerned. Other guidelines, e.g., those of the European Atomic Community on maximally permissible doses of radiation primarily serve as the protector of the public from the dangers of nuclear energy.

In general it may be said that on all levels those influence factors and activities relating to environmental protection motivated by the desire to protect the public from the dangers of modern technology have substantially gained in importance over purely trade-policy-related issues. One example of this is the fundamental recognition of the pay-as-you-pollute principle in environmental protection by the EEC. The member countries of the EEC have agreed on the principle that whoever excessively exploits

Table 8.1 Safety legislation: a survey

Target	Protection from danger	Legal basis Liability	Sanction(s)
Safety in vehicular traffic			
Road traffic	Art. 6 StVG; StVZO	Art. 7 ff StVG	Art. 21 ff. StVG; 315 b StGB
Air traffic	Art. 1 ff. LuftVG	Art. 33 LuftVG	Art. 58 ff. LuftVG 315 StGB
Rail traffic	Art. 3 Allg. EisenbahnG	Art. 1 Liability Law	Art 315 StGB
Safety in the home			
Buildings	State (Länder) building codes	Art. 823,836-838 BGB	Art. 323 St.GB
Furniture and fixtures		Art. 823 I, II BGB	
Safety at the place of work			
Buildings	State (Länder) building codes	Art. 823, 836–838 BGB	Art. 323 StGB
Human beings (Industrial safety)	Art. 618 BGB	Art. 618 III BGB	Art. 710 RVO
	Art. 120 ff. GewO + Regulations governing Industrial safety	Art. 823, II BGB	Art. 147, 148 GewO
	Art. 10,11 AtomEnL. +	Art. 25 ff Atomic Energy Law or: AtomEnL	Art. 81 Radiation Protection Regulation; Art. 45 ff At.En.L.
	Regulation for the Protection from radiation		Art. 311 ff. StGB

Technical devices and installations	Art. 3 Law Governing the Safety of Appliances		
	Art. 1 ff Fed. Emission Law + Regulations	Art. 823, II BGB	
	Art. 24 ff. GewO + Regulations	Art. 2 Liability Law	Art. 143, 148 GewO 45 ff AtomEnL, 311c ff. StGB
	Art. 1 ff. AtomEnL + Regulations	Art. 25 ff. AtomEnL	
Public safety	Art. 1 ff. Fed.Emiss.Law + Regulations thereto	Art. 823 I,II BGB	
	Art. 24 ff GewO + Regulations	Art. 2 Liability Law	Art. 143, 148 GewO Art. 45 AtomEnL; 311 c ff. StGB
	Art. 1 ff. AtomEnL + Regulations	Art. 25 ff. AtomEnL	

Legend:

BGB	= Bürgerliches Gesetzbuch	(Civil Code;)
StGB	= Strafgesetzbuch	(Penal Code;)
LuftVG	= Luftverkehrsgesetz	(Air Traffic Law;)
StVG	= Straßenverkehrsgesetz	(Road Traffic Law;)
StVZO	= Straßenverkehrszulassungsordnung	(Road Transport Licensing Regulation;)
GewO	= Gewerbeordnung	(Industrial Code;)
BImschG	= Bundes-Immissionsschutz-Gesetz	(Federal Emission Law;)
Allg. EisenbahnG	= Allgemeines Eisenbahngesetz	(General Law Governing the Operation of Railroads;)
RVO	= Reichsversicherungsordnung	(Reich Insurance Code.)

or pollutes the natural environment, i.e., the air and water, for his own advantage should also bear the responsibility for his actions, i.e., defray the necessary expenses. The wording of the pay-as-you-pollute regulation, however, is still controversial as far as its details are concerned. In particular, this rule does not force the individual countries to pass laws of a certain nature, e.g., to determine the amount of liability for polluting air or water. Nevertheless, the public is becoming increasingly aware of the necessity of a joint reaction toward any hazards or dangers affecting the citizens of several countries simultaneously.

C. Protection of the Individual. In every industrialized nation with a democratic constitution, the state (the government) no longer considers

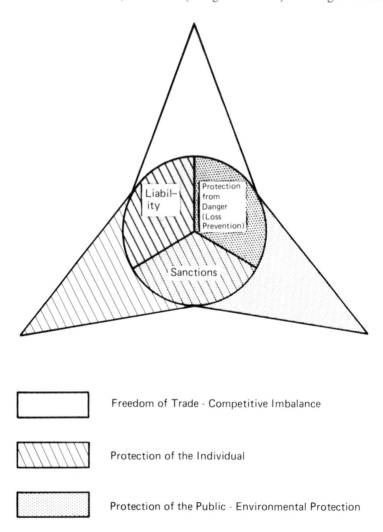

Figure 8.2 Factors influencing safety legislation.

itself solely the guarantor of public safety and law and order, but also as indirectly responsible for safeguarding the life, health, and property of each citizen. The activities summarized under the heading of consumer protection are a direct expression of this development. The greater portion of legislative activities directed towards consumer protection is designed to protect the consumer from health hazards arising from improperly manufactured products. Parallel developments in all member countries of the EEC have, in the meantime, produced a draft for guidelines which are intended to render the various national legislations more uniform, and perhaps stricter. We will discuss this draft later. However, the climax of this development, which attempts to safeguard the protection of the individual by the government as far as is humanly possible, has not yet been attained.

Figure 8.2 demonstrates in a simplified way the main goals of international and supranational activities as well as those parts of the national legislation which are essentially influenced by it.

8.2 The Goals of Safety Legislation

8.2.1 The Maintenance of Law and Order (Administrative Fines)—Guaranteeing the Ethical Minimum (Penal Code)

An element of order is indispensable, and safety legislation is no exception. People can be forced to obey specified rules and regulations if any contravention against these will lead to punishment. Whenever these rules are obeyed out of fear of punishment, a certain gain in safety will have been achieved. In this manner, safety is a direct result of an effective deterrent.

The legislature distinguishes between misdemeanor and criminal offenses. Misdemeanors are contraventions against rules and regulations which, e.g., in the interest of maintaining a safe environment in the workplace, cannot be tolerated but which, on the other hand, do not bear the stigma of criminal behavior. Therefore, the punishment for such violations will generally be a fine and not a criminal penalty. One pertinent example is Art. 710 of the Reich Insurance Code in which violations of regulations passed by professional associations and designed to prevent accidents may be punishable by fines not exceeding DM 20,000.

Specific violations of safety regulations, because of the severity of their possible consequences, can no longer be regarded as mere misdemeanors. These violations not only constitute contraventions against set rules which are necessary for the maintenance of safety and order, but they also endanger the foundations of peaceful coexistence which the legislation guarantees by making them subject to criminal prosecution as a form of "ethical minimum." The criminal acts which must be classified as part of safety legislation can chiefly be found in Art. 310b ff. Penal Code. There, these violations have been classified as criminal acts mainly because of the danger they pose to the public. As a typical example of safety-related penal legislation, we may mention Art. 323 Penal Code, because it specifically refers to the "accepted and recognized rules of technology," where importance to safety legislation will be discussed in detail later on.

> Art. 323 Sec. 1: Anyone violating the generally accepted and recognized rules of technology during the planning, supervision, erection, or wrecking of a building and thereby endangering the life and health of another person will be liable to a prison term not exceeding five years, or a fine.
>
> Sec. 2: Likewise, any person who, in the course of exercising his profession or trade in the planning, supervision, or completion of a project which includes the addition of technical installations to the project or to make changes in the technical installation contained therein, violates the generally accepted rules of technology and thereby endangers the health or the life of another person, shall be liable to punishment.

8.2.2 Settlement of Claims Through Liability Standards

Penalties and fines as a part of safety legislation are of limited value. Invariably only a small part of the violations actually committed will be covered by the rules and regulations and an even lesser percentage will lead to punishment.

Most of all, imposing a fine or penalizing the perpetrator will be of little use to the victim once an accident has actually occurred. Besides the (in the interest of the public) necessary condemnation of a violation of the rules of coexistence, there must be an individual settlement of the damage between the perpetrator and the victim (occasioner and sufferer of the damage). Liability standards that determine a liability on the part of the perpetrator (occasioner) and the victim (sufferer) for the damage caused by the former, like penal statutes or administrative fines, will act as a deterrent. Any person faced with the risk of being held responsible for any damage caused by technical products will, in most cases, strive to avoid such damages or loss as far as possible. Whenever the liability risk must be insured, the insurance carrier (underwriter), through the wording of the insurance policy, will exert his influence to avoid damages during the erection and operation of a technical plant. Thus liability standards should be regarded as an integral part of safety legislation.

A. The Principle of Strict (Direct) Liability. Liability standards may be regarded as the oldest form of safety legislation. At the beginning of the industrial revolution, there were no regulations in existence that covered the erection and operation of a technical plant because technical development was one jump ahead of the law. There were only general laws prohibiting the individual from doing harm to another person's life or property through negligence. Whoever violated these laws was automatically forced to make restitution—apart from being liable to punishment. Thus restitution for loss or damage incurred marked the beginning of safety legislation.

The pivotal liability clause in the present legislation of the Federal Republic of Germany is Art. 823 Civil Code (Bürgerliches Gesetzbuch or BGB):

> Para 1: Whoever deliberately and negligently injures the life, health, freedom, property, or any other civil right of another person shall be liable to make restitution for any damage resulting from such action.
>
> Para. 2: The same liability applies to anyone violating a law designed for the protection of another person . . . .

Art. 823, Para. 1 protects these civil rights, yet it fails to protect a person's property from purely financial losses. Conversely, Art. 823 Civil Code (BGB) basically offers protection from financial losses.

Provisions of the type Art. 823 Civil Code, e.g., penalties or fines, presuppose negligence or fault on the part of the acting party. There must be sufficient evidence as to the acting person having been at fault in a certain (but not in any other) way. Within the framework of safety legislation, the principle of criminal negligence incorporates a number of grave disadvantages. The standard itself, of course, can be interpreted very widely and, basically, may encompass any and all forms of human behavior. The result of this will be clear and unequivocal laws. The crucial question, however, is which standard should be applied in order to assess the actual degree of negligence. Within the scope of Art. 823, Para. 2 Civil Code (BGB), for the sphere of application of Art. 823 Para. 1, such a mandatory behavioral standard is often contained in the "protective law." Within the sphere of application for Art. 823, Para. 1 Civil Code there is a complete lack of legal standards. This means that the legislation would have to develop a voluminous catalog of behavioral standards, any violation of which would constitute grounds for an accusation of criminal negligence. Technical standards and regulations play an important part in the development of behavioral standards defining that which is regarded as reasonable and necessary. More of this later on.

B. Absolute (Strict) Liability as an Example of Modern Safety Legislation.
The progress of industrialization in the 19th century proved the inadequacy of the principle of criminal negligence. Liability was always attached to the person actively involved and, in an environment where work was allotted to a number of people, this was often the wrong person. Thus, e.g., the person operating a steam engine was held liable whenever an explosion occurred, and not the entrepreneur or owner, as the person actually responsible toward third parties, because the latter was always in a position

to point to a lack of direct responsibility. The premise of "fault" thus often excluded any liability because the damage was caused by a not-yet-matured technology and could not be attributed to anyone in particular. The state of technology was not yet sufficiently developed so that the cause of loss or damage could be foreseen or prevented. In other words, no one in particular had to assume responsibility for the increased danger inherent in modern technology.

To pursue the example further, liability in the case of the boiler operator fell mostly on the financially weakest person. He was exposed to a liability risk which threatened his livelihood and he had no way of influencing this situation effectively. For these reasons, the Reich Liability Law of 1871 established strict or direct liability of the owners and operators of railroads and certain other technical installations. This direct liability no longer insisted on negligence as a liability-establishing element, solely used the increased danger emanating from the operation of a technical installation as a starting point, and imposed a general liability on the owner for any loss or damage caused by the technical installation itself. As compensation for its nondependence on fault or negligence, this type of liability is limited in its extent.

With the Reich Liability Law, legislation created a basic example of future-oriented liability standards which meet the requirements of an increased hazard potential brought about by modern technology. In the following decades, Articles 7 ff Traffic Law, 33 ff Air Traffic Law, and 25 ff Atomic Energy Law were published. They are further examples of direct or strict liability of great importance to safety engineering.

C. The Necessity of Discretionary Liability Limitation in the Case of Strict Liability. Strict or direct liability is a no-fault liability. However, it is much more than a mere contingent liability which would solely require a causative connection between the operation of a technical product and the occurrence of the damage in order to be able to establish liability. Strict or direct liability which is all-encompassing is restricted by assessing points of view which, besides the already-mentioned upper limit, include, in general:

—the necessity of being able to control the technical product or the ability to influence it;
—the actual impossibility of preventing an accident;
—the exclusion of operationally atypical dangers as well as acts of God;
—taking into consideration the actions and behavior of the victim.

We shall illustrate this by means of an example. Strict or direct liability, according to Art. 7 ff. Road Traffic Law, was introduced because the legislature recognized the fact that it was impossible for the average person to remove himself from the added dangers of motor vehicle traffic. Any direct liability on the part of the vehicle owner who is basically responsible will be excluded in such cases where the vehicle has been used without his knowledge and consent; furthermore, no liability is attached to the latter if the accident was caused by an unavoidable occurrence neither attributable to a defect in design nor to a malfunction of any of the vehicle's devices. Here the concepts of "controllability of the risk" and "exclusion of operation-unrelated hazards and acts of God" have been fully considered. Finally, Art. 9 Road Traffic Law reduces the amount of liability if the victim is partially at fault. Direct or strict liability—and this is obvious from the aforementioned liability-limiting aspects—does not condemn modern technology in principle and does not attempt to reduce it by means of unavoidable or mandatory liability. Strict liability, despite recognizing the dangers connected with modern technology, maintains a basically positive attitude toward technology. It is nothing more than a balancing force between the interests of the victim, protecting him from risks being forced on him, and the interests of the operator or owner who does not want to assume any liability if he is not at fault or wants such liability limited to a predictable amount and only to an extent which reflects the increased danger actually attributable to the operation of the technical installation in question.

D. Product Liability: From Liability Arising from Negligence Via Strict Liability to Contingent Liability. We will use the example of product liability in order to demonstrate how safety legislation can develop further and may endanger the limits of a proper and practical indemnification or even transcend them. The concept of "product liability"—also called producer or manufacturer liability—describes the common law responsibility on the part of the manufacturer of defective mass-produced products for any health hazards to the user. Product liability has become an issue because of the industrial mass production of technical products for an anonymous market. The function of the distributor changed from that of a guarantor of freedom from defects of the product to that of a mere distributor or agent of the manufacturer. Another contributing factor within the drive to expand the functions of the government was its growing awareness of its obligation to protect the health and the property of the individual citizen in his capacity as a consumer.

In the Federal Republic of Germany, product liability in its present form is less the result of legislative activities than that of precedents established by the courts. The essential basis is Art. 823 Para. 1 Civil Code (BGB). Without changing the wording of the law, the legislation developed a host of mandatory behavioral patterns which characterize product liability. The manufacturer must test the product for defects during development, design production phases, and even after it has been marketed, and, if necessary, make the required changes. The manufacturer's obligation to provide information on the product as well as a far-reaching reversal of the burden of proof have changed the practical application of product liability in the consumer's favor. However, the principle of liability for acts of negligence determined in Art. 823 Para. 1 Civil Code (BGB) is still valid, despite the progress towards consumer protection. The result of this is that the manufacturer will not be held liable, e.g., for "mavericks or outliers" which simply cannot be avoided, even with the utmost care.

According to the suggestions of the EEC commission for the adaptation of rules and regulations covering defective products, certain changes will be forthcoming. According to these suggestions, product liability is to be converted from a purely negligence-related liability to strict or direct liability in such a way that the manufacturer will be liable even in such cases where marketed merchandise had to be regarded as free from defects and the defects were discovered only later. The idea of holding the manufacturer responsible for defective mass products, whereby it is immaterial whether he is directly at fault or not, suggests itself because, in everyday life, the user cannot escape endangerment by certain products. Legislation in the Federal Republic of Germany still adheres to the principle of liability through negligence. However, responsibility of the manufacturer to exercise reasonable care, and the requirements for an adequate settlement (recovery) have been increased to such a degree that it is almost safe to speak of strict liability. Seen in this light, the legislature's introduction of strict liability would solely constitute the standardized perpetuation of the development already anticipated by the legislation. However, it may well be that, with the realization of the suggested guidelines of the EEC, those boundaries would be crossed which even a legislature is constrained to observe in the interest of a proper settlement based on the principle of equality.

The EEC guideline proposal seems to perceive liability-determining endangerment in the manufacture of the product, for it does not distinguish between dangerous and less dangerous products. A social compulsion to accept the assumed endangerment by any type of product is regarded as inevitable. Liability in cases resulting in death or damage to a person's health is only theoretically limited by the maximum amount a person can recover; in reality it is practically equivalent to unlimited liability. This, however, is in direct contradiction to the limitation of liability as compensation for dropping the negligence. Above all, the proposed ten-year validity period of the liability neglects the principle that the risk involved must be controllable in order to establish liability due to negligence or fault. After all, the manufacturer can only influence the product to a certain point, namely to the time when he markets it. Any liability will be voided whenever the manufacturer can prove that the product was free from defects. Whenever a product causes damage after having been marketed, the manufacturer must prove that initially the product was free from defect. In many cases, this will be practically impossible. As a consequence, the manufacturer will have to make restitution for any damage. Thus strict liability would become pure contingent liability which is alien to the present legislation.

8.3 The Prevention of Damage Through Governmental Protection Measures

8.3.1 The Role of the Administration in Safety Legislation

Penalty, fine, and liability are designed to create a safe environment through their deterrent effect. If an accident should happen despite this deterrent effect, it will be settled financially, at least, through the liability of the perpetrator.

The deterrent effect of threatened sanctions has often been doubted. Restitution (indemnity, penalty, or fine once an accident has occurred) are merely inadequate reactions on the part of the legal system. They cannot undo what has been done. However, more importantly and meaningfully than subsequent sanctions or inadequate indemnification would be appropriate action for the prevention of accidents. Such realizations are by no means new. In the early part of this century, when numerous boiler explosions pointed to the danger inherent in the newly introduced steam engines, attempts were made through rules and regulations governing design, construction, and operation to reduce the danger of boiler explosions to an absolute minimum. In the Central European countries, the public turned to the authorities for regulatory action. The reasons for this reside in the concept of the responsibilities of government during the liberalistic period of enlightenment of the late 18th century and also in the structural differences of centrally administered authoritarian states.

Protection of the public from hazards or danger was simply and generally regarded as "police business," which would cover practically the entire interior administration of the country in question. Part II Title 17 Art. 10 of the Prussian Common Law of 1794 contained the following "classical" wording:

> It is the task of the police to institute the necessary measures for maintenance of (public) safety, law, and order and to protect the Public or the individual from any danger.

Basically, there has not been any change to this day. The following is an excerpt from the directives for the Regulatory Authorities, the present name for the former Administrative Police, namely Clause #2 of the North Rhine–Westphalian Reorganization Law of March 25, 1980:

> Art. 1 Para. 1: The Regulatory Authorities are charged with the task of actively averting any danger to the public safety or (Law and) order (Aversion of Danger) . . .
> Art. 14 Para. 1: The Regulatory Authorities are empowered to institute the necessary measures in order to avert any danger to public safety or (Law and) Order in individual cases.

As a result of the steam boiler explosions during the first half of the 19th century, the government felt challenged to avert danger and protect the public. The very first regulation covering the use of steam boilers, the Prussian "All-Highest Order on Council dated January 1, 1831 concerning the installation and operation of steam engines" exhibits the typical characteristics of almost all subsequent rules and regulations:

—licensing required;
—regulation of the licensing process;
—public disclosure of applications before granting a license;
—administrative penalties for erecting or operating the plant without a license or in a manner inconsistent with the scope of the license;
—authorization to remove an unlicensed or unlicensable plant.

A later "instruction" for the implementation of this directive dated October 31, 1831 [8-3] contains several supplemental procedural directives and, above all, technical details with regard to the erection and construction of steam boilers. The technical specifications must be met in order to render the installation licensable "from the police's point of view."

The example of safety legislation for the purpose of governmental activities for the aversion of danger presented here has not been changed to this day. It is characterized by two seemingly directly opposed basic elements (fundamentals), namely, on one hand, the restraint on the government which obligates it to intervene only if absolutely necessary, and, on the other hand, the perfectionism which motivates it to prescribe even the smallest technical details.

8.3.2 "Private" Protection from Danger (Insurance, Supervisory Bodies)

Protection from danger on the basis of laws and by officials is by no means the only adequate form of loss prevention.

In Great Britain, the first rules and regulations governing the design and operation of steam engines were not introduced by legislation but by associations of steam engine operators which had been formed for the purpose of mutual support and assistance in case of loss or damage, but which, however, were also intended to prevent such accidents from happening. Their primary objective was not exclusively the indemnification for losses suffered by third parties, but rather the prevention of losses to themselves. At a later date, the principle of loss prevention was also included in the form of insurance contracts for the coverage of the liability risk involved. The insurance companies, which were eminently interested in minimizing the insurance risk involved, attempted to achieve this goal through rules and regulations designed to reduce the danger of accidents, whereby any violation of these rules and regulations might cancel the insurance coverage altogether. To this day, this type of loss prevention plays a very important role in the Anglo-Saxon countries.

A somewhat comparable solution to this problem is the promulgation of accident prevention regulations in accordance with Articles 708, 709 Reich Insurance Code by the entrepreneurs organized in professional or trade associations. However—and herein resides an essential difference—membership in these professional associations is mandatory. Moreover, these professional associations are semi-official organizations with comparable authority.

In this connection mention should be made of the activities of the so-called Technische Überwachungsvereine (Technical Supervisory Associations). From 1866 on, they were formed as associations of operators of steam boilers with the purpose of offering help and assistance to their members in the construction and operation of installations with a view toward avoiding boiler explosions. This turned out to be such an extremely successful form of cooperation that, later on, operators who had their boilers periodically checked by association experts were exempt from simultaneous inspections by government officials. This setup has remained basically unchanged. Private operators render governmental protection from danger superfluous by taking matters into their own hands. The state is relieved of some of its obligations.

8.3.3 The Principles of Governmental Protection Measures Against Danger

Safety legislation cannot legislate danger to man and his environment out of existence. For this, we would have to forego the services of technology. Safety legislation needs a differentiating standard which it can use for its own orientation, with regard to form as well as content. In a constitutional country, such a standard can never be found outside the law. It should rather be taken from the basic laws which govern the relationship between the state and its citizens. In the Federal Republic of Germany, those standards significant for its safety legislation can be found in Articles 1, 2, 12, and 14 of the Constitution. Due to their eminent importance, they deserve special mention in this context:

> Art. 1 Para. 1: The dignity of the individual is inviolable. To respect and to protect it is the responsibility of all governmental authority...
> Art. 2 Para 1: All persons have the right to freely develop their personalities provided this does not violate the civil rights of others or contravene the constitutional law or the moral code. Para. 2: Each individual enjoys the right of life and to physical integrity. The freedom of the individual is inviolable. Any intervention of these basic rights must be based on an existing law.
> Art. 12 Para. 1: All German nationals enjoy the liberty of freely choosing their professions, places of work, and places of training. The practicing of an individual's profession may be regulated by law or on the basis of a particular law in force.

Art. 14 Para 1: Property and right of inheritance are guaranteed. The
contents thereof as well as possible limitations shall be governed by the law.

Without wishing to enter into constitutional details, the following is
becoming clear: The freedoms and rights guaranteed in the Articles 2, 12,
and 14 have their limits when the identical freedoms and rights of others
are being touched, especially where the limitless guaranteed right of the
individual to life and physical integrity is concerned. Thus, the life and
health of the individual are of greater importance than his general
freedom of action.

Despite the unconditional priority given to the protection of life and
health, a general ban on all technology would be basically unconstitutional.
Article 2, Para. 1 of the Constitution guarantees each individual the right
to freely develop his or her personality and thus also the right to freely
develop and operate technical devices. The freedom of choice of a
profession guaranteed in Art. 12 underscores this. Moreover, Art. 14,
concerning the right to property, also guarantees the right of individual
exploitation of a new technical development.

As far as safety legislation is concerned, this means the following:
Opposing interests have to be weighed, general freedom of action should
be limited to the least possible degree, and that which is necessary in the
case of technical installations will be determined by the degree of risk
involved. The smaller the risk, the wider the freedom of action on the part
of the manufacturer or operator. The greater the risk, the more the rights
and interests of others will be affected which, in turn, means that the
government must reserve for itself the right to intervene. As explained in
Chapter 2, the degree of risk is determined by the degree of probability of
loss or damage, and by the extent of the damage to be expected. Even
losses or damage with a lesser degree of probability may thus necessitate
extensive preventive measures if the possible damage is of a considerable
magnitude.

8.3.4 Graduations of Governmental Intervention

The legislature has taken into consideration the necessity for relativity in
the field of loss prevention by graduating the possibility for intervention on
the part of the authorities according to the installation's danger potential
(hazard potential) or its intended use.

In the case of relatively danger-free activities, the legislation waives the
right to demand a license or permit prior to the practice of such activities
and resigns itself to the possibility of retroactive intervention. However,
the possibility of intervention may even include a ban on these activities.
Thus, e.g., taking up a trade is basically free. If, however, the tradesman
or artisan, through his activity, poses a danger to the public, the authorities
may enjoin him from pursuing his activity further. This is commonly called
a "revocable license or permit" (license whereby the issuing authority

reserves the right to revoke if the necessity arises). The law governing the safety of appliances is based on the identical principle. In Art. 3, Para. 1 it spells out the following:

> The manufacturer or importer (marketer) of technical tools may only market or exhibit such tools if, according to the generally accepted rules of technology as well as the existing work safety and accident prevention regulations, they have been designed in such a way that the user or third parties, when putting them to their intended use, are protected against all hazards to life and health to the degree permitted by the type of application intended.

Thus the technical tools must meet "the accepted Rules of Technology." However, the manufacturer is under no obligation to prove this to the authorities before marketing the technical tool in question. On the contrary, authorities are limited to a subsequent intervention once the danger of an appliance has been discovered. In many cases, taking up a profession or other activity requires notification of the authorities. In its mildest form, such notification is designed to afford the authorities a more comprehensive knowledge of the project in question. In most cases, however, the activity may only be resumed after a certain period of time has elapsed and the authorities have not intervened within a certain grace period. At this point, legislation has now entered the next phase and has placed the prior step in the place of subsequent control. One no longer alludes to a "permit subject to later revocation," but of a "ban requiring notification." One example is the law covering construction where insignificant buildings or structures do not require a building permit yet must be reported before construction begins. Should the project fail to meet the rules and regulations of the building code, the authorities have the right to intervene.

In the next phase, the resumption of an activity is not only dependent on merely reporting that activity, but on a permit issued by the authorities. Legally this is referred to as a "ban with the reservation to grant a permit." It will be imposed any time the hazard potential of the installation or activity is of a magnitude which requires that, prior to issuing a permit, it does not pose a danger to the public. Within the scope of technical safety legislation and environmental protection, the "ban providing for a later permit" is the most frequently used form of governmental protection of the public from danger. As an example, the building laws can be invoked in such cases where a building permit is mandatory; also within the framework of the road traffic laws in the form of mandatory registration of vehicles and the licensing of drivers; finally, the obligation to obtain a permit for installations according to Art. 24 Industrial Code or according to the provisions of the federal law on emmissions. The feature which the ban with the proviso of notification and a later permit have in common is that the obligation to report an intended activity to the authorities or the obligation to obtain a permit for it prohibits the intended activity only temporarily until a governmental inspection has been concluded, however, it does not impose a ban on the activity itself.

There are activities where a hazard potential is of such a magnitude that legislation, in principle, regards them as a danger to the public and prohibits them by imposing a ban, one however, which may be overcome by specific and exactly defined exceptions. Legally speaking, this is a "ban involving a proviso for later dispensation."

Further examples—without entering into details—are the mandatory permits required for the acquisition of a firearm or for the handling of dangerous explosives.

The difference between reservation to grant a license and a reservation of dispensation has been described by the Federal Supreme Court (Federal Constitutional Court) as follows [8-4]:

> A reservation to grant a license constituted under the rule of law is endowed with the task of bringing in the authorities at an early stage in order to conduct a preventive examination under circumstances or for procedures which, according to experience, often entail breaches of existing rules or violations. The statutory obligation to apply for a permit thus does not mean that the activity requiring a permit or license is prohibited per se but only that any legal execution of said activity may only be started once the legality of the project has been tested within the framework of a regulated process. The legal importance of such a permit thus resides in the fact that the preliminary ban which has been temporarily imposed on the legal execution is now being lifted.

The permit or license itself is thus nothing but a mere certificate of nonobjection.

> The ban containing the reservation of a later dispensation does not regard the suppressed activity as one which is permitted in principle and affording the possibility of prohibiting illegal conduct. On the contrary, it looks on such activity as prohibited in principle, whereby there is a "chance" of obtaining an exemption from said ban. The permit or license issued no longer only means that there are no legal obstacles in the path of the project, but also gives permission for the aforementioned activity for the first time. The permit or license does not just constitute a formula meant for the preventive supervision of the legal execution of a normally nonprohibited activity, but contains the lifting of a repressive ban on an objective right. Thus it is the material prerequisite for the right as such. It is the permit or license which, for the first time, is justified in a constitutive manner.

An eminent administrative jurist has formulated this in a much simpler way [8-5]:

> While a permit or license—figuratively speaking—lifts a barrier blocking a thoroughfare, the dispensation or exemption from a ban just gives a person the right to climb over a fence.

The difference between a reservation to grant a permit or license and the reservation to grant a dispensation may sound like a very theoretical one. However, it is of considerable practical import. The activity which may be taken up under a license granted by the authorities is basically a legal one. Its execution is covered by the freedom of action guaranteed by constitutional law. Thus, in general, the applicant has a legal right to the permit or license; he must be granted this license if all legal prerequisites have been met. The same goes for the field of technical safety legislation. The building permit, the trade permit or license, the emission-related license, all of them must be granted whenever the planned technical device or installation meets the pertinent legal prerequisites and standards. The type of activity, however, which may only be undertaken on the basis of a governmental (dispensatory) permit or license, is basically an unlawful one. Its execution is socially damaging and thus is moving outside the right to the freedom of action guaranteed by constitutional law. Nobody can be entitled to a license for the execution of a basically unlawful activity. The granting of a dispensation must thus remain at the discretion of the issuing authority. Even if all requirements are being met, the permit or license applied for may be denied.

However, it would be a mistake to assume that a ban with the reservation of granting a permit and a ban with the reservation of granting a dispensation is a normatively exact description of a definition. In individual laws, the borderlines are often not clearly defined and sometimes there may be a dispute as to which form of ban is being referred to.

According to Art. 7 Para. 2 No. 3 Atomic Energy Law, the licensing of an installation "is permissible only if the required precautions against loss have been taken." The draft of the Atomic Energy Law which, upon a suggestion by the Bundesrat (Upper House) had been changed, still had the following wording: "a license shall be granted whenever . . . ." While the draft was based on a right to a license if the requirements were fulfilled, this is no longer the case in the wording which has now become the law. Therefore a legal claim to the granting of a license under the Atomic Energy Law no longer exists; the authorities have been endowed with the right to discretionary denial.

In principle, a discretionary denial is only possible within the framework of a ban with reservation to grant dispensation, whereby the activity in question is regarded as socially hazardous (or a danger to the public) and thus is relegated to a position outside the freedom of action guaranteed by constitutional law. The discretionary powers to deny licenses on the part of the authorities could, therefore, give rise to the assumption that the erection and operation of a nuclear plant constitute activities which are banned in principle because they pose a hazard to the public. In its resolution of August 8, 1978 [8-6], the Federal Constitutional Court did not rule along these lines. It qualified Art. 7 Para. 2 Atomic Energy Law as a ban with the reservation of granting a license; however, simultaneously it ruled as constitutional the discretionary powers of the authorities.

8.3.5 Precautionary Measures Against Risks and Protection from Danger

"Classical" loss prevention was only able to counteract activities where the degree of danger was known as measurable. There was no way of actively combatting possibly existing but not yet recognized dangers inherent in new types of activities. The activity itself, however, could not be banned because there was no proof that it actually posed a danger. The risk of damage had to be left to subsequent indemnification under the provisions of a strict liability, if extant.

Faced with the hazard potential of modern large-scale technology, the legislature no longer accepted this; it now also attempted to prevent endangerment not covered by the conventional type of loss prevention. This has now been done by the law covering (industrial) emissions and the Atomic Energy Law. According to Art. 5 Federal Emission Control Law, installations requiring licensing must be built and operated in such a way that:

1. Environmental pollution and other hazards cannot create considerable disadvantages and considerable inconvenience for the public and the immediate neighborhood.
2. Precautions are being taken against environmental pollution, especially by state-of-the-art emission control measures.

According to Art. 1, the purpose of the Atomic Energy Law is the following:

> ... to protect life, health, and property from the dangers of nuclear energy . . . .

According to Art. 7 Para 2. No. 3 Atomic Energy Law, a license may be granted only when:

> ... precautionary measures dictated by the state of science and technology against possible damages caused by the erection and operation of the installation have been instituted.

Both laws distinguish between a seemingly categorically demanded loss prevention (protection from danger) and precautionary measures against damage—precaution against possible risks—which are mandatory only within the scope of the state of the art (Emission Control Law) or the state of science and technology (Atomic Energy Law). The question is how to draw a line between loss prevention (protection from danger) and precaution against risks (or contingencies).

As for the conventional type of loss prevention (protection from danger), there is no doubt that risks which are extremely unlikely to occur and which would only cause inconsiderable damage, or where any damage was unlikely to occur even without any precautionary measures, do not constitute endangerment to be averted in a law-enforcement-related sense, yet constitute only a suspected danger which cannot be averted. With a view to the considerable danger potential inherent in modern large-scale technology, there is the question of whether the limit should not be extended and whether relatively insignificant occurrences of losses, which, however, may lead to catastrophic results, should be allocated to loss prevention (protection of the public from danger) and thus be subjected to stricter regulations than those of precautionary measures against risks. In the literature and in legislation these questions remain controversial; "prevalent opinions" cannot yet be discerned. Three positions, however, can be defined at this time:

(a) The "classical" concept of danger as defined by police law is being retained. Extremely improbable occurrences of loss, independent of the extent of the potential consequences, do not create an actual danger and should be accepted as a "residual risk."

(b) The nature and extent of precautionary measures against risks and the prevention of losses are determined by a standard of practical reasoning and are dictated by the precept of proportionality. Once recognized, any risk must be minimized to a degree where any loss is "practically" excluded.

(c) Even extremely improbable occurrences of loss must be effectively forestalled, where the loss potential is of a considerable magnitude. Risks of danger which either have not yet been recognized, or are not recognizable, should be allocated to precautionary measures against risks.

In atomic energy legislation, the consequences of various attitudes of the Courts as to whether rupture protection would be required for the reactor pressure vessel has become evident. The Administrative Court of Freiburg, in its judgment of March 14, 1977 [8-7] has affirmed the necessity of rupture protection. The following is an excerpt from the judgment:

According to the concurring views of the chamber, the premise for a permit or license standardized in Art. 7 Para. 2 No. 3 is lacking, namely that precautionary measures according to the current state of science and technology against possible loss or damage through the operation of the plant be instituted because of pressurized water reactor of the Nuclear Plant South may only be built with so-called rupture protection which, however, is not provided for in the plans. . . . Such rupture protection . . . prevents the reactor pressure vessel—of which, unlike all other components of safety-technology-related importance, there is only one—from bursting, whereby pieces of the ruptured steel may penetrate the reactor containment like projectiles and thereby, very rapidly, a vast amount of radioactivity in the reactor core may be set free. Even if the chain of loss occurrences described above is highly unlikely to happen, the nature and the extent of the consequent loss expressly caused by it are of such a magnitude that this risk, in accordance with the standards set by the Chamber with regard to the required precautionary measures, cannot be regarded as a negligibly small contingency. Therefore, it should not be classified (qualified) as a so-called residual risk but should be taken into consideration by including a rupture protection device into the design.

Conversely, and several days later, the Administrative Court of Würzburg, in its judgment of March 25, 1977 [8-8], came to a different conclusion. As far as the necessity of including a rupture protection device is concerned, the judgment states the following:

> ... there is no legal claim on the part of the plaintiff to insist on the inclusion of a rupture protection device. The pressurized water reactor used at the installation at G., affords the required degree of protection according to the state of science and technology and according to Art. 7 Para. 2 No. 3. However ... a catastrophic failure of the pressurized water vessel cannot entirely be excluded, even given today's state of the art in science and technology. Nevertheless, the Court ... is convinced that this does not constitute a violation of the plaintiffs rights, because a rupturing of the reactor pressure vessel—and this is the decisive point with a view to the above (fundamental) representations—is, as far as one can judge, improbable; protection against a rupturing of the pressurized water vessel can also be achieved in suitable ways other than the inclusion of a rupture protection device.

These two judgments, arrived at almost simultaneously, produced diametrically opposed results in dealing with a central problem of technical safety. At the same time, they demonstrate the currently prevalent degree of legal uncertainty (unsureness) with regards to the licensing process for an atomic installation. Efforts to achieve changes in this direction will be dealt with in the final portion of this book.

The Emission Protection Law also makes provisions for certain contingencies. The extent as well as the standards of these provisions are as controversial as those found in Atomic Energy Legislation. We will revert to this topic during a discussion of the legal quality of technical standards. In connection with protection against risk or contingencies, the new Directive for the Prevention of Incidents [8-9] deserves special mention. It is a directive governing the implementation of the Federal Emission Protection Law which, above all, is intended to prevent ecological disasters in the field of chemistry and to safeguard effective protective or preventive measures during incidents. The directive is intended to guarantee the advent of an intensive and ceaseless check on all safety-related matters in industrial plants handling dangerous chemical substances. The nucleus of this directive is the safety analysis provided for in Art. 7, whereby the complete safety technology and organization of the entire technical plant must be disclosed. The Directive for the Prevention of Incidents—for the very first time in history—makes a formal attempt to investigate the multitude of imaginable causes of incidents and to establish a breakdown of these causes according to their individual hazard potential in order to gain from this measure a safety concept for the erection and operation of an installation. Projective safety analysis of the type demonstrated in the Directive for the Prevention of Incidents transcends the conventional type of loss prevention by far and simultaneously designates (constitutes) one of the paths of development in safety legislation of the future.

The noteworthy result is that safety legislation, in certain subdomains, has been expanded by risk or damage prevention measures which have advanced the front line of safety legislation for the protection of man and his environment much more effectively than would have been possible by loss prevention in the conventional sense of the word.

8.3.6 The Importance of Administrative Procedures

Loss prevention must not be limited to the minimization of risks inherent in technical installations by establishing rules and regulations for their planning, construction, and operation. One might also consider the approach of achieving material safety through the corresponding creation of the administrative process governing the licensing of an installation. This is less astounding than it may sound.

The concept of "danger or hazard" has not only an objective but a subjective side to it. Here is one example: Danger is what is perceived as danger. If a certain danger or hazard is subjectively greater than in reality, fear and excessive precautionary measures will be the result. Whenever the danger or hazard is objectively greater than perceived subjectively, man will tend to become careless and negligent. One example for the differing subjective evaluation of various risks is the current dispute over the licensing of nuclear installations. It demonstrates clearly that the very real dangers of street traffic which every citizen experiences daily are obviously regarded as less of a hazard than the theoretical and less likely hazards inherent in nuclear energy installations. Taking into consideration the fact that, in a democratically organized society, it is ultimately the voter who decides which type of hazard he is willing to accept and which not, the next thing to do is to attempt to align the objectively extant hazards in technical installations with the degree of subjective assessment of this danger by the individual citizen. The enlightened and informed citizen whose interests have been integrated into a transparent and constitutionally organized administrative process will much sooner eschew excessive assessments of danger than his uninformed neighbor. The result would be a gain in internal legitimacy of the administrative process, and in many cases a gain in safety for the individual citizen, because the existing hazard or danger is recognized in its actual magnitude.

In the United States the above-mentioned path, especially with regard to the licensing of nuclear plants, had been chosen from the beginning. In the Federal Republic of Germany, the participation of the individual citizen in the administrative procedure, besides providing protection for all concerned, serves—above all—the purpose of providing comprehensive information to the authorities.

In the United States, in a much more formalized administrative process with much greater participation on the part of the citizen, the life and health of the individual is protected as much by the formal input of their vital interests as by material directives governing the erection and operation of the installation. The impact of legal supervision by the courts is thus essentially less significant than in the Federal Republic of Germany.

The U.S. administrative procedure has its origin in the democratic and liberal tradition of that country, whose citizens prefer to take care of their own protection rather than looking to the government and the authorities for salvation. In the Federal Republic of Germany this attitude is not yet strongly developed. Based on tradition, the protection of the individual is still primarily regarded as the domain of the government and the authorities. Only gradually has the individual begun to recognize that a fair and constitutionally organized administrative process also serves for the protection of the civil rights of the citizen and contains the possibility of factual integration, a possibility which should not be underestimated. The Constitutional Court, however, has always been aware of the importance of procedural rules. In its resolution of December 19, 1978 [8-10], it seized the opportunity once more of determining the following:

> ... that protection by fundamental rights should largely be effected by the regulation of procedures and that, therefore, fundamental rights do not solely influence the entire material law but also procedural law as far as the latter is of any importance for an effective protection through fundamental rights ... The fundamental right of Art. 2 Para. 2 of the Constitutional Law likewise influences the application or implementation of the directive governing governmental and judicial procedures in the licensing of nuclear installations whereby their first and foremost purpose is the protection of life and health from the hazards inherent in nuclear energy.

The administrative procedure—and this should be especially noted at this point—serves for the protection of the public to the same degree as the rules and regulations written within the framework of loss prevention and governing the erection and operation of a technical installation.

8.3.7 Interaction of Sanction, Liability, and Loss Prevention

The concepts of sanction, liability, and loss prevention (protection from danger) do not figure within the boundaries of safety legislation as unconnected or unrelated parts. On the contrary, they are linked together by means of multiple and varied interrelationships and thus form a regulatory network designed to guarantee the protection of the individual to a maximal degree. Whenever existing rules and regulations concerning the construction and operation of a technical installation are ignored, the authorities may intervene and force the operator to run the installation according to the law or, if necessary, close the installation down. Furthermore, a violation of the operating regulations may constitute a statutory offense, usually in such cases where the violation endangers, or causes damage to, human lives or substantial property. One good example is Art. 323 Civil Code (BGB) referred to above, in which reference to the "generally accepted rules of technology" builds a bridge reaching over to the building codes which, likewise, demand adherence to these rules and regulations.

Whenever a violation of the rules and regulations causes damage, this constitutes an offense according to Art. 823 Para. 2 Civil Code (BGB), that is, if negligence can be proven and the regulations violated come under the heading of "protective laws." This will clear the way for indemnification.

The same general "panoramic" view also permits a different type of approach. A statutory offense in the domain of ecology (Art. 311c ff. Penal Code (StGB)) resulting in actual damage will always simultaneously constitute a violation of a liability standard involving mandatory indemnification and will force the authorities to intervene.

The interlinking of liability, sanction, and loss prevention (protection from danger) is a necessary measure. A legal system confined to subsequent indemnification and sanctions would be just as ineffective and inadequate as preventive protection from danger which, in the case of failure, is not supplemented by the obligation to provide indemnification and, in the case of violation, is not empowered to impose the proper sanctions.

8.4 Safety Legislation and Technical Standards

8.4.1 The Need for Technical Standards

Safety science and safety technology are dynamic, geared to continuous development and radical changes. The development is characterized by more and more new findings and the effort to find profitable applications for this knowledge in daily practice.

Technical progress is faster than progress in other areas of life. Thousands of new technical products appear annually on the market. In the highly developed industrialized countries, half the labor force is engaged in manufacturing or selling products which were completely unknown at the turn of the century.

The law is unable to keep up with this pace. As an evolved historical value system, it appears to be rather static. It does not absorb every instance of technical progress for its own sake, but weighs its inherent disadvantages against its advantages. The law cannot react immediately to every technical innovation and must therefore work with generalized formulations to keep the gap from becoming excessively large. Lacunae must be filled by technical standards and rules.

Technical developments create perpetually greater needs for standards, legal as well as technical. One hundred years ago there wasn't a single section in any law dealing with the motor vehicle. At present no less than 20 laws and 76 regulations with roughly 1200 sections deal with motor vehicles and vehicular traffic. Twenty years ago there wasn't a single federal regulation in the field of nuclear energy. Now there are hundreds of such sections. Parallel to this development, the volume of technical standardization has increased at the same ratio. The VDE compendium of regulations, for example, which essentially deals with safety requirements for electrical installation parts and equipment, increased from 1950 to 1974 from roughly 2500 pages to more than 6000 pages. Of roughly 500 DIN (German Industrial Standards) standards and drafts with safety technology determinations, roughly 65% were written between 1960 and 1967, roughly 25% between 1945 and 1960, and about 10% prior to 1945. Since technical safety standards are written only when there is a corresponding requirement, the above data permit conclusions concerning the increased complexity and danger of technological products as well as the increased requirement for regulations.

Technical standards can be drafted more rapidly and changed with greater ease than legal standards. They can be adjusted to technical developments more rapidly than the rules of law. They are therefore indispensable as an intermediary between law and technology as well as the pathway to subsequent legislation.

It would be ideal if the operator or owner of technical equipment were able to determine from statutory regulations how to act in order to avoid exposure to liability claims and penalties. It would be useful if perusal of laws could serve to show how the planned technical product or technical installation should be executed to avoid objections by the authorities upon assembly or in operation. Statutory regulations do not live up to this ideal. This is primarily due to liability provisions, insofar as they assume fault. They do not include any kind of precise description of required action. How the accusation of fault can be avoided must therefore be found elsewhere.

Section 323 of the Penal Code, cited previously, is not noteworthy for the desired precision: it provides penalties for endangering man as a result of violation of "generally accepted rules of technology." What these rules of technology are, where they can be found, or what their significance is, cannot be determined from the law.

The above law governing safety of instruments also refers, in Section 3, to "generally accepted rules of technology." Apparently the law does not specify the required conduct in the field of technology with the precision required to make it possible to view legal standards of safety legislation alone as adequate instructions on conduct of operations. The reason for the inadequacy of the rule of law is the rapid pace of technical development. For the legislator this pace makes it impossible to attempt regulation of technical progress by means of laws. Since parliamentary bodies lack knowledge of the subject matter, the legislator would be strained beyond his limits. Insofar as safety legislation—as in the form of statutes governing liability and penalties—is not primarily an instruction manual for designers and operators, this lack is less serious. Fault is a suitable way of including criteria outside the law, for example, in the form of technical standards for the concretization of obligations governing operation and conduct. If, however, safety legislation, as in the area of danger of risk prevention, is intended to be and must be instructions for operation, this possibility does not exist. It is up to the legislator to specify the required conduct. In order to keep up with technical developments to some extent, at least, the legislator resorts to the use of standards with varying liability degrees, ranging from general to specific technical data. The basic means for determining what is permitted and what is not in generally binding form is the law. It determines the consequences of violating the rights of others. Only the law can interfere with constitutionally guaranteed freedom of action. The law must delineate general compatibility as a boundary of freedom of action with adequate accuracy. Since, in the area of technical safety legislation, the legislator is less and less able to do so, the law can merely outline general protection aims. The more specific form of legal conditions must be set forth in regulations which rank below the law. An example is Section 24 of the Industrial Code which in fact represents no more than a comprehensive authorization to issue specific regulations.

The more specific formulation will be by "administrative orders" which the appropriate agencies will be authorized to issue for enforcement of the law once the legislature has created the authorization, sufficiently specific with respect to substance and extent, required for such action. Administrative orders are legal standards as binding as the law. Manufacturers and operators are bound by them to the same extent as the authorities and the courts. The difficulties faced by those who issue the administrative orders are, however, hardly less serious than those faced by the legislator. They, too, find it necessary to either overload the regulation with technical specifications or to opt for escape into unspecific legal concepts which incorporate rules of technology or the state of the art in differing form into the regulation. Examples of this appear in regulations governing installations which require supervision according to Section 24 of the Industrial Code or the construction regulations of the States of the Federal Republic.

The level at which specific technical data generally appear first is that of so-called "administrative regulations." There it is frequently stated that requirements of the law or of the regulation will be considered to have been complied with when the technical product corresponds to the individual specifications of the administrative regulation. Administrative regulations are standards for internal use, binding only on the administration, and are not directly mandatory for manufacturers or operators. They may, however, become indirectly mandatory.

8.4.2 Technical Standards and the Legal System

The law, administrative orders, and administrative regulations constitute the forms of safety legislation by rule of law. From the judicial point of view, they could be the most essential.

For the practicing engineer technical regulations and standards are more important because he has to deal with them on a daily basis. This is due to the large number of standards which are combined in the various regulation manuals of private institutions. Compared to the wealth of existing technical standards, in the safety technology area the legal system could still almost be called an understandable whole. This is due, among other things, to the fact that the required specific technical instructions are found first on the level of the technical standard. Finally, technical standards concretize the substance of statutory regulations with readily applicable precision. The pulsebeat of technical safety law is apparent in technical standards. Rules of law, created by the legislator following the procedure prescribed by the constitution, are rules provided with absolute validity claims, applicable and enforceable rules governing human responsibility. This does not hold true for technical standards. Technical regulations are not established and decided by the legislator, but are worked out and set forth by those affiliated with professional associations; the engineers. The institutions concerned have no legislative powers. For this reason technical regulations at first do not have legal but actual validity. They originate as the record of the experiences of technicians in a specific area. Following new experiences and new insights, the rules are changed. Thus, technical regulations are not drafted and decided on in the manner of laws, but develop from practice and are continually refined.

Though technical regulations are not rules of law, they are not without legal significance. The greater the extent of knowledge and experience included in a technical standard, the greater the inherent claim of the standard to recognition expected not only from the professional circles concerned, but beyond this, it gives a more detailed description of that which in laws and regulations is merely outlined by rules of technology or similar formulations. If, however, in the standardization procedure, in addition to technical expertise, "public interest" is also brought into play and is sufficiently stressed as well as the economic interests of the circles concerned, there can be no objection to technical standard recognition by the legal system. The legal system has, in many ways, heeded the appeal emanating from these standards for the legal adoption of these standards.

The substance of technical regulations can be "incorporated" into laws or decrees, that is, included verbatim. Here the technical standard is no longer recognizable as such, but becomes an integral part of the law or decree which includes it. The advantage of this procedure is obvious: the legal position is clear and the regulations are reported in the appropriate official publication. There are no problems of validity: operators, manufacturers, and the authorities are bound to observe laws and regulations. These advantages are outweighed, however, by major disadvantages, namely that incorporation is not suitable when new technical areas are covered by laws. If one wished to turn all technology law—even if this meant only safety law—into laws including all the rules of technology, the volume of legislation could no longer be measured in pages but would have to be measured in tons of paper, and not only during the first adoption into law but continuously, since every change in the state of the art would mean a change in the law or the regulation and the change would have to be published. The legislator would be strained beyond practical limits. Wherever this method is still followed, it is due to historical tradition. At the inception of technical development, legislation still had the ambition, certainly welcome from constitutional viewpoints, to regulate technical particulars as well. In view of the rapid pace of technical development, at present this procedure means that absorption of standards into laws is always years behind the state of technology.

The above reservations are not applicable to the same extent to the incorporation of technical standards into administrative regulations. Administrative regulations need not be published and can therefore be adjusted more rapidly to new developments and findings. Sometimes administrative regulations, for example "technical instructions for air purity maintenance" in emission protection law, are technical standards in a different guise. The advantage of flexibility is offset, however, by the apparent lack of authority. Inclusion of technical standards into administrative regulations is at first only an acceptance by the administration. Acceptance by the legislator is not the case; for the courts, technical standards do not become binding by having been included into administrative regulations.

The legislator or the author of decrees can refer in the text of the regulation to specific technical standards, either by way of referring to the valid version of a standard or to a specific standard by citing a precisely designated version, giving date of issue and place of publication. The first form is referred to as a flexible reference. It is generally considered insufficient because it represents a disguised shift of legislative authority to private standardization. The reference to a standard in a specific version is called a firm reference. It is admissible because by means of the reference the legislator cites a specific standard and only that standard. As a result of the firm reference, the contents of the technical standard referred to practically becomes a part of the legal ruling using the reference and thereby acquires the same authority. Firm references are frequently used. The advantage over the incorporation method is that the text of the law or regulation remains free of technical specifications. This gives rise to the question of whether the substance of a technical standard which has become part of a rule of law due to such reference would not have to be published in the same manner. However, the demand that the referenced technical standard should also be published in the official legal publication would undo the advantage of the reference method: whether a technical standard is published as part of a rule of law in the text of the statute or as an appendix makes no difference with respect to volume. It is therefore considered sufficient that the standard as such is published, accessible to everybody, and that the place where it can be found is indicated in the referencing rule of law. A disadvantage remains: every change in the technical standard must be followed by a corresponding change in the legal instrument and must be published.

Finally, the legislator can, by means of one or several unspecific legal concepts in the form of a general clause, give a general description of the protection aim of the regulation, leaving the concretization to administrative regulations which in turn refer to technical standards or which are technical standards in their own right. The content of the general clause usually reflects the hazard potential of the technical installation. In addition there will be a statement about the rate at which the actual state of knowledge must be taken into account, which in turn depends on the familiar and, above all, the unknown hazards of the technical product. Essentially there are three levels with varying import which the Federal Constitutional Court described as follows in its decision dated August 8, 1978 [8-11]:

> The law may, as for example in Section 3 Para. 1 of the law governing technical equipment (machinery protection law), refer to *generally accepted rules of technology*. In the case of this type of linkage of law and technology, the authorities and the courts can limit their activity: they can ascertain the prevailing views among technical practitioners to determine whether the technical equipment concerned may or may not be released for operation. There is a disadvantage to this solution: the legal system, using the criteria of generally accepted rules of technology, is continuously lagging behind technical developments. This is avoided if the law refers to the *state of the art* (as, for example, in Section 5 No. 2 of the Federal Emission Protection Law). The legal yardstick for that which is permitted or indicated is thereby shifted to the front of the technical development, since general recognition and practical satisfaction alone are not decisive for the state of the art. In the case of the state-of-the-art formula, the determination and evaluation of decisive facts, however, becomes more difficult for the authorities and the courts. They have to enter into the differences of opinions held by engineers

to ascertain what is technically necessary, suitable, appropriate, and avoidable. Section 7 II No. 3 of the Atomic Energy Law finally goes one step further by applying the *state of science and technology*. By referring to the state of science, the legislator exerts even greater constraint to make sure that technical regulations keep up with scientific and technical developments. Those precautions against loss which are considered to be required by the most recent scientific findings must be taken. In the event that such precaution cannot be effected, the permit may not be granted; the required precautions are thus not limited by that which is technically feasible.

The Federal Constitutional Court has been cited so extensively because the decision is of importance in another connection as well. The "state of science" as mentioned need not be considered; it will not yet be included in regulations. The general clauses "generally accepted rules of technology" and "state of the art" remain. Concerning the use of the first formula, the instrument of safety law was mentioned as an example.

An administrative regulation pertaining to this law lists the technical standards, and the authority assumes that compliance with these standards means compliance with statutory requirements, namely, that the instrument complies with the "generally accepted rules of technology." The great advantage of this procedure is its flexibility, which makes it possible to take technical progress reflected in a change of the standard into account without having to change the law. Neither law nor regulation need be changed since they contain no reference to the technical standard. The administrative regulation need not be changed either, since it can refer to the standard in whatever its actual version without encountering the reservations which apply to a statutory optional reference.

The "general clause method" has its limitations, for example, when it is doubtful whether and when certain technical standards do or do not represent a "generally accepted rule of technology" or the "state of the art." Until the decision by the Federal Constitutional Court referred to above, this problem was hardly considered significant. This is no longer true. Due to the distinction made by the Federal Constitutional Court, it may become necessary to analyze technical standards to see whether they represent merely "generally accepted rules of technology" or have gone further and represent the "state of the art." This will not be simple and will lead to conflicts. At the same time, a questionable classification is introduced. To use an old railroad image, the "generally accepted rules of technology" can be viewed as the wooden seat carriage, the "state of the art" as the leather seat carriage, and the "state of the art and technology" as the velvet seat carriage. The buyer acquiring a household appliance with the instrument safety (GS) symbol confirming compliance of the instrument with the instrument safety law would be highly surprised to find that he had not acquired the appliance with highest degree of safety.

On the other hand, the legislative attempt to guarantee every consumer the optimal attainable instrument safety would lead to unattainable legal insecurity. The more the yardstick of the legally permissible and indicated is shifted to the front of technical development, the less will there be of technical standards which can be used for concretization. It would be up to the courts to equalize this deficit and they would have to determine in every individual case what is required with respect to safety technology without being able to adequately use firm criteria. If, in addition, the "state of science" were to be considered, the results would be as unpredictable as—and this is clearly shown by the examples given—is the case in the procedures surrounding nuclear power plant permits at present. The hazard potential of a nuclear power plant is, however, incomparably greater than that of a household appliance and therefore the legal insecurity in the latter case is, for this narrow area of technology, acceptable. A higher degree of legal insecurity would presumably mean waiving optimal hazard protection. This cannot possibly be desirable. At present the general clause method is, in spite of its drawbacks, probably the best method of linking technical standards with legal standards. Its advantage above all is the possibility of quick reaction to changes in technical development without the need for major legal change.

A technical regulation which has not become generally binding is not therefore negligible; it is the standardization reflection of practical experiences gathered by experts in a specific field. It combines, particularly when we are dealing with a "generally accepted rule of technology," the long term experiences of the vast majority of experts. Therefore, a technical rule has the advantage of the assumption of correctness and completeness and is compulsory in the sense that the quality of the underlying expertise and the extent of experiences contributed set measures for that which is appropriate and required.

In the event that the experts agree that an existing standard is outdated, it loses its effectiveness as a reference. In such a case, deviation from the standard is possible. This means that the intrinsic authority of a technical rule provides it with indirect binding force which even the courts cannot ignore. The definition of an offense pursuant to Section 323 of the Penal Code will generally have been satisfied when one of the "generally accepted rules of technology" to which reference is made have been violated. The situation in the area of liability regulations is similar. There, moreover, technical rules have a double effect: First, they concretize, for example, within the context of manufacturer liability, obligations of action developed by legal decision concerning Section 823 Para. 1 of the Civil Code. Technical regulations could have the same effect via Section 823 Para. 2 if technical rules could be used via the "protection law." Second, technical rules are of significance in connection with fault, since they can concretize the "required operational care." Thus, in the example of manufacturer liability, in the event that the user of a technical instrument suffers damage, the manufacturer will generally not be accused of fault if the instrument was manufactured according to technical standards. Matters will be different, of course, once it is determined that a standard is outdated. This, however, would have to be proved by the one who suffered the loss. On the other hand, the manufacturer of an instrument is not unconditionally bound to follow the standards: in principle, they are not mandatory. If, however, there is a loss in such a case, the manufacturer will have to prove that the damage-causing instrument shows the same degree of safety as one manufactured according to standards. The first impression would then show that he is at fault.

In the area of risk protection, matters are similar with respect to results. When an administrative regulation refers to technical rules or if the administrative regulation is such a rule in substance, binding force is attained at first only for the authority applying the law. However, the administration is obliged to treat all citizens equally. This means that the citizen must be able to assume that the law will be interpreted and applied in the manner provided by the administrative regulation. In case of reference to a technical regulation in an administrative regulation, the regulation becomes indirectly mandatory via the equal treatment statute.

One example of this is the frequently quoted instrument safety law. An administrative regulation of the law names numerous technical rules and standards and at the same time states that the requirements of the law will be considered complied with when the instrument corresponds to the standards named. If the instrument is produced according to standards, the manufacturer will be certain that the authorities will not object to the instrument. The manufacturer may also deviate from the standards. He is expressly permitted to do so. The instrument safety law is not intended to hinder technical development. The manufacturer must, however, prove that his instrument is as safe as the one produced according to standards.

Another example of the indirect binding force of technical standards is in the area of emission protection rights. In an administrative regulation accompanying the Federal Emission Protection Law, the "Technical Guidelines for Pure Air Maintenance" of 1974, boundary values are given which the operator of a power plant, for example, must prove to have observed. If they are observed, the agency considers the legal requirements fulfilled. Nevertheless, the Higher Administrative Court in Münster, in its decision dated July 7, 1976 [8-12] responded to the action brought by a resident in the area where a power plant was seeking approval by rescinding the permit, though the values were lower than those specified by the technical guidelines on air. The court did not feel bound by the values set forth in an administrative regulation and viewed the hazard of harmful environmental effects as a reality:

> The Senate ... arrived at the conviction that the emission values specified in the technical air guidelines do not have the weight of rigid absolute boundaries but should be viewed as demarcations which designate a not precisely known transition range between harmful and harmless environmental effects or which—in other words—can be compared to warnings posted

before or in unsafe terrain and which in this sense are in the nature of reference values or guidelines.

As a result of this interpretation, every permit application procedure could include, irrespective of the question of whether technical air guideline values are observed, an examination to ascertain whether harmful environmental effects can be expected. For practical purposes, the results cannot be predicted, the loss of certainty with respect to legal decisions is obvious. For this reason the Federal Administrative Court rescinded the decision by the Higher Administrative Court in Münster in rendering its decision of February 17, 1978 [8-13]:

> The emission values set forth in the technical guidelines governing air correspond, according to the determinations of the Court of Appeals, to the requirements provided in Sections 48 No. 1, 51 Federal Emission Protection Law for the type and manner of determination of emission values. They thereby reflect the experiences available and the state of scientific knowledge on the potential of certain noxious materials in the production of the impairment as stated in Section 3 Para. 1 of the Federal Emission Protection Law available at the time that the technical guidelines for air were promulgated and can therefore—subject to new and better findings—serve as a basis for legal decisions by way of an anticipated expert opinion.

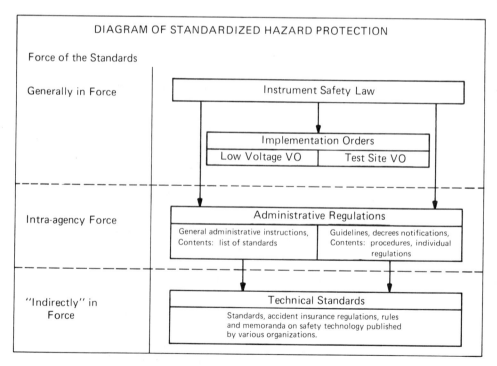

Figure 8.3 Diagram of standardized hazard protection.

Thus technical standards and administrative regulations are indirectly binding, depending on the extent to which they reflect the state of technology or, in the case of emission protection rights, the state of knowledge concerning specific emission hazards. The effectiveness of the technical guidelines on air as anticipatory expert opinion is therefore at all times subject to new and better findings.

Notwithstanding the uncontestable advantages of the general clause method, it has a major disadvantage: there is considerable lack of regulation and the final decision is passed on to the courts.

Interrelationships between legal and technical standards can be illustrated by means of the example of the instrument safety law as shown in Figure 8.3.

8.5 Correlations Between the Purposes and the Effect of Safety Rules

8.5.1 Technical Standardization

Technical standards are operating instructions for the practicing technician. To enable him to comply with this task, the technician must adjust these technical requirements meticulously and down to the last detail. General guidelines for design implementation of technical requirements are not included in standards. In this respect the design engineer is free to proceed as he sees fit. The safety standard is a rule governing particulars and is essentially characterized by its functional relation. Technical safety standards are not limited to the function which can be readily seen in their name, but have a series of additional functions which overlap to some extent. These will be briefly enumerated.

Technical safety regulations are a standardization example. Only the uniformity of specific technical processes or requirements will create a common criterion and thereby the foundation for science and practice in order to be able to make comparisons. Technical rules facilitate efficiency. Once there is uniformity of technical requirements, the designer and manufacturer can search for uniform solutions of similar technical problems and expect acceptance of these solutions. Only uniform technical products will create, in broad areas of our highly technical world, the premise for orderly operation. Technical standards often affect everyday life, ranging from dimensions of writing paper to window sizes, from the shape of a wall outlet to the bricks for a building.

The guarantee function must be cited as an important standardization task. Rules of technology, provided they have an orderly evolution, offer the participants and the consumer a certain degree of security. They guarantee that a given conduct is practical and responsible. For this reason quality or test symbols, for example, the GS symbol, can be awarded when specific technical standards have been fulfilled. The guarantee function of the technical standard stems from its intrinsic authority, i.e., from the strength of conviction due to the input of expertise and the participating interests.

The guarantee force of the technical standard also places the latter in the position of assuming higher functions with legal implications. In the first place, the protection function of the safety standard should be mentioned. Since the standard determines specific requirements, it attempts to translate the protection of all who could be exposed to the hazards of the technical product into reality in the best possible manner. In this sense the standard, and this is part of its nature, in its statement and power of conviction is constantly subjected to new and better findings. To the same extent as new safety-technical knowledge develops and becomes established, the existing technical standard loses the claim to optimal realization of the desired protection.

There is a close connection between the above functions and the force of the technical standard in creating legal certainty. This is the natural and necessary consequence which the legal system draws from the power of conviction of a technical standard when and to the extent to which the standard fulfills its guarantee and protection functions. As has been shown, technical standards can attain direct or indirect force in various ways. The technical standard can create legal certainty when the one who observes the standard can proceed with the assumption—putting it in simple terms—that nothing can happen to him in court. It is this import of technical standards which elevates them *de facto* to the level of legal standards for the practicing technician.

Above all, the legal certainty created by a technical standard will be noted by the consumer. The regulating bodies, however, will try not only to establish a technical consensus but also to give due regard to prevailing economic and public interest in such a way that the technical standard can be viewed as having the function of providing satisfaction. Belief in the force of conviction of "sedimented expertise" is limited, as evidenced by the current controversy surrounding nuclear power plant permits. Similar events can be observed in other areas of modern large-scale technology. In this connection technical regulations acquire considerable significance. In the final analysis they determine how the intended protection of the citizen will appear in concrete terms. Therefore, in terms of intention and substance, though they have a technical and natural science character, they do not exist in a vacuum. Technical regulations do not deal with technical aspects only; they must also show consideration for economic and judicial aspects.

Only during recent years has satisfaction been viewed as a significant and necessary function of a technical regulation. This is a noteworthy contradiction of the general understanding of technical standards which are given far greater social significance than the authors of the standards intended. With increasing social significance, technical standards can no longer be viewed as merely object- and function-related operating instructions but must also be viewed as value-related maxims. Thus the boundary between technical and legal standards becomes fluid.

8.5.2 The Law

In contrast to technical standards, legal standards are not maxims primarily following laws of nature and the technical purposes to be attained, but standardizing rules of action which aim for order in social co-existence and follow a value system conceived as static. Value-related design of social existence is the primary function of legal standards, while it is only in recent times, and by no means generally, that technical standards have taken on a similar function. Nevertheless, there are subareas where the functions run parallel.

An essential purpose of the law is to create legal certainty. This has been the function of the law ever since the inception of an established legal system. This is incompatible with continuous changes in the law. The citizen who must observe the laws must be certain that laws in effect today will be valid tomorrow. Otherwise actions related to the future would no longer be possible. Technical standards can also fulfill these functions, but only as a secondary measure and rather incidentally. They are subject to continuous reservations with respect to changes. Thus their significance in terms of certainty of the law can extend only over the period during which the technical standard represents the actual state of knowledge.

A next task, which can be fulfilled only by rules of law, is the evaluation of interests. In the area of technical safety law, this will be the evaluation of the benefits versus the hazards of technical progress. The particular difficulty in interest evaluation in the area of safety is the requirement that the legal system must be receptive to technical developments and findings while also providing satisfaction to constitutional demands for clarity, certainty, and dependability. This is achieved by means of the sequence described, namely general laws dealing with the aim of protection, implementation regulations, and concrete administrative regulations with technical standards.

The interest evaluation between benefits and hazards in the narrower sense should be distinguished from the guarantee obligation of the state which is fulfilled by rules of law. This means that the responsibility of the state with respect to protection of important legal rights, such as life and health, is comprehensive and indivisible. There are additional aspects, for example, humaneness in the work place. Interest evaluation would not be sufficiently precise to outline the function of rules of law since the need for the three-pronged safety legislation, the conjunction of liability, sanction, and hazard protection, would not be stressed with sufficient clarity.

Due to the mandate to preserve human dignity, a legal system which would be limited to subsequent compensation of damages and to sanctions would be as inadequate as preventive hazard protection which, in the event of failure, would not provide for the obligation to furnish equalizing damage compensation and which would not be equipped with sanctions for cases of violation. All three substantive elements of safety legislation complement one another and in their totality fulfill the guarantee obligation of the state.

It has been characteristic of the functions described thus far that varying and contradictory interests are not really denied, but that there is unanimity on the view that the benefits of technical progress generally outweigh the hazards and therefore the legal system should, in principle, be receptive to technical progress. The usefulness of technical progress as such has not been questioned. As a result of new forms of modern large-scale technology, this has changed. The controversies surrounding nuclear power plant permits clearly show that among some segments of the population, belief in the overwhelming benefits of technical progress has yielded to mistrust. The consensus prevalent earlier has been shattered. Not only is the maximum possible safety of large-scale technical installations demanded, but their necessity and benefits are being questioned more frequently. Such an attitude, contrary to the views held until now, cannot remain without impact on the legal system. In this way the scope of interest evaluation becomes far broader. In the balance are the rights of man to quality of life, clean air, and pure water, tolerable noise levels, and absence of exhaust gas as well as the needs of industry and job security, and in the final analysis, the need to secure the existence and welfare of an industrial society. This is no longer evaluation of interests but integration of interests. Such fusion of contrasting contradictory concerns and positions can be the function of the legal system only, since through this fusion process the law develops its substance and purpose.

The process of the evolution of the law does not take place in a legal vacuum: the abrogation of unconstitutional legal regulations is part of the daily work of the Federal Constitutional Court. The constitution, in particular the civil rights set forth therein, are the legal criteria which must serve to orient the process of the evolution of laws. First there is the obligation of all state power, which includes the legislator, to respect and support the dignity of man. The universal right to life and physical safety has equal rank. In the creation of laws which respond to the commandment to protect the dignity, life, and health of man to the same extent as they respect his freedom of action, the legal system attempts to fulfill its primary and actual function of creating material justice.

This is beyond the scope of technical standards, although by their function of providing satisfaction, they may contribute to material justice. The criteria, the replies to the question of what is just, can be determined only by the legal system. Rules of law determine to what extent technical regulations can contribute to material justice.

8.6 The Limits of Safety Standardization

8.6.1 Regulatory Limits

It is the aim of the legislator to create codification which standardizes a legal sector without contradictions and according to uniform principles, as expressed, for example, in labor legislation or social legislation codes. Insofar as liability regulations and individual penalty regulations are described as parts of safety legislation, such codification exists in the Civil Code and the Penal Code.

There is no such codification in the regulations governing hazard protection, not even a beginning in that direction. For this reason the request for comprehensive technology legislation, a "technical safety legal code" has been frequently voiced. The request is understandable. There are good reasons for it, since the legislator is not free in his evaluations. He must follow criteria which are basically set forth in the constitution and thereby make their imprint on the law. In the area of hazard protection, it is above all the general welfare which the legislator must translate into reality. Analogously, individual freedom of action encounters its limit in the general compatibility concept. If the limits of freedom dictated by common welfare were clearly demonstrated, every sensible citizen would find such limits understandable. He would also accept a body of legislation which regulates the obligations of the citizen to the state in a comprehensible manner. These principles are barely perceptible in legislative practice. The consequences could well be that individual rules of law will no longer be viewed as law in the sense of a system of responsibilities following uniform values.

As far as possible, legislation must be simple, clear, understandable, and guided by uniform principles in order to fulfill its function of formulating the law. In addition, safety legislation must permit ready adjustment to technical progress. This is easier said than done. Particularly in the area of safety, the complaint about the flood of standards is as vocal as the call for providing legal form for numerous matters, the call for new laws—and for good reasons. Technical safety regulations, if they are to have practical success, must enter into particulars. If they meet these requirements they either hinder progress or are swept away by progress unless they are developed further in time. Continuous rescission and change of technical regulations by the legislator or author of regulations would be the only method of dealing with the safety problems of technical progress. The legislator, however, even if he wanted to, would not be in a position to do so. In spite of all his legislative energy, the image of rapid technical development with standardization trailing behind still resembles the race between the hare and the tortoise.

The idea of a single comprehensive law on technology may have to be waived. It would either be devoid of practical substance or, if it did include substantive regulations, would be obsolete before final promulgation. However, the principles should be adhered to. Legislative procedures, for example, which follow the general clause method will be determining factors in the future as well.

8.6.2 Constitutional Limits

A. Reservation of the Law. To a great extent, technical safety law uses unspecific legal concepts in the form of general clauses. In addition, standards which set forth in concrete terms how life and health are to be protected from the hazards of technical progress are often merely technical standards which cannot claim any direct force on their own. This could constitute a violation of the principle of the general reservations of the law.

The reservation of the law states that in specified areas the questions left open by the constitution may not be decided by the executive by means of legal measure or administrative action but are subject to formal law and therefore require a decision by the parliament. Reservation of the law thus entails parliamentary reservation. The executive—and naturally a private standardization body all the more—may not be active in the area subject to reservation or only when empowered by law. In the administration of justice, in particular by the Constitutional Court, the principle of reservation of the law is described in such a way as to make it incumbent on the legislator to make all essential decisions in the fundamental standardization areas.

In the area of technical safety law, a violation of the reservation of the law was affirmed for the first time in a decision by the Higher Administrative Court in Münster dated August 18, 1977 [8-14], which found that Section 7 Para. 1 and Para. 2 of the Atomic Energy Law violated the constitution. In a decision dated August 8, 1978 [8-15] on the subject of the decision by the Higher Administrative Court in Münster, the Federal Constitutional Court ruled in principle and by way of guideline on the constitutional context of technical safety law. This decision deserves to be quoted due to statements with significance which, for all practical purposes, is equivalent to the law:

> The decision in principle with respect to standardization, for or against legal admissibility of peaceful use of nuclear energy on the sovereign territory of the German Federal Republic is, due to its far-reaching consequences for the citizens, in particular for their freedom and equality, for general conditions of life and due to the necessarily connected type and intensity of

the regulation, a basic and fundamental decision in the sense of the reservation of the law. Only the legislator is qualified to make it . . . .

Within this context he has regulated all fundamental and basic questions of permits in Section 7 I, II of the Atomic Energy Law. He has established, in terms of standards, the conditions governing installations in the sense of Section 7 I Atomic Energy Law, including fast breeder reactors which may be constructed, operated, owned, or essentially changed; he has made this determination, as will be shown, with sufficient specificity. Thereby the constitutional requirements stemming in the present situation from the reservation of the law are complied with.

The Federal Constitutional Court was supposed to decide merely on the constitutionality of the Atomic Energy Law. The decision wll not be overinterpreted by the deduction that the present regulation of the technical safety law in general is not defeated by the principle of reservation of the law. The use of technical standards and their legal significance appears to be secure to the extent practiced at present. The requirements set forth by the Federal Constitutional Court calling for "substance" of legal regulations are met by existing authorization foundations.

B. Certainty of the Law. The numerous, general-clause-type uncertain legal concepts of technical safety law could—in particular, because they are used in the basic formal laws—violate the constitutional mandate calling for sufficient certainty of the law.

This mandate used to serve the purpose of securing the liberty of the citizen. It is therefore directly linked to the above-mentioned principle of the reservation of the law. Action against the citizen would be permissible only on the basis of a law, namely a law which makes the time, the manner and kind, and intensity of government intervention sufficiently predictable. Aside from this focal area, the mandate calling for sufficient certainty of the law also has a democratic function at present. It is designed to separate executive and legislative areas. The purpose is that, on the one hand, parliament makes the key political decision and does not transfer any global authorization of whatever kind to the executive; on the other hand, the purpose is to keep the legislator from becoming lost in regulation of specifics for which he may lack the necessary expertise and thereby fail in the basic political function of legislation and democratic control.

With respect to the Atomic Energy Law and thus all of technical safety law, the Federal Constitutional Court, by its decision of August 8, 1978 [8-15] approved the far-reaching use of uncertain legal concepts:

Nor does Section 7 I, II of the Atomic Energy Law violate the constitutional requirement calling for sufficient certainty of the law. Section 7 I, II of the Atomic Energy Law uses, to a large extent, uncertain legal concepts. According to customary administration of justice, the use of uncertain legal concepts is, in principle not objectionable with respect to constitutional legal provisions. In reply to the question of which certainty requirements in particular must be complied with, the special aspects of every object to be regulated as well as the intensity of regulation should be considered. Lower requirements should be made especially in the case of multiform factualities or when it is expected that actual conditions will change rapidly. By means of these criteria the result is:

... There is sufficient certainty in Section 7 II No. 3 of the Atomic Energy Law. This regulation is in the area of technical safety law. In the event that the legislator intends to establish legal regulations in this area in order to avoid jeopardizing the individual or the general public, he is faced by special problems inherent in the nature of the object to be regulated. Even in those rare instances where a technical state of knowledge and development seems already completed, it is not possible, as a rule, to determine all specific safety requirements the individual installations and objects must meet, due to the many-faceted and complex problems of technical questions and processes. In areas such as that of the peaceful use of nuclear energy, where innovations must be expected continuously due to rapid technical development, there is an additional element to be considered: if the legislator had decided on regulation of specifics he would have to update them continuously to keep up with the state of the art.

For the promulgation of such standardization, as will keep pace with knowledge and developments in science and technology and will have the force of law, the legislator basically has several options.

... When faced with the question of whether to use uncertain legal concepts or draft a text entering into specifics, the legislator has leeway in his formulation, and considerations of practical handling may also influence his decision. In the case of Section 7 II No. 3 Atomic Energy Law, there are good reasons for the use of uncertain legal concepts included in the section. In the formulation of Section 7 II No. 3 Atomic Energy Law, providing for future changes serves to allow for dynamic protection of constitutional rights. It assists in optimal realization of the protection aim of Section 1 No. II of the Atomic Energy Law. Firm legal determination of a specific safety standard by means of formulation of rigid rules would by contrast, were it possible to provide, inhibit rather than promote technical development and the proper securing of constitutional rights. It would mean a step back at the expense of safety. It would also mean a misunderstanding of the mandate of certainty if that was the obligation intended for the legislator. Certainty requirements are indeed also intended to guarantee legal decision certainty; this, however, cannot be implemented in the case of every object to be regulated to the same extent. A measure of legal uncertainty, which over the course of time will to some extent be reduced by executive decrees, by administrative practice and by legal decisions, must be accepted, at any rate, wherever the legislator would otherwise be constrained to either adopt regulations unfit for practical application or to abstain completely from adoption of any regulation, and both options would, in the final analysis, be at the expense of the protection of constitutional rights.

This quotation has been extensive since it is not only convincing, but also establishes guidelines for technical safety legislation in the future. The decision is convincing, above all, by its restraint and by the attempt to find an appropriate compromise between that which is necessary and that which is possible.

C. The Basic Freedom Priority. Thus far the question considered has been whether the present system of technical safety law exceeds constitutional limitations. The Constitutional Court has ruled that it does not, by showing that only a flexible safety law open to future changes can guarantee optimal protection. On the other hand, a question arises concerning the limits of such safety legislation which is bound to intervene in the constitutionally guaranteed rights of the citizen to freedom and equality. It has been stated elsewhere that as a consequence of the general freedom of action and profession as well as the safeguarding of property rights granted by the constitution, there is a claim to approval of a technical installation once legal conditions have been fulfilled. The contrary will apply in the area of socially harmful activity which is outside the constitutionally protected freedom of action area and therefore need not be accepted.

With respect to atomic energy law, the authorities may refuse a permit even when all legal requirements have been met. This is unusual since the Atomic Energy Law recognizes the construction and operation of a nuclear energy plant as an activity permissible in principle. The operator would therefore have a legal claim to a permit once the conditions are met. According to the decision by the Constitutional Court [8-15], there are no constitutional grounds for objections to a negative judgment by the authorities. The reasons for this, according to the Constitutional Court, are the risk and hazards, not fully clear at the time of the deliberations on the Atomic Energy Law, which could result from the use and application of fissionable materials. In view of the high danger potential of the installations requiring a permit according to Section 7 Para. 1 of the Atomic Energy Law, no objections can be raised if the legislature exercises particular care by providing the executive with the right to refuse a permit which should basically be granted, when special and unforeseen circumstances dictate that such an action is necessary.

In spite of the refusal power bestowed on the authorities, granting a permit should be the rule once legal conditions have been met and refusal should be the exception. Thus the freedom priority of the Basic Law will have been satisfied. The legislator would not have been permitted to make refusal the rule and approval the exception.

By way of summary, the following should be stated: the manufacture, construction, and operation of technical installations should, as a rule, be viewed as an activity protected by constitutional guarantees of freedom. The authorities may intervene only when legal requirements have not been complied with or when binding regulations have been violated. Action by the authorities will confine the "violator" to the area within the boundaries. If, however, legal conditions have been met, the authorities must accept the activity and grant permits applied for unless they have been legally granted refusal power for the exceptional case.

8.7 Extralegal Limits

8.7.1 The Calculated Risk Inherent in Technical Civilization

Every year many thousands die on the roads as victims of vehicular traffic accidents. This loss of life could be avoided if driving of motor vehicles were banned. There is no one who would consider this seriously. On the contrary, density of vehicular traffic increases every year and thus the danger of traffic increases as well. Safety science attempts to develop methods to provide optimal safety for the system "motor vehicle–man– environment" without wanting to do away with the basic cause for the death of man, motor vehicle traffic.

The reason why vehicular traffic is not banned is obvious. Nobody can or wishes to turn back the clock by returning to the age of the horse-drawn carriage. That would mean that entire segments of the economy which provide work and sustenance for millions would lose the basis for their existence. The standard of living would drop. The economic consequences for the individual as well as society would be immeasurable.

Society is ready to pay for technical progress in the form of motor vehicle traffic with thousands of lives. This is not a matter in which the legal system can simply stand aside. By regulating the consequences and effects of technical civilization and by attempting to minimize them in numerous ways, it legally acknowledges hazards of technology and limited risk as a matter of principle. Technical safety law is not concerned with the creation of a completely safe world, but is designed to evaluate the degree of danger man should live with. What is involved here, and this deserves to be stated, is a genuine judicial function. That which, to use the example of the Atomic Energy Law, is "required" with respect to hazard protection or risk prevention is to be determined exclusively by the legal system. The unavoidability of a danger situation for engineering reasons does not constitute an impediment to the requirement postulate for countermeasures. In such an instance, the permit may not be granted. To quote the Constitutional Court [8-15]:

> The required precaution provisions are not limited by that which at present is technically feasible.

8.7.2 The Human Knowledge Potential

Though required hazard protection and risk prevention may not be limited by presently available technical possibilities, there is, however, a limit to hazard protection as a matter of course: human knowledge. An enemy who is not known cannot be fought. Only by paying the price of waiving any technical progress could risks of technology, which are not yet even identified as such, be excluded. That is not the intention of the legal system. In this regard, the Federal Constitutional Court [8-15] advanced the criterion of "practical good sense":

> Section 7 I, II of the Atomic Energy Law, however, agrees to the granting of permits even when it cannot be completely excluded that in the future, due to the construction or operation of the installation, loss may appear. As mentioned before, the regulation accepts residual risk . . . .
>
> In the event that the legislator wants to evaluate the possibility of future losses due to construction or operation of an installation or due to a technical process, he depends to a great extent on conclusions from observations of past actual events for the relative frequency of occurrence and on an analogous course of analogous events in the future; in the absence of adequate data from experience, he is limited to conclusions drawn from simulated events. Empirical knowledge of this type, even after it has been condensed into the formulation of natural science law, will, as long as human experience has not been terminated, remain merely approximate knowledge which does not provide complete certainty but is subject to correction by every new experience and will therefore always remain merely at the most recent level of possible unrefuted error. In view of his obligation to provide protection, to demand of the legislator a regulatory system which would exclude threats to constitutional rights with absolute certainty, threats which could possibly arise from technical installation permits or from their operation, would mean misunderstanding the limits of human knowledge potential and would to a great extent ban any government permit for the utilization of technology. Therefore, for the purpose of formulation of the social order, there must be acquiescence in evaluations guided by practical good sense.
>
> Insofar as losses involving life, health, and material goods are concerned, the legislator, by means of the principles set forth in Section 1 No. 2 and Section 7 II of the Atomic Energy Law on optimal hazard protection and risk prevention, established a criterion which allows the granting of permits only when, according to the state of science and technology, it seems *practically* impossible that such events involving loss will come about. Uncertainties on the far side of this threshold of practical good sense are due to the limits of human knowledge; they are inevitable and therefore must be accepted by all citizens as socially equitable burdens.

8.8 Trends in the Development of Safety Legislation

At present it is hardly possible to foresee how safety legislation will develop in the long term. At this time it is basically possible to discern three trends which determine the actual discussion of safety legislation and which have three aims: intensification of legal provisions, limitation of examination by judges, and greater consideration of technical-scientific expertise.

Legal uncertainty in the area of modern technology is not acceptable. It is not possible to have decisions on permits for technical installations with all the inherent consequences *de facto* made by courts with no political accountability nor sufficient expertise. The legislative and the executive branches are subject to political responsibility which they are not allowed to avoid. This holds true even in view of the decision by the Constitutional Court which accepts the legal uncertainty in the Atomic Energy Law on constitutional grounds. Greater legal security is required and must be possible, though absolute legal certainty may neither be possible nor desirable. It is therefore absolutely necessary that technical safety law should, at least in subareas, become intensified by legal concretization or by binding legal reference to standards to a greater extent than before, insofar as this is possible without detriment to the safety standard which must always be up to date.

To the extent to which the legislator needs to use general clauses which are concretized by use of technical regulations, one should consider attributing greater force to decisions made by the authorities in permit proceedings by limiting the courts, voluntarily or by law, to ascertaining whether there is misuse when they review decisions by the authorities.

Comparable self-imposed limitations on judges and statutory limitations already exist. Violation of Article 19 Para. 4 of the Basic Law which provides for due process of law for anyone against legal violation by public authorities has not yet been determined in this context. The Constitutional Court [8-15] is specific in leaving this path open:

> Here, in this connection, it can remain open, where, with regard to Art. 19 IV of the Basic Law, in the evaluation of technical standards for estimation of future loss possibilities, the limits of the judge's review obligations are situated and whether the courts may not limit themselves to ascertain whether, in the case of knowledge gaps and uncertainties in areas of natural science and technical determinations and judgments, the limits of the resulting "bandwidth" have been respected.

How the limitation of review by judges might be handled remains an open question. Possibly, for the sake of balance, the administrative process would have to be developed according to the American pattern, would have to become more formalized and equipped with judicial elements. The procedure should be an object-related democratic process rather than mere instructions of the authorities.

The courts could then, in essence, limit their function to procedural supervision.

Finally, there has been some thought given to strengthening the significance of the expert level. A start has been made. Thus, in the 1960s it was proposed to make technical standards binding by creating panels with special authorizations, panels which would be permitted to create technical regulations with force equal to that of rules of law. In substance these would be expert subparliaments. Models from the Anglo-American legal system were described. In this connection the Royal Commission, Research Parliament, and Science Court should be mentioned. The notion of a technology commission is along these lines. All these models could be accommodated in the expanded administrative procedure outlined above. "In the process of developing political objectives, material law, supported by the parties concerned, should be formulated within certain limits. The courts could then, ..."

Thus far there is hardly any chance that these plans will become reality. The difficulties rest in the far-reaching constitutional changes this would require, but that is not the most important element. In view of the largely lost unanimity with respect to the area of modern large-scale technology concerning the benefits of technical progress, it is hard to believe that institutionalized expertise could capture the confidence which would provide it with the authority to create technical law on its own. For analogous reasons, plans for an expanded administrative procedure may be possible to carry out only in the long term. Plans of this kind require an enlightened citizen with a sense of responsibility and readiness for compromise. In addition the citizen must always be aware that in a democracy it is he who determines how much technical safety he is to have.

In conclusion, technical safety law must, as a matter of principle, follow the direction of the best possible hazard protection. New and better findings of safety science concerning the type and extent of hazards in dealing with matters of technology change the creative leeway of the legislative, executive, and judicial branches. They also formulate legally admissible risk. Safety legislation follows at varying intervals in the wake of scientific findings. The gap must remain as small as possible. On these grounds the legal system is in close relation of safety science and, in some special areas, even linked with it.

Chapter 9

Social Aspects of Safety Science

Clinging to the familiar is part of man's general instinct for self-preservation and protection, an instinct evident in all societies. An old Spanish proverb tells us that "The bad we know is better than the good we do not know." The desire for stability in life is, however, in conflict with the desire for progress. This desire is so strong that in the life of the individual, as well as in the history of nations, it is the motivating force which leads to the rejection of the established order for the sake of the search for the new. As long as the goal of completely satisfying living conditions has not been attained, the two demands, for preservation and progress, will be compatible only when the order of the society is open and dynamic and keeps the way free for progress for the sake of its own stability.

By means of theoretical systems analysis, the effects of a new technology on society and the environment can be ascertained prior to its realization, and possible dangers which can be expected, together with the benefits, can be identified. This type of procedure, known as "technology assessment," which will be explained in Section 9.1, has its limitations since it is not based on its own specific scientific methodology. However, in view of the serious and, under some circumstances, irreversible effects of a technical innovation, from case to case technology assessment cannot be waived.

In evaluations of operational and economic aspects, technology assessment will be useful for necessary corrections with a view to stress on the environment or the like. When technology assessment is used at an early stage of technical development—in particular, for evaluation of safety as well as economic consequences—then these are, strictly speaking, not necessarily competing aspects. Nevertheless, this type of conflicting individual interests will be evident in the social structure. They can be merely identified by technology assessment; their settlement and subordination to the general welfare are subject to political decision and thus cannot be effected either by science or by other free forces within society. The decision by the state will be made with due regard for society's willingness to assume risks. This will be subject to further comment in Section 9.2.

There are tendencies to use the actual reaction of society as expressed in acceptance or resistance as standardization measure of the readiness to assume risks and of technology acceptance. With respect to method, these tendencies are directly comparable with the efforts of the utilitarian school of the last century: enumeration of effective action in catalogues and statistical-empirical determination of the laws governing this effectiveness. It must be agreed that, within specific boundaries of that which is "normal," such attempts are definitely useful and they can contribute to the factualization of a risk-acceptance dispute. These attempts, however, misunderstand the spontaneous and often irrational core of human decision making and this is often evidenced by a suprisingly high or low level of risk-acceptance readiness.

The state should assume an altogether more rational risk behavior. We should take the position that the state will have to expect acceptance of fair risks. Since the choice is not between "risk" and "no risk," because risk-free conditions cannot be realized in this world, the demand for zero risk must be rejected as an irrational demand. The measure of protection to be required and the risk for which there can be expected agreement can be determined only in concrete cases according to the principle of a means-and-ends ratio. This opens a broad field which requires legal regulations. In this connection, to a certain extent, waiver of a quantifying standard of admissible risks will not be possible. The designer and user of dangerous technology must be furnished with clear goals established by numerical data with respect to risk reduction. In Section 9.2, two admissibility criteria for technical risks will be discussed in this sense. The instrument used to render technological hazards transparent for the general public will be introduced in the form of a risk register showing all risks in plane-related form. This is suitable as a planning tool for improvement of existing conditions as well as for the planning of new technical installations.

9.1 Technology Assessment

9.1.1 Concept and Aim

The phrase "technology assessment" was coined in the United States in the 1960s. It is understood as an analytical procedure designed to estimate the effects of a specific technology on society and the environment in a comprehensive manner. This involves the determination of negative as well as positive effects in the short term and long term within the context of possibilities.

In the past, industrial development of new technologies was promoted primarily by private economic interests and was marked by the ideas of elitist elements. As a result, not all interests of the population were fully considered in specific cases. By way of example, the manufacture of detergents should be mentioned, which undoubtedly was profitable for the manufacturer and useful for the consumer. However, the hard tensides used in the 1950s in the manufacture of detergents resulted in surface foam formation on water which was detrimental to water sports and also created problems for the drinking-water supply. In addition, eutrophication, in particular, in the case of standing or slowly flowing waters, due also to the phosphate contents of detergents, required countermeasures. Here the remedy was the use of readily biodegradable anionic tensides and sodium-aluminum silicates to replace the phosphates. In the German Federal Republic, the legislature was able to exert considerable influence on the direction taken by the development of laundry detergents by means of the Detergent Law (1961) and the Laundry Detergent Law (1975), promulgated in the interest of effective water protection.

One can assume that in all major industrialized nations the legislative authorities will exercise comparable control functions in the public interest. As the tendency to voice criticism becomes stronger within the population, the decision-making process does not take place exclusively within the sphere of influence of the political parties but is at times influenced by vocal organized minorities. The measures taken in the example of laundry detergent have the character of consequential improvements. Such measures, however, are only conditionally in keeping with the premises of technology assessment. As a rule, an attempt should be made to evaluate economic, political, and social repercussions of technical developments in advance and to use appropriate measures to guide such developments in the desired direction. Not all that is technically feasible is also desirable. In 1937 the possible consequences of a multifaceted developing technology were examined by order of a U.S. Congressional Committee for the first time [9-1]. The report made at that time predicted with astounding accuracy, among other things, the development of television and its influences on society and industry. The influence of the automobile, street traffic, and aviation on the economic and social mobility of the population must be confirmed now as having been accurately appraised. Due to agricultural mechanization, a population shift from rural areas to industrial centers was predicted and there, in turn, accelerated effects of automation on the economic structure became evident. Had there been due consideration for the research results at the time, several negative effects could have been offset, but the start of World War II imposed different priorities.

It was only in the 1960s that the technology assessment discussion started again in the United States, undoubtedly because of intensified development and use of new findings and technologies. The Committee on Science and Astronautics of the House of Representatives of the U.S. Congress created the Subcommittee on Science, Research, and Development in 1963 and commissioned research, in particular, on the negative side effects of technology application. The concept of technology assessment developed from these studies [9-2].

The term "technology assessment" (TA) was used and defined for the first time in a report made in 1967 on the consequences and secondary effects of technical innovations. Though V. T. Coates [9-3] writes in 1974 that TA can and should be applied in the private as well as the public sector, realistic thinking will see the main responsibility, the main interest vested in the state as representative of society. The ground rules of a free economy are determined by competition which permits self-imposed restrictions only in rare cases. On the other hand, by means of legislation, government activities, and public subsidies, government influence is so strong that it can stimulate technological developments.

It should be the goal of TA to assist the decision-maker in arriving at a decision and propose possible alternatives beyond the mere yes/no evaluation. The decision-making aid extends beyond the purely technical sector to management and organization as well as to the area of health and legislation. In dealing with the latter instances, some authors state a preference for general evaluation of consequences or action analysis over a comprehensive technology–consequence evaluation which, in their view, would be too narrow in substance. It is the pluralistic task of TA within the context of safety science to identify and evaluate the potential effect of new technologies on the environment, on laws, morals, culture, and the economy.

In principle there are three different types of effects which should be determined within the context of the analysis:

—intended or desired effects;
—undesirable effects, that is, negative side effects;
—possible but uncertain negative effects.

The first type corresponds directly to the aim of the technology and the connected favorable side effects. If one were to use the example of nuclear technology to show the differences, the advantage of an alternate energy source should be mentioned first, together with practically inexhaustible yield under certain circumstances (breeder reactors). The incidental production of radioactive isotopes, which can be applied in various ways in chemistry, medicine, sterilization, measuring technology, etc., is also noted as a positive effect.

Undesirable effects, that is, negative side effects, must be considered in the case of radioactive fission products, in particular in the case of those with long life, which for whatever reason can no longer be applied to a useful purpose.

Finally, a third area of effects must be considered, among them mutations resulting from radioactive radiation effects of the smallest dosages as well as the frequently discussed consequences of major reactor accidents.

9.1.2 Areas of Application and Classification

TA analyses, executed as well as contemplated, can be assigned to varying concept classes according to different criteria. Examples will be used to provide a clear understanding of the system. At first, classification according to subject-related grouping is obvious: raw materials and energy sources, production technology, transportation and traffic, communications technology, humanization of the work environment, health protection, city planning and construction, defense, education and training systems, and "social inventions." It is evident that the concept "technology" is very broad and that it includes areas which are not safety-science subjects. Connections do, however, exist where they are not suspected, for example, an exhaust chimney in the vicinity of a skyscraper to be erected later or the consequences of ground subsidence and ground-water-level lowering on structures.

In principle, the approach of a TA study can be applied to all areas. If differentiation is according to purpose, which means according to the problem, the technology, or the project, the result is a TA typology which makes a distinction between problem-induced, technology-induced, and project-induced TA studies.

Problem-induced TA studies are directed toward acute or foreseeable problems in society, for example, the CO_2 enrichment of the atmosphere or ozone layer destruction in the stratosphere by fluorine-chlorine hydrocarbons and the resulting consequences. Here the research leads to consideration of so-called macroalternatives, which lead away from the original problem-solution.

Technology-induced TA studies deal with the varied consequential effects of applications of technology. Here society is confronted with possible effects and must decide whether and which of the solutions indicated it will, if necessary, be willing to adopt.

Finally there are project-induced TA studies which, as a rule, are based on a concrete case. In this connection, questions of location of airports, power plants, and the like should be recalled.

The studies thus classified are not unrelated. However, the problem-induced TA study is a contradiction of the basic idea of technology-consequence evaluation, at least when it deals with acute problems which the analysis should have prevented. However, one should not forget that a development could proceed in a way which was not foreseen or was considered improbable. It would then be appropriate to return to the TA studies at suitable intervals to check earlier decisions in the context of new findings, data, and time frames. This means that the study must be updated, which in turn means that it is within an additional aspect of the category of reaction-project TA studies. This category is based on a time-related subdivision. L. H. Mayo [9-4] draws a distinction between reaction and project analysis, namely whether the TA study is made before or after the application of the technology, and finally the reaction-project analysis as a combined long-term investigation which continues beyond the planning stage and includes the actual consequences. The investigation of eutrophication of water by fertilizers is an example of reaction analysis; the investigation of massive use of supersonic aircraft in private aviation is an example of project analysis.

For the purpose of additional experience with technology evaluation methods, retrospective analysis of historical cases is undertaken as well.

9.1.3 Methodology

Various attempts have been made to find a generally valid methodology for the execution of effect evaluation. The multiplicity of the questions, however, casts doubts on the success of such efforts. In general the results are simple procedure instructions, possibly in the form of a more or less roughly structured sequence diagram. Here, as a rule, criticism of system analysis begins, since analysis of complex systems cannot be executed adequately according to preset diagrams, especially since partial results influence the continuation. Literature frequently refers to the research results of MITRE Corporation dating back to 1971 [9-5], for example, which are comprehensive and therefore highly informative.

In principle the following steps are suggested by way of a sequence:

—definition and limits of the task;
—description of the technology complex to be evaluated (main, complementary, and alternative technologies);
—data and fact inquiry;
—identification of possible effects and spheres of action;
—consideration of possible action options from the vantage point of maximum benefits to society;
—consideration of exogenous influences on the application and effects of the technology concerned;
—conclusions and possible recommendations for the decision makers.

Not infrequently reactions, i.e., feedback will make it necessary to view the sequence diagrams as iterative processes. Figure 9.1 shows the sequence diagram selected by E. Jochem in 1976 [9-6] for the investigation of motorization and its effects.

In general the sequence indicated for the execution of a concrete TA analysis is not very helpful. In 1978 H. Paschen, K. Gresser, and F. Conrad [9-7], on the other hand, had undertaken to set up principles, i.e.,

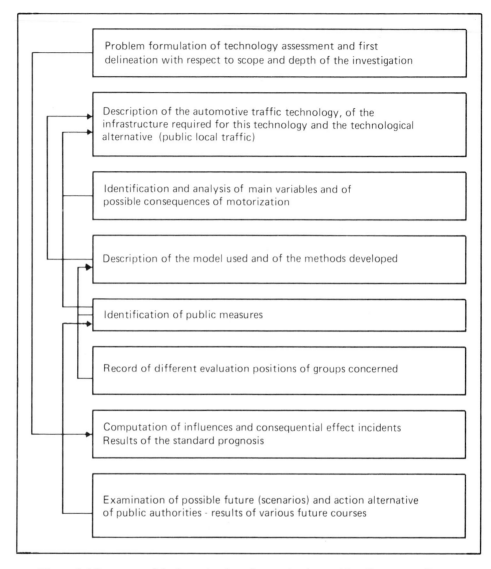

Figure 9.1 Sequence of the investigation of motorization and its effects according to E. Jochem [9-6].

guidelines to be observed in the planning and execution of the investigations:

—The project process must be rendered transparent step by step so that all who use it are able to reason through these steps.
—The public, and in particular those concerned, must be kept informed during the course of the investigation of intermediate results and

decisions as well as the reasons thereof at an early stage and in generally understandable terms.

—A maximum active participation in the TA process, at least on the part of those most directly affected, must be sought, since the situation-specific knowledge is distributed over all those who participate and are affected.

—TA processes are to be understood as argumentative processes where every answer raises new questions and the positions of the parties participating or affected can differ according to location to a considerable degree.

—Individual aspects must not be viewed separately. Therefore TA investigations must be approached in the manner of interdisciplinary cooperation.

—TA studies can be fairly voluminous and costly. The first approach should evaluate with results, which in the end may not be more informative, before the decision concerning costly optimization or simulation models is made.

—Viewed in their entirety—and this results from what has been stated above—TA processes should be organized as transparent, participant-oriented, and argumentative processes, with tools basically known and merely requiring adjustment as well as those which remain to be developed.

With respect to the tools, MITRE Corporation [9-8] provides a list based on an analysis of American studies, which include the following work methods:

—(computer) simulations,
—matrix techniques,
—trend projections,
—conventional statistical methods,
—problem-oriented target analyses,
—cost-effectiveness analyses,
—operations-research processes,
—behavioral science experiments and observations,
—group consensus processes, including the Delphi technique,
—historical analogies.

H. Paschen *et al.* [9-7] add

—cross-impact analyses,
—morphological methods,
—scenario writing,
—relevance tree, relevance matrix process.

The method finally adopted is up to the judgment of the study group involved. None of the methods are TA specific but are borrowed from other fields of endeavor. The individual choice will be determined by the expectations, the costs, and the value of the part problem concerned. The study of various TA investigations may yield tactical indications for the solution of special tasks and may provide ideas. A uniform method will, however, hardly be found.

9.1.4 Boundary Problems

Technology-consequence assessment is not new in principle. Investigations of this type were made before the term "technology assessment" was coined; but compared to those in the past, the scope of the investigations has expanded. At present the thrust of the studies is aimed at the unintended indirect effects of technology applications which may possibly become apparent only in the long term and which may cast doubts on the benefit of intended "primary" effects.

To this extent the inquiry has changed, as well as those who make the inquiry and decide on the boundary. Compared with feasibility studies, cost-effectiveness analyses, etc., which have technical and economic motivations, in TA studies social and ecological considerations dominate at present. The data and fact survey must be as comprehensive as possible, and the determination of effects must be comparable. There are no boundaries set for the extent of the survey, unless financial, personal, and time-related considerations require a limitation of the study. This boundary, however, amounts to an evaluation which reveals the individual position dictated by interests. It should be decided upon in consensus with the social groups affected so that the results of the study meet with general recognition.

Boundary decisions are exceedingly critical. They must be checked and documented with care since they could have considerable effect on the statement value of the study. The MITRE study [9-5] explored various possibilities and discussed specifically reducing the number of social groups "impacted," the limitation of the categories of consequences to be investigated, waiver of higher-level consequential effects, reduction of the time span in the case of future projections or waiver of accuracy of results in the case of solely qualitative evaluation.

In this connection independent analysts will be viewed with reservations when it comes to neutrality and objectivity, since they, too, are members of society and influenced by society. They will act under this influence, though perhaps unwittingly. The danger of bias is greater, however, in those representing interests, who as members of an investigative team may, in situations of conflict, be inclined to suppress formulations of questions or omit certain aspects, considering them nonessential, in order to guide the study in a desired direction. This means that the fairness dispute, where participants and those affected present their arguments, gains in significance. It undoubtedly may prove practical and sensible

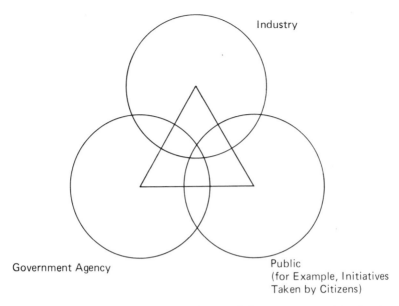

Industry

Government Agency

Public
(for Example, Initiatives
Taken by Citizens)

Figure 9.2 Equilibrated relationship of forces of the partners participating in the TA (the circle diameters correspond to the publicly stated arguments). Source: D. Altenpohl [9-9].

during the later course of the TA process to reflect on the boundary or, after a certain time span, to rework the study with due consideration to the time element of external influences. This knowledge, however, results as a matter of course and is not so much due to the position of the analyst.

Thus far the majority of TA studies have dealt with ecological effects in which the civil population has been particularly interested, and this position has been supported by the state. Early and wide participation by the public is desirable as well as helpful in dealing with boundary problems. D. Altenpohl [9-9] considers the ideal case to be an equilibrated relationship of forces among partners in technology assessment, and he provides an illustration (Figure 9.2). The diameter of the circles symbolizes the extent of publicly stated arguments. The equilibrium is frequently disturbed. Here it is uncertain whether Altenpohl's special representation (Figure 9.3) accurately reflects the facts in the example "nuclear power plant" or whether the magnitude ratios and intersections should have been selected differently. The illustration can, however, be useful as a clarification of the statement of principle.

Apprehension with respect to possible competitive disadvantages or of revelation of state secrets may speak in favor of restraint in public reports of arguments or in the release of available data and facts. Thus, under certain circumstances, there are bound to be incongruous demarcations, i.e., misevaluations, unless it is possible to change the procedure by

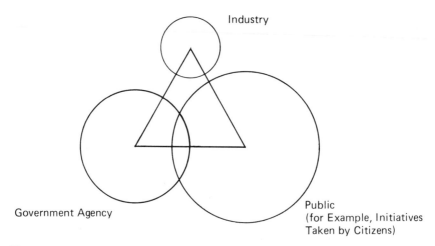

Figure 9.3 Disequilibrium in a TA study due to domination by public authority (example: construction of nuclear power plants). [9-9].

convincing arguments and motivation. In the event that specific information is actually unavailable, it should, if possible, be obtained in the context of separate research programs if it would serve to provide essential support for the TA study statement. In practice there is invariably a constraint to effect demarcation. If, however, there is a great readiness to give in to this pressure, little will be left of the technology assessment proper.

9.1.5 Effectiveness

Worldwide preoccupation with technology consequence assessment leads to the conclusion that it is viewed as a strategy resulting from a general need. Though it is not possible to demonstrate a general methodology, current investigations, as compared to methods of the past, show a more systematic approach in the way technological effects are comprehended and evaluated. Since the solution of difficulties in the method is found only from case to case, this leads to the possibility of adjustment which permits doing greater justice to individual tasks; but, nevertheless, there has been no lack of criticism of the TA concept.

As for this criticism, it is sometimes claimed that TA studies inhibit technological development and its motivation. This statement, however, is accurate only when the innovative forces risk being lost in sociological or ecological "dead-end streets." Indications of alternatives and choice of actions can frequently be considered particularly positive and stimulating.

TA studies are intended to facilitate good decisions. They include collection, analysis, and structuring of information as a basis for entrepreneurial or political decisions. The difficulties involved in the collection of data and facts and the conflict situations possibly arising in the course of the evaluation are undoubtedly weaknesses of the concept, in particular when expectations are too great. In the purely technological area, the attainable degree of accuracy is relatively high. Sociological and ecological subjects, however, require greater caution due to the absence of quantification.

Readiness to serve on the part of those concerned and the multidisciplinary composition of the research group offer the best guarantee for objective results and recommendations for action. Political pressure, which is present as a matter of course in the majority of TA studies, accounts for the circumstance that results of the study are always exposed to criticism and attacks in areas of controversy, such as the economy, the state, and society.

Technology assessment deals with interaction within man–machine–environment–systems and its time linkage in a manner quite different from other disciplines. Viewed in the context of the short and medium terms, social effects and reactions which deserve particular emphasis in safety science are dealt with. There are, however, often difficulties with doing justice to the problem, since connections are frequently unknown and corresponding "measuring concepts" are not extant. It is hoped, however, that the desired information basis can be created by means of social indicators, but this development is only beginning.

In several fields, technology assessment already provides satisfactory results. There are uncertainties, however, with respect to future projections, the same uncertainties which are part of any prognosis. Reliability depends essentially on the degree of accuracy of the basic information. Statement certainty is reduced according to the number of nonquantifiable effects of social areas which must be included. Here, in particular, development processes unintentionally slip easily out of control, since functional connections are little known. Extrapolation and trend evaluations are subject to time-related influences of various types which generally require representation of large-scale TA projects in the form of "continuous updating." The question of institutionalizing technology assessment is under discussion in many countries. The United States has been (and presumably still is) the only country to create a special TA institution for its legislative branch by establishing the Office of Technology Assessment (OTA) [9-10]. Thus far, other countries have reached only a rudimentary stage.

9.1.6 Example of a TA Study

The considerations which lead to technology assessment in a concrete case will be illustrated by the example of "energy sources under the ocean." They are based on studies supported by the American Office of Exploratory Research and Problem Assessment in the Research Applications Directorate of the National Science Foundation (NSF).

Within the scope of the RANN program (Research Applied to National Needs—Grant Number GI-29942), the opening up of energy sources in the outer continental shelf of the U.S. was investigated. The results were published under the title "Energy under the Oceans: A Technology Assessment of Outer Continental Shelf Oil and Gas Operations," 1973 [9-11], [9-12].

The technology consequences subject to evaluation are designated as OCS technologies (Outer Continental Shelf). Basically the problem was examined by eight experts in an interdisciplinary study group within a period of 20 months. The participants were engineers, physicists, biologists, a policy expert, and a lawyer. An Oversight Committee including seven experts from outside organizations supported the work of the group.

In its first stage, the TA study was set up as a technology-initiated analysis, and alternative procedures were examined in the sense of a cost-effectiveness analysis. On the basis of data and facts collected on the subject of possible energy supply alternatives and subsequent evaluation of such data and facts, it was concluded that no substantial change could be expected by the end of the 1980s. By 1985 oil and gas would have to cover 70% of U.S. energy needs. Coal, terrestrial heat, solar energy, hydrodynamic power, breeder reactors, nuclear fusion, etc., could not yet replace oil and gas to any noteworthy extent, if at all. Even intensified research would hardly serve to solve existing problems.

These conclusions motivated a change in the definition of the task which had made the second TA study stage appear to be a problem-initiated analysis. Now the determination of negative effects of the opening up of the outer continental shelf, and finally the exploration of the causes, moved into the foreground of the endeavor. In this case too, it should be possible to counter negative developments and to eliminate drawbacks by utilizing the knowledge and findings which had been acquired.

Intensified exploitation of gas and oil required, on the one hand, larger investments by companies and, on the other hand, promised profits as well. Risks to be expected in connection with regional employment and economic development prospects had to be calculated. Possible effects on the environment are not difficult to imagine now, in view of the accident in the Ekofisk field in the North Sea and various incidents with supertankers during the recent past. Effects on fishing and tourism may be mentioned in this connection, and legal aspects of national and international scope are certainly not the least of the problems.

Data and facts serving as basis for the analysis and recommendations were culled from literature or acquired from inquiries addressed to experts. Of particular significance was a supplementary cost-effectiveness analysis of three alternatives with respect to national self-sufficiency and security. Here the opening up of the outer continental shelf was contrasted with supply via the trans-Alaska pipeline system and imports. The team conducting the investigation arrived at the conclusion that intensified exploitation of the United States outer continental shelf is preferable to the two above-mentioned supply possibilities. Physical technology problems could be solved by application of familiar techniques, and it was indeed possible to eliminate points of weakness by suitable application of existing equipment and devices. The majority of changes had to be effected in the area of social technology (rules, regulations, provisions, etc.). A large number of concrete recommendations pointed the way to socially optimal solutions of problems.

9.1.7 Critical Evaluation of TA

It would certainly be wrong to introduce critical evaluation of TA by the question of whether it makes sense. The sense and purpose are uncontested. However, the question of whether TA is always feasible in individual cases and whether it also leads to useful answers with dependable statement values is not inappropriate. TA does not make decisions, but finds potential problems, provides data for evaluation, and shows weak points and possible alternative solutions. This, of course, should be done with the maximum possible objectivity and with scientific neutrality, though this is easier said than done.

For reasons of practicability, an in-depth analysis of the consequences of a technology requires selection according to importance. The determination is up to the judgment of the compiler and therefore inevitably introduces subjective evaluation into the report. Linkage of effects of several innovations is undoubtedly possible. Selective views of effects or complete neglect of these effects reveal the views of the analyst and possibly the interests of his employer.

The manner in which the problems are defined and alternatives shown already indicates the direction for subsequent work. Critical groups among the public fear that the technical-scientific elite offers false advice. Therefore, from the very start, their concept of values must be compared with the views of those directly concerned. Due to this aspect, representatives of underrepresented groups should follow the progress of the investigations with a critical attitude. The experts do not consider this type of involvement feasible, since TA deals especially with complex processes for which, as a rule, the layman lacks the necessary understanding and insight. In addition, only well known facts and special interests related to the present would be introduced into the discussion. The main concerns of TA analysts, however, are long-range consequences and detection of unexpected effects. Legislators who have the power of decision are, as a matter of course, exposed to political pressure from public groups and are subject to controversial influences. They should not be made even more insecure, but should receive the expert advice which will place them in a position that will enable them to fulfill their political function.

Who should be viewed as representative within society or empowered to represent the interests of minorities? TA results are not free of value judgments. Vociferous minorities, financially powerful economic cartels, or organized majorities are readily suspected of using the conflict of interests to assert their views. Results should also be viewed against the social and cultural background of society.

Experiences during the recent past show that difficult tasks cannot be solved without the social aspect. Multidisciplinary and interpersonal handling makes TA results appear "objectivated." An advantage which should not be underestimated with respect to this systematic analytical procedure, as compared to pure assumptions and fears, is the investigation which can be *post facto* executed in detail on the basis of the indicated assumptions, selective decisions, and value positions. This does not exclude that separate evaluations can, in the case of the same initial position but fundamentally differing conceptions, lead to differing results.

In view of the difficulties inherent in TA, the objection of the inhibition of technological development by TA can be rejected as unfounded. As a rule, what occurs is not obstruction of a technology, but possibly more cost-effective or ecologically better solutions.

9.2 Risk Acceptance

9.2.1 The Irrationality of Human Risk Behavior

The evaluation of human risk reaction towards technology raises the question of defensible risks which can be accepted in the course of utilization. For some time this has been the subject of intensive discussion on the part of those active in the humanities, theology, and the engineering sciences. "Defensible" is intended to mean that developments, conditions, and decisions can be justified by those responsible vis à vis society in such a way that refutation of criticism and predominant agreement can be expected.

In a pluralistic society the obvious idea is not to refer for justification to recognized principles of ethics or religion, but to establish a connection with guidelines expressed in the actual conduct of society and to prove that they are observed. In other words; standards according to which an activity can be accepted or rejected should be obtained from the empirical regularity governing the conduct of man, which the social sciences are seeking to detect. In concrete terms this means that standards of acceptable or no longer acceptable risks would have to be derived from the empirical laws on which human readiness to accept risk is based.

Such attempts at statistical expression of the patterns of acts of will in individual and social life have existed since Bentham [9-13], Quételet [9-14], [9-15], Buckle [9-16], Wagner [9-17], and other researchers of the last century, in particular those of the utilitarian school. These statisticians aim to create a catalogue-type compilation of basic motives of man and to make an empirical determination of the laws governing formation of intent, thus—to use the words of Quételet—to work out a physiology of human society.

If this "physiology" were available, it would actually be possible to exactly predict, in the most varied cases of actions vis à vis society, in particular in the imputation of risks, acceptance or rejection of those risks. It would be possible to almost demand them as a physiological reaction of the social body in the sense of a so-called "objectivation of decision processes." A reference to morality and law would hardly be necessary.

It should not be denied that attempts to rationalize individual and social opinion molding—in particular attitudes toward risks—in this manner are successful within certain limits. The "normal" motives in those motivation-catalogues, such as health, welfare, esteem, rest, etc. for the individual, for the family, etc., are within certain limits the principal motives which opinion molding would like to follow with sensible rules. This will hold true all the more for humanly simpler motives and simpler practical conditions and rules for giving satisfaction. It will be possible to state with certainty that in the past man, in dealing with his technology, intuitively and without specific safety science argument, asked for the measure of safety for life and limb which corresponded to the measure of safety with respect to "natural" hazards in his environment. Compared to present demands, the safety level was low. However, the unwritten rule assumed, without articulation of this thought, that hazardous technology affected the life of man by influences deemed positive, that it was advantageous in particular with respect to economics.

Nor is it surprising to find that man's readiness to accept risks in the case of voluntary participation in dangerous technical activities is greater than in the case of risks imposed by the state or by society—at least as long as the understanding of "readiness to accept risks" does not include submission to coercion. A great measure of confidence in one's skill or luck is responsible for this behavior and almost all human beings are inspired by such feelings in freely chosen endeavors, implying better chances than those stated by objective statistics.

There is a question here, however, namely whether the greater readiness to accept risks when the activity is voluntary actually stems from such self-deception. This is claimed by the philosophical theory of the so-called ethical intellectualism and in particular the utilitarian school. For the most part these self-deceptions are possibly superficial, psychological rationalizations of the pleasure derived from free choice.

In the event that the state or society imposes risks on the individual, this joy of freedom motivating readiness to accept risk is missing and more so with increasing superiority and anonymity of the imposition. This point will have to be looked at more closely later on.

On the basis of statistical data on risk of death in the United States, C. Starr [9-18] tried to provide a quantitative view off the difference in readiness in the case of voluntarily accepted as compared to imposed risks.

In his model ("revealed reference model") he measures risk readiness by retrospective data on actual risks in the past, which he understands as a result of an adjustment process of technical risk and risk readiness in the course of the maturation of technologies. According to these investigations covering the United States, in the case of free activity such as participation in road or air traffic, risk readiness as high as the overall mean of the natural death rate must be expected. According to Starr, it drops to one thousandth of the value in the case of nonvoluntary participation in technical systems, even if, as in the case of an electricity network, it is hardly possible not to be connected.

In his model C. Starr saw a socio-psychological instrument for the evaluation of new risks by means of exact data of the past on the basis of the following premises:

—Due to the already mentioned adjustment process of risk and risk readiness, long-term statistics on fatal accidents reflect, in the objective risk values, the attitude of the population toward those risks in the past.
—The retrospectively ascertained social attitudes with respect to "old" dangers can be extrapolated into the future to predict attitudes toward "new" dangers with an objectively equal extent of risk.

It was found, however, that application of the model method to data of the past shows model-contrary inconsistencies in the population's attitude toward risks. This was explained by means of the difference between objective risk, which is the deciding quantity in Starr's model, and perceived risk. Neither does the risk perception remain constant for the "old" risks nor can the difference between objective risk and subjective risk perception be neglected in the case of "new risks". Though the "revealed reference model" is therefore subject to criticism as an acceptance-prediction model, the risk-acceptance comparison which can be obtained with the help of this model is suited as an argumentum ad hominem in public discussion.

The problem of the difference between objective risk and subjective risk perception is avoided by the so-called psychometric models which aim to directly record subjective evaluations as a basis for decision-making processes. The "expressed preference method" [9-10], which uses opinion polls, did, however, show the change in the way the population evaluates technical developments.

The term "quality of life" was coined as a concept comprising the human aspects of technology evaluation. It is, however, not yet provided with sufficient substance to serve as an instrument for specific qualitative or quantitative evaluation of new technology and its risks. Hierarchies of requirements and hazards for profit and loss evaluations in view of humane aspects were established especially in connection with Maslow's [9-20] motivation and personality theory. In the sequence of their assumed importance the following hazards were placed in the hierarchy: danger of premature death, danger of illness and invalidity, curtailment of basic life expectancy, premature exhaustion of natural resources, prejudice to social security, prejudice to monetary status, prejudice to human bonds, prejudice to egocentric factors, and prejudice to self-fulfillment [9-21]. There is an effort to set up more specific danger and requirement catalogues for evaluation analyses and to show their sequence in the form of a diagram scale.

The latter (used only as examples) contemporary decision-theory attempts at benefit-risk evaluation apparently refer back to the above-mentioned efforts, directly comparable, with respect to method, to those of the utilitarian school of the last century: enumeration of effective motives (requirements, hazards) in the form of catalogues and statistical, empirical determination of the laws governing their efficacy (arrangement in hierarchies, in scales).

Here it must again be pointed out that the benefit of all these efforts to create objectivity in evaluating decisions on socio-empirical foundations must be acknowledged for a specific validity range of that which is "normal" with limitations with respect to space, time, and substance. On the other hand, the difficulties (inconsistent risk behavior, the difference between objective risk and risk perception, the difference between risks assumed voluntarily and involuntarily, rapid change of risk evaluation, etc.) indicate more serious problems than developmental flaws not yet corrected by scientific progress. As contemporary critics of the utilitarian school already objected, it did not allow for the spontaneous, to a large extent irrational nature of human decisions: "What is one to do about the millions of cases which prove that man knowingly, meaning with full comprehension of true advantages, ignores the latter and follows a different way, toward risks, into uncertainty, not forced to do so by anyone or anything" (Dostoevski [9-22], compare also [9-23]). Not even the basic assumption, that risks are accepted according to risk/benefit ratios, to be determined, for the sake of advantages, which can be catalogued proves right; sometimes risks are taken for their own sake. That this behavior is not limited to rare, abnormal personalities is confirmed by the example of widespread enthusiasm for the motorcycle which shows no objective advantages over driving a car, but is certainly more dangerous. The reason for this enthusiasm, which at times endangers life, is given more or less unanimously by the mostly younger riders: the "marvelous feeling of freedom." Man wishes to be free and to prove to himself that he is free.

For this reason a "systematic analysis or a consistent theory of risk acceptance" in the exact sense of a rational system of humans, respectively social motivation and readiness to take risks, is not merely "wanting completely thus far" [9-24]. It cannot be expected either from future progress of empirical social sciences or social psychology—neither as a computation system for decision-making nor as a yardstick system for "democratic" justification of demands to take risks—nor as an instruction and guidance instrument of society. The theory of such risk/benefit motivation systems must fail—aside from the reasons already given—because the motives (benefit, advantages, disadvantages) not only have not been comprehended in catalogues, but are not comprehensible. For "advantage" frequently means nothing else than tautologically "that which is approved of one's free will.*

The originally intended plan was to obtain a system of guidelines for "ideology-free", "democratic" justification from the empirics of social conditions. If beyond this, on the basis of such a theory-attempt it were attempted to eliminate the irrational in man by means of enlightening him on his "true advantages" and by accustoming or finally compelling him to promote them, the results would be destructive. Rationalized reality leads to dreadful boredom and stimulates within society abnormal desires which serve to destroy the enlightened rational structure. The rapid change of social preferences which creates such confusion for empirical social psychology is then the least of the evils.

9.2.2 Government Standardization as a Moment of Rationality

Government authority can neither be merely the executive agency of the irrational inclination toward risk-behavior actually present in society nor feel called upon to eliminate the irrational in society. On the other hand, to fulfill its function it must protect citizens from risks, sometimes with and sometimes against their will, and must altogether adopt rational attitudes toward risk. This holds true to a special degree in the case of a democratic, constitutional state where governmental authority must furnish political justification to democratic institutions and where administrative actions are subject to review by the courts.

These two contradictory trends can be reconciled only by the state granting the individual citizen as much freedom as possible. He must also be permitted to take objective, freely accepted risks. In this sense, for example, the German constitution limits "free personality development" only insofar as it violates the rights of others and is in conflict with the constitutional order or the law of morality (Article 2 Para. 1).

With respect to freely accepted risks, which do not threaten the rights of others, nowadays the state remains largely within the limits of Steinbuch's [9-24] "three dimensions of acceptance of (technical) risks":

—reduction of safety hazards of the means used (technical dimension),

---

* Compare again the apt passage from Dostoevski [9-22]:
"One's own, freely willed and free decision, one's own though most peculiar whim, one's imagination though sometimes stimulated to the degree of madness,—that is the self-same overlooked, the same most advantageous advantage which consistently causes all systems and theories to go to the devil."

—enlightening concerning existing dangers in the sense of correction of risk perception (educational dimension),
—sharpening of a sense of responsibility (ethical dimension).

Social protection (health insurance, social insurance) is preserved to a maximum extent if in freely accepted risks the case of damage happens to occur. In view of its protective function from more or less unavoidable risks, the state will have no choice but to follow the premise of acceptance behavior which enables society to deal with these risks on the level of the spirit. For behavior toward the risk the following helpful gradation following W. Stoll [9-25] could be devised:

—Less than one fatality annually per 100 million persons (for example death due to fall of a meteorite): The danger is not felt consciously.
—One fatality annually per 1 million persons (for example, death due to lightning): accidents of this type occur within the information range not expanded by the mass media. The threat is shifted aside, rather than reduced, by means of advice based on proverbs.
—One fatality annually per 10,000 persons (for example, due to fire, traffic accident, capital crime etc.: To counter these threats familiar to everyone in everyday life, the citizen demands organized, primarily governmental protection (police, fire fighters, etc.).
—One fatality annually per 100 persons (for example, due to illness): Risks of this type are countered rationally by personal measures and irrationally by suppression.

Efforts made by the state to reduce risks can either be geared more toward reduction of social costs or more toward avoidance of major personal suffering by the individual. In the first instance, for example, major funds will be used to combat illnesses like colds with serious effects on economic activities; in the second instance, to combat multiple sclerosis which is comparatively insignificant with respect to social costs. A clearcut *a priori* decision is not always possible in the individual case, as will be shown later. The real problem—and this case arises for risks due to technical progress—is the boundary of the free action sphere of one party against the protection claim of the other. For a long time the state operated along the above-mentioned unwritten rule of society which demanded the same measure of protection from harmful side effects of industrial activity as it believed to have from natural threats. Rapid overall progress of technology and technical civilization has reduced natural threats in our geographical area to a previously unimaginable degree, yet at the same time it has created new types of threats, some of major proportions. The atomic bombs over Hiroshima and Nagasaki have become symbolic in this sense for many people. Beyond this the progress of scientific knowledge has, after all, brought about the awareness of long-range dangers, so that, for example, the question of whether a large share of cancer cases may be the long-range effect of exposing the environment to new chemical substances can be formulated.

On the grounds of the heightened awareness of the risks brought about by technical civilization and probably also due to a higher instinct, the demand was formulated to the effect that the protection claim of one party had absolute and unlimited priority over the free action possibility of the other and that the imputation of risks was not permissible. It was stated, for example, that determination of risk boundary values was in violation of Art. 2, Para. 2 of the Constitution which grants everyone the right to life and physical inviolability.

State Standardization Based on Ethics and Law. This statement calls for reflection on ethics and law to which it lays claim. In fact, if there were a choice between "risk" and "no risk", then it would not be clear why one party should have to accept a risk and pay for the activities of the other. This, however, is just not the case. Rather, a choice must be made between one risk and another; risk-free conditions cannot be created. Undoubtedly it would have been possible to do without technical civilization and its risks—and this could still be done. But at what a price! No one will want to claim that the risks for every individual would be reduced by this decision. The many "alternative" schools of thought, apparently based on the idea of a "gentle" technology, which would be comfortable as well as free of risk or at least incomparably less risk-laden than the present technology, are, obviously, wishful thinking. In reality man wants the benefits of technical progress. The demand appears to be justified. This holds true in particular with respect to the Third World. With its increasing population it can hope for continued existence only if the industrialized countries provide the basis by means of modern technology.

Stated in abstract concepts of ethics (compare, for example, R. Lauth [9-26]): state and society could not exist in a legal relationship only, where the good are protected from the bad by prohibitions and threats of punishment. Between all citizens there must be an action relationship, namely, the freedom to follow goals and produce the means, which while serving various purposes, may first promote the self-interest of the individual but at the same time benefit the community as well. The legal relationship must not be allowed to throttle the action relationship.

Thus the claim by the individual that the state provide basic freedom from risk impositions must be rejected, and not only because the claim cannot always be complied with technically. Even where this is possible in some points, it must be refused on moral grounds because of the connection with risks to others or to the community as a whole. In fact the German Constitution expressly permits legal intervention in the rights guaranteed by Article 2—at least as long as they are not affected in their essence (Article 19, Para. 2).

A classical example of the right to risk imputation claimed by the state and of the altogether surprising risk acceptance by society is compulsory military service.* This example also shows clearly the fundamental difference between state imposition of *fatality risk*, no matter how great, or a *death sentence* set for a specific day and to be implemented without fail. If the existence of states depended on this sacrifice by man again and again, the struggle of all against all would be preferred to the endurance of this torment.

Thus, if the demand for zero risk must be rejected as a nonsensical, absolute demand in view of technical-practical conditions prevailing in this world, while the basic claim of the citizen to protection from dangers remains, the measure of protection, and the imposed risk can be determined only in concrete instances according to the principle of the relationship between means and ends. The limitations of our abilities only permit that we transform into reality some of that which is feasible in view of outside, factual reality; in this case the moral law commands selection of the best possible alternative. Between equally good or, if the occasion arises, between only probably equally good means and ends there is free choice. What will determine our choice is morally a matter of indifference [9-26].

With this demand for safety to the extent possible, and for risk impositions only to the extent required for inevitable purposes ethics is to be satisfied. The Constitution does not attempt to make a more specific determination of the boundary between the inviolable right to life and physical safety and the surrounding area which is open to legal intervention by introduction of additional general concepts.

The following should be pointed out for the sake of completeness: in principle there is a claim to uniform distribution of the expected risk-burden to all citizens. However, for moral and inevitable pragmatic reasons (women and children, respectively, important public servants, etc.) this principle can be restricted or become impracticable due to the force of circumstances. This is demonstrated by the previously cited example of military service. In the case of technical risks it does not, as a matter of course, seem ethically objectionable if, for example, the neighbors or employees of a major technical plant are expected to bear a higher risk—quite aside from a higher acceptance-expectation in the case of freely assumed risks.

State Standardization Concretized by Technical Language. Reflections on ethics and legal dogma generally lead to mere verbal requirements in

---

* This matter is not affected by the fact that the Constitution does not impose this obligation without limitations. The right to refuse military service is linked to moral scruples and not to concern for one's welfare.

conjunction with the principle of the ratio of ends and means. The latter imposes the need for a hierarchical order of means and ends in free technical development and safety with respect to its side effects. This opens a broad field between "no danger" and "too dangerous," between "no risk" and "risk too high," which requires positive legal regulation.

Positive regulation of risk, which can or cannot be expected to be assumed, can, as long as quantification of risk is waived, consist more or less only of the assignment of authorization for arbitrary decision by one authority or another: either a range of action is given to the administrative authority for this broad field, or it is regulated by unspecific legal concepts with compliance subject to full examination by the courts, or the lawgiver decides individual cases by unequivocal regulation.

If the decision is to be objectivized and placed on a general and definite basis, it will not be possible to waive a quantifying standard of the risks up to a permissible degree. Mere classification of risks is an inadequate guarantee of equal treatment. The laws on which technical planning and construction of individual aggregates are based are expressed in the exact language of mathematics. The designer and user of dangerous technology must be provided with clear goals set forth in figures.

The preceding chapters have shown that present-day science has acquired or can acquire the ability to furnish quantitative determinations of risks to life, health, and material goods. Thus a general setting of risk-reduction aims (technical dimension) as well as correction of erroneous risk perception (educational dimension) becomes possible.

By means of analytical-mathematical or empirical methods, results which provide absolute data on the risks of a technical installation or a technical system can be obtained. The statement of absolute risk data could, however, under some circumstances represent an insufficient indication. The data, for example, on accidental death risk per person and per working hour in mining, namely 0.0000003, represent accurate but not very helpful knowledge. If, however, it is known that the risk of death while working in a mine is 30 times greater than in the case of professional activity with the health service [9-27], then on the basis of this relationship meaningful decisions on the social or industrial level can be made.

Here the necessity to introduce a measure related to a basic level of human risk becomes apparent. This will be done subsequently. The ratio of existing risks, or of those to be accepted, to this quantity, which means the relative risk with the dimension of a figure, appears better suited for communication and as a basis for decision.

As remains to be shown, there are serious differences in the risk level of various areas of life and work. Government standardization must aim for reduction and uniformity within the range of possibilities which have their limits in economics and the state of knowledge. On the basis of risk quantification the following strategies for government standardization are possible and they can also be linked:

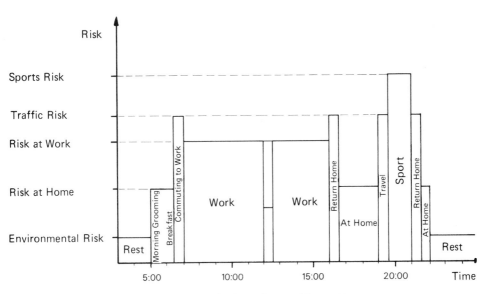

Figure 9.4 Typical employee risk profile.

1. It must be the aim of all political safety efforts to design living conditions in such a way that the additional risk due to technology for the predominant part of citizens does not exceed a determined absolute basic risk R_{ag}. To guarantee this, the first step will be to ascertain actual risk load of the inhabitants of a region, in the vicinity of a technical installation, etc. The risks run by the individual according to life style can be shown, for example, by a risk profile plotted over the time axis. Figure 9.4 shows a typical employee risk profile. The annual mean risk of the person concerned depends partly on the extents of objective risk (ordinate values) of the various risk sources and areas (home, environment, traffic, workplace, sport) and partly on personal exposure times to these individual risk sources (abscissa segments: shift duration at work, travel duration in street traffic, etc.). The exposure times given within a population can be gathered from so-called time budget analyses. When the annual mean risk is weighted with the statistical distribution of these exposure times in the population concerned, a distribution function of the annual mean risk is obtained. Figure 9.5 shows the typical course of such a distribution function.

This immediately shows how high the share will be of those of the population observed, whose annual mean risk due to technology is smaller or greater than the absolute basic risk R_{ag}. In the same manner the load due to a single risk source, as for example a concrete technical installation, can be ascertained from comparison of the two annual mean risk-distribution functions—once with, once without the installation concerned—and thereby an admissibility criterion can be obtained.

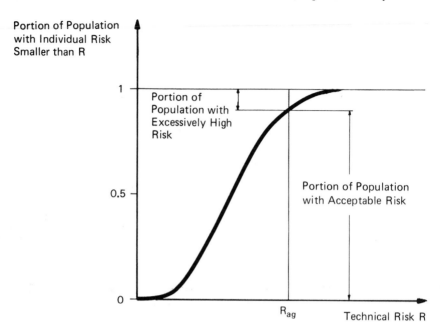

Portion of Population
with Individual Risk
Smaller than R

1

Portion of
Population with
Excessively High
Risk

0.5

Portion of Population
with Acceptable Risk

0

R_{ag} Technical Risk R

Figure 9.5 Typical course of an annual mean risk distribution function.

2.   One strategy which is oriented less toward personal life style and its
exposure times but more toward a single technical installation and the risk
field which is created in its vicinity and which can be registered in an
inventory, is the following: For the admissibility criterion for an installation
it is stated that the "impressed" risk field $R(r)$ caused by the installation in
its vicinity (see Chapter 2) is everywhere less than the determined,
maximum admissible risk R_{adm}.

$$R(r) \le R_{adm} \text{ for all locations } r. \tag{9-1}$$

It seems necessary, however, to weigh possible incidents with a particularly
extensive loss level more strongly. The threshold value for major incidents
in our geographical area can be considered to be about 100 fatalities. In the
event that a major technical installation causes this threshold value to be
exceeded under certain circumstances, then the admissible death risk for
the individual should be reduced in a ratio inverse to the number of those
possibly affected. This means that in the case of an expected loss of 1000
fatalities the admissible risk value may be lowered to only $R_{adm}/10$, of
10,000 fatalities $R_{adm}/100$ and so on (Figure 9.6). In this overall formulation the procedure was as though the individual death risk for every person
within the population considered was the same. In the vicinity of technical
installations the individual risk R is, however, as a rule, a function of the
site r.

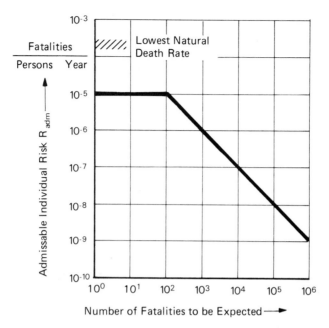

Figure 9.6 Admissible individual fatality risk depending on the number of expected fatalities.

In addition to the limiting condition of Equation (9-1) concerning individual risk, therefore, a second condition for the population risk R_P

$$R_P = \int R(r) \cdot B(r) \cdot df \leq 100 \text{ persons} \cdot R_{adm} \qquad (9\text{-}2)$$

will have to be added (see Chapter 2). It integrates the (impressed) risk $R(r)$ weighted with the population density $B(r)$ over the vicinity of the installation and compares this integral risk with 100 times the admissible risk R_{adm}. In the exceptional case that in the vicinity of an installation the considered risk is the same for all inhabitants there, this limiting condition for the integral risk will lead back to the overall formulation.

In the event that the limiting value for individual risk or for population risk in the interior or in the vicinity of a technical installation is not maintained, it must be improved by system conservation measures such as maintenance and repair or by measures altering the system. The type of measure taken depends essentially on economic aspects.

3. A third way is limited to separately giving actual values of the risk of installations, fields of activity, etc., and to establish a ratio. Cost-effectiveness analysis can then show which risks in this group can and should be reduced next. Into this, semi-empirical fictitious equations are entered, for example, the equation of the radiation dosage of 1 man-rem

with property damage of $1000 in the USNRC Regulatory Guide 1.110 [9-28].

This strategy has the advantage that it leads to direct risk reduction at comparatively low cost—even if the risk is possibly already below an admissible boundary value. The problem, however, of defining the benefits and costs entered into the analysis cannot be solved without arbitrariness. On the other hand, the attempt which may be necessary under certain circumstances to express the loss due to fatal accident in terms of money [9-29], [9-30], with its uninhibited calculating character ("Dollar per fatality") may not be a means toward stimulation of public acceptance.

9.2.3 Basic Risk Values, Admissible Risks

The ethical admissibility of risk impositions by the state is based on the premise that social conditions without any risk are impossible to create. It is necessary to demand the best possible, indeed, without wanting to foresee every single human situation, certainly in view of the various hazards to the individual and not only in view of social costs.

The actual risk in various areas of life and work is easily stated on the basis of statistical surveys, for example, in Table 9.1.

In this connection it should be noted that the available statistical data must be viewed with caution since thus far no uniform system to determine technical risks in areas of life and economic sectors is available. To create order in this area is undoubtedly an important project.

The greatly varying risks in the areas listed in Table 9.1 provoke the question of whether there is a base measure of risk which would serve to

Table 9.1 Risk values per person and exposure hour for fatal work accidents, 1977 [9-27]

Professional activity	Risk values
Mining	$30 \cdot 10^{-8}$
Traffic	$30 \cdot 10^{-8}$
Construction	$20 \cdot 10^{-8}$
Nonmetallic minerals	$15 \cdot 10^{-8}$
Gas and water	$4 \cdot 10^{-8}$
Iron and metal	$7 \cdot 10^{-8}$
Wood and pulp	$4 \cdot 10^{-8}$
Food and beverages	$6 \cdot 10^{-8}$
Paper and print	$5 \cdot 10^{-8}$
Electronics, precision mechanics, optics	$4 \cdot 10^{-8}$
Chemistry	$3 \cdot 10^{-8}$
Trade, money, and insurance services	$5 \cdot 10^{-8}$
Textile and leather	$2 \cdot 10^{-8}$
Health service	$1 \cdot 10^{-8}$
Average value	$8 \cdot 10^{-8}$

evaluate the situation represented here with respect to social risk acceptance.

For a long time we had proposed that the natural death risk be used as a guideline. It is practically equal in all industrialized countries, changes only slowly with time, and reflects the civilizational effects of technology. The death rate depending on age shows after deduction of "unnatural" causes of death: accidents, congenital deformities and immaturity, in the 5–15 age group a broad minimum. This, the lowest natural death risk seems suited for an "absolute base measure" of technical risk with respect to death. In the Federal Republic of Germany its value is

$$R_{ag} = 2 \times 10^{-4} \text{ dead per person and year}$$
$$= 2 \times 10^{-8} \text{ dead per person and hour.} \qquad (9\text{-}3)$$

In the long term mean there are per one accidental death 20 accident injury cases with permanent physical disability and about 200 accident injury cases with temporary disability, in terms of magnitude. Therefore, starting with this premise, the risk base value for permanent physical disability should be set one, and for temporary physical disability two powers of ten higher than the absolute base measure. In the case of material and financial assets, the setting of a risk base value is not required. Here economic evaluation will determine to what extent capital should be invested in safety and protection measures or reserved for possible damage compensation. The special case of material goods with high ideal value requiring special protection will not be discussed here.

The "admissible base risk" R_g for hazard due to a single technical installation cannot, however, be identical to the absolute base risk. It must be assumed that in many instances man is threatened simultaneously by several technical installations. In view of this aspect the admissible base risk R_g, accepted for a single technical installation will have to be lower.

Our proposal for admissible base risk by a single technical installation is as follows:

$$R_g = 10^{-5} \text{ dead/person and year,} \qquad (9\text{-}4a)$$
$$R_g = 10^{-4} \text{ permanently disabled/person and year,} \qquad (9\text{-}4b)$$
$$R_g = 10^{-3} \text{ temporarily disabled/person and year.} \qquad (9\text{-}4c)$$

An "admissible base risk" R_g is also assigned to the major life areas such as work, traffic, household, and leisure time and is about one-third of the "absolute base risk" R_{ag}:

$$R_g = 1 \times 10^{-4} \text{ dead per person and year,}$$
$$R_g = 1 \times 10^{-8} \text{ per person and hour.} \qquad (9\text{-}5)$$

Once the base risk-value system has been established, what remains is to establish a relationship to present actual-risk-values as are shown in Table 9.1. Viewing the presently known risk values, it is immediately obvious

Table 9.2 Relative risks R^* according to areas

Area	R^*
Work	1.4
Transport	2.4
Household and leisure time	1.7

that safety efforts need not be made uniformly along a wide front. Applied to the broad areas such as work, transport, household, and leisure time, the relative risks R^* as actual risk and admissible base risk quotients show this priority particularly clearly; they are given in Table 9.2. Therefore, in view of present conditions, work must be concentrated on the areas of transport and household.

For insertion in the proposed Equations (9-1) and (9-2),

$$R(r) \leq R_{adm} \text{ for all locations } r,$$

$$\int R(r) \cdot B(r) \cdot df \leq R_{adm} \cdot 100 \text{ persons}$$

as criteria for risks which can be imposed in connection with a technical installation, finally values of admissible risk R_{adm} above the admissible base risk R_g, dynamic with respect to time and varying according to sectors must be established:

$$R_{adm} = X \times R_g (X \geq 1). \tag{9-6}$$

Table 9.3 Configuration of risk base values

Areas: life/work	abs. risk-base-value R_{ag} [accident/ year inhabitants]	adm. risk-base-value R_g [accid./year. inhab.] death	perm. disability	temp. disability	adm. risk R_{adm}
Traffic Household and leisure time Work	$2 \cdot 10^{-4}$	10^{-4}	10^{-3}	10^{-2}	
Mining Traffic Construction Nonmetallic minerals Gas and water Iron and metal Wood and pulp Food and beverages etc.	$2 \cdot 10^{-4}$	10^{-5}	10^{-4}	10^{-3}	$R_{adm} = X \cdot R_g$ $(X \geq 1)$

In the determination of sector-dependent factors X cost-benefit aspects must be observed. In the lead, however, must be the desire to approximate R_{adm} to R_g; i.e., X to value 1, as quickly as possible. As a general rule, that will have to take place gradually. In the event that the R_{adm}-value remains too far from the R_g-value, a solution by means of alternative technologies with lower risk will have to be considered.

Expert commissions should be given the task of establishing admissible risk values R_{adm} and of updating them according to the state of science and economic feasibility. Table 9.3 shows a review of the proposed system of risk base values and admissible risks.

9.2.4 Risk Inventory

The possibilities of risk limitation and reduction by means of standards, as dealt with thus far, assume that the risk can be described quantitatively. Observations used in this connection start with the premise of a single technical installation. It is assumed that the risk to be assigned to the installation can be stated from knowledge of the installation structure, the processes taking place inside, and site conditions in the installation surroundings. The risk in the surroundings concerned will be due to the combined action of the individual technical installations.

The task here will be to make sure that the risk to man will be as low as possible in the surroundings of technical plants and that it will remain below the proposed limits. At the same time, interest in maximum effectiveness is dictated by economic aspects. There is no point in exaggerated reduction of risk due to a single installation involving high costs, if risks due to others are permitted to remain comparatively high. On the contrary, it is necessary to include all the risks generated in an area and to represent them in a suitable manner. This can be done by means of a risk inventory. If necessary, risks connected with different types of loss or impairment should be shown separately.

Individual risks $R(r)$ dependent on local conditions can be assigned to any technical installation concerned. The actual risks to which individual persons are exposed in the surroundings of technical installations depend on how often and how long these persons are at that location. To exclude this influence one will refer, in the sense of the "impressed risk" according to Chapter 2, for practical purposes to a hypothetical person staying permanently at the location concerned. Thereby the inventory-like representation of the impressed individual risk will furnish a representation of the characteristics of a technical installation at a specific location, unconnected with the behavior of the persons concerned.

As a rule, individual risks are reduced with increasing distance from the technical installation. Their location dependence can be represented by Iso-risk lines. The impressed risk field is usually not a symmetrical circle around the installation site. If, for example, the risk is due to emission of harmful substances into the atmosphere, the field will depend to a large extent on meteorological conditions.

The impressed individual risk due to several technical installations emerges from the superposition of the actions of the individual installations. In the case of incident considerations it is frequently possible to assume that the risks are additively superposed. The assumption here is that simultaneous appearance of these incidents at several installations need not be considered because of the low probability of the individual incident. In this case we obtain the resulting impressed individual risk R_S by N installations:

$$R_S(r) = \sum_{n=1}^{N} R_n(r). \tag{9-7}$$

Figure 9.7 schematically shows the superposition of impressed risk fields of three installations in varying locations vis à vis each other. The impressed risk included and represented in its location dependence in the risk inventory can be evaluated directly by means of the suggested admissibility criterion of Equation (9-1). But it can also be easily subjected to the second admissibility criterion given in Equation (9-2), which limits the population risk when it is weighted with the population density $B(r)$ and integrated over the area included in the inventory.

It is possible to enter not only existing technical installations in the risk inventory, but planned installations as well. It can be determined how risk stress will be changed by planned installations and how by suitable location selection high risk values can be avoided. The risk inventory together with the proposed admissibility criteria thus becomes a planning instrument and an objective source of information for persons living in the risk area.

9.2.5 The Obligation

In Section 9.2.2 the basic authority of the state to adopt regulations which include risks for an undetermined circle of persons concerned was affirmed on ethical and legal grounds. Notwithstanding all the objective legal reasons, preponderant agreement by society and the heeding of criticism voiced by public opinion within society cannot be waived in the long range.

It is necessary to have the agreement of public opinion to a rational policy toward technical risks and to be aware of the mechanisms important in shaping public opinion. In order to reach a broad consensus within society on important questions of technical safety policy, it is above all necessary to free the risk concept of its negative connotation and to stress that it is to be understood exclusively as a concept of mathematical-natural science in the technology discussion.

Scientific education and the dialogue within society and with government bodies cannot replace the final decision by authorized government agencies nor bring it about as a direct consequence. It must be borne in mind that

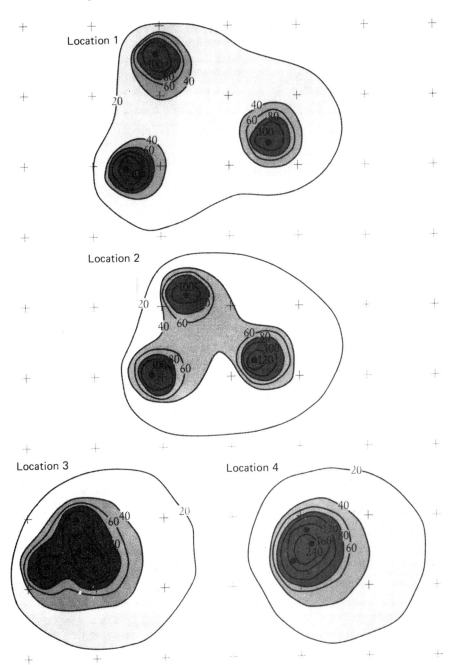

Figure 9.7 Superposition of "impressed risks" for varying locations of three hazard sources in relation to each other.

excessively long discussions and perpetually renewed educational efforts on the part of scientific circles will become self-defeating after some time. If the state delays a decision in a specific direction for too long, this will be bound to provide the opposition argument with weight which in some circumstances may not be justified. Groups within society which lack the ability, time, or inclination to follow scientific arguments concerning a technology and its risks are led to the general assumption that where there is smoke, there is fire.

However, those competent to form an opinion also must become insecure when they find their well founded and sincere conviction represented in society as an extreme party opinion which government agencies hesitate to espouse. After a certain time the substantively sensible view of the problem can appear narrow, and it becomes fashionable to become "critical" of a new technology and the mastery of its risks and to make the pending decision more difficult.

The following aspect, however, is even more weighty: Starting with the fact that one and the same opinion can be formed for varying reasons, it is possible that during the course of the exchange of views extraneous "political" considerations insinuate themselves into the originally honest motivations. The contested question becomes a mere crystallization focus of political fronts. This process of substitution of motives in social controversy has been described in a classical manner by Ch. Péguy [9-31].

Before a difference of opinion started with honest motives finally becomes poisoned, government bodies must make use of their authority and decide whether the technical risks can be imposed and on the regulation pertaining to them. The readiness of citizens to respond in the affirmative to determined government action in favor of such decisions is not simply identical with risk acceptance; as a rule, it is higher.

Man's fear of technology has frequently been described as the fear of something unfathomable, apparently overwhelming. It has been said that in our society individual minds know more and have greater ability than we can collectively transform into desired shapes of life [9-32]. It should be added that man has, for the first time, become fearful of invisible long-term dangers. The pace of the technical progress of our time has led, via suspicion and apprehension, to fear and resistance and has created problems for the state.

To maintain or restore confidence in technology the principle of conservative burden-of-proof distribution should be recalled. This principle tells us that until there is proof to the contrary, the reasonableness of existing conditions, of valid standards, and of decisions previously made must be trusted. Technical progress may be inhibited only if this burden of proof can be borne convincingly. Otherwise we would impair the welfare of man in whose nature technology is inherent.

References

Chapter 2

[2-1] dtv-Lexikon der Physik. Bd. 2, München 1970
[2-2] Wallot, J.: Größengleichungen, Einheiten, Dimensionen. Leipzig 1954
[2-3] Kuhlmann, A.: Alptraum Technik? Verlag Hoppenstedt und Verlag TÜV Rheinland, Köln 1977

Chapter 3

[3-1] Wiener, Norbert: Kybernetik. Econ, Düsseldorf und Wien 1963
[3-2] Dubach, Paul: Die Kybernetik als theoretische, praktische und interdisziplinäre Wissenschaft (Eine Einführung). Industrielle Organisation 38 (69) Nr. 7, S. 275 ff.
[3-3] Ashby, W. Ross: Einführung in die Kybernetik. Suhrkamp, Frankfurt 1974
[3-4] Föllinger, Otto: Regelungstechnik. Elitera, Berlin 1978
[3-5] Simulation technischer Systeme. Hrsg.: A. Schöne. München, Wien, Hauser 74—76, Bd. 1-3
[3-6] Koelle, H.H.u.a.:Systemtechnik. Vorlesungsmanuskript, Institut für Luft- und Raumfahrt
[3-7] Grundlagen und Anwendung der Systemtechnik. VDI-Berichte 262, Düsseldorf 1976
[3-8] Systemtechnik, Grundlagen und Anwendung. Hrsg.: G. Ropohl, München, Wien, Hauser 1975
[3-9] Graf, Henning, Stange: Formeln und Tabellen der mathematischen Statistik. Springer 1966, 2. Auflage
[3-10] Stange, L.: Angewandte Statistik. Erster Teil, Eindimensionale Probleme, Springer 1970
[3-11] Stange, L.: Angewandte Statistik. Zweiter Teil: Mehrdimensionale Probleme, Springer 1971
[3-12] Sachs, L.: Angewandte Statistik, Springer, 4. Auflage, 1974
[3-13] Überla, K.: Faktorenanalyse, Springer 1968
[3-14] Gordon, G.: Systemsimulation, Oldenbourg 1972
[3-15] Stachowiak, H.: Allgemeine Modelltheorie, Springer 1973
[3-16] Zienkiewicz, QC.: Methode der finiten Elemente, Hauser, München 1975

[3-17] Bathe, K-J. u. Wilson, E.L.: Numerical Methods in Finite Element Analysis. Prentice-Hall. Inc., Englewood Cliffs, New Jersey

[3-18] Schwarz, H.R.: Methode der finiten Elemente. Teubner Studienbücher, Stuttgart 1980

[3-19] Marsal, D.: Die numerische Lösung partieller Differentialgleichungen in Wissenschaft und Technik. Bibliographisches Institut AG. Zürich 1976

[3-20] Herman, H.H.: Modern Factor Analysis. University of Chicago Press. Chicago, London 1960

[3-21] von Bertalanffy, L.: Zu einer allgemeinen Systemlehre. Biologia Generalis 1/49. S. 114–129

Chapter 4

[4-1] Gaede, K.-W.: Zuverlässigkeit, Mathematische Modelle. Hanser, München, Wien 1977

[4-2] Höfle-Isphording, V.: Zuverlässigkeitsrechnung, Einführung in ihre Methoden. Springer 1978

[4-3] Schneeweiss, W.: Zuverlässigkeitstheorie. Springer 1973

[4-4] Koslow, Uschakow: Handbuch zur Berechnung der Zuverlässigkeit für Ingenieure. Hanser 1979

[4-5] Fischbach, Groß, Ott: Entscheidungstabellen. IVI Köln 1975

[4-6] Thurner, Reinhold: Entscheidungstabellen. VDI Düsseldorf 1972

[4-7] Erbesdohler/Heinemann/Mey: Entscheidungstabellentechnik. Springer 1976

[4-8] Reichelt, C.: Rechnerische Ermittlung der Kenngrößen der Weibullverteilung. Fortschr.-Ber. VDI-Z Reihe 1 Nr. 56. VDI Düsseldorf 1978

[4-9] DIN 25448, Ausfalleffektanalyse. Juni 1980, Beuth Verlag GmbH, Berlin 30 und Köln 1

[4-10] DIN 25424, Fehlerbaumanalyse, Methode und Bildzeichen. Juni 1977, Beuth Verlag GmbH, Berlin 30 und Köln 1

[4-11] DIN 25419, Störfallablaufanalyse
Teil 1: Störfallablaufdiagramm, Methode und Bildzeichen, Juni 1977
Teil 2: Auswertung des Störfallablaufdiagramms mit Hilfe der Wahrscheinlichkeitsrechnung, Februar 1979
Beuth Verlag GmbH, Berlin 30 und Köln 1

[4-12] TÜV Rheinland: Zentralabteilung Angewandte Elektrotechnik und Steuerungstechnik, Berechnung der Ausfallwahrscheinlichkeit (Ausfallrichtungsanalyse) des Silometers FMC280Z, 27. 09. 1979

[4-13] TÜV Rheinland: Zentralabteilung Steuerungs- und Bahntechnik, Gutachten zur Funktions- und Sicherheitsschreibung der Demonstrationsanlage für Magnet-Schwebetechnik zur IVA 1979 in Hamburg. Bericht Nr.: 947:B 79/100, Köln, den 15. 02. 1979

[4-14] TÜV Rheinland: ZUPRO, Zuverlässigkeitsprogramm zur Errechnung von Unverfügbarkeiten und Ausfallwahrscheinlichkeiten von komplexen Systemen. Januar 1974, K.R. Hartung, Fachbereich Kerntechnik

[4-15] TÜV Rheinland: AP1, Handhabung, Input-Beschreibung und Anwendung des analytischen Zuverlässigkeitsprogramms AP1 zur Errechnung der

Unverfügbarkeiten und Ausfallwahrscheinlichkeiten von komplexen System, Februar 1975, K.R. Hartung, Fachbereich Kerntechnik

[4-16] Deutsche Risikostudie Kernkraftwerke: Eine Untersuchung zu dem durch Störfälle in Kernkraftwerken verursachten Risiko. Copyright: Der Bundesminister für Forschung und Technologie, Bonn 1979, Verlag TÜV Rheinland GmbH, Köln

[4-17] Hammer, W.: Handbook of System and Product Safety, Prentice-Hall, Inc. Englewood Cliffs, N.J., London, 1972

[4-18] Balfanz, H.P.: Sicherheitsanalyse-Plan, Anwendung verschiedener sicherheits- und Zuverlässigkeitsanalysen zum richtigen Zeitpunkt und zu speziellen Problemen, IRS- W -2, Köln, 1972

[4-19] Knoglinger, H.: Sicherheit. VDI-Lehrgang ,,Technische Zuverlässigkeit". Vortrag Nr.: 3687

[4-20] Lawley, H.G.: Operability Study and Hazard Analysis. Chemical Engineering Progress 70 (1974), No. 4, p. 45

[4-21] Lawley, H.G.: Size up Plant Hazard this Way. Hydrocarbon Process, 55 (1976), 247

[4-22] Grein, W. und Braubach, H.O.: Abnahme und Überwachung der Anlagen. Dechema Monographien, Vol. 88: Das Sicherheitskonzept für die Chemische Technik. Verlag Chemie, Weinheim, 1980, 149 f.

[4-23] Schuster, K.: Anlagenbeurteilung aus sicherheitstechnischer Sicht—dargestellt am Beispiel einer chemischen Produktionsanlage—Vortrag auf dem Druckbehälter-Seminar des TÜV Rheinland, Köln, 11. 12. 1979

[4-24] Ullrich, H.: Checklisten zur Prüfung von Fließbildern. (Z-Chemie-Technik 2 1973), 110

[4-25] Chemical Industry Safety and Health Council of the Chemical Industries Association Limited: A Guide Hazard and Operability Studies London, Alembic House, 1977
deutsch: Der Störfall im chemischen Betrieb. Herausgeber: Berufsgenossenschaft der chemischen Industrie, Heidelberg

[4-26] Krass, Kittel, Uhde: Pipeline- Technik, Band 3, S. 243. TÜV Rheinland 1979

[4-27] Verordnung über Druckbehälter, Druckgasanlagen und Füllanlagen (DruckbehV) vom 27. 02. 1980 (BGBl. I S. 184), Carl Heymanns Verlag KG, Köln

[4-28] Technische Regeln Druckbehälter (TRB)

[4-29] Unfallverhütungsvorschrift ,,Druckbehälter" UVV VBG 17 vom 01. 04. 1965. Hauptverband der gewerblichen Berufsgenossenschaften. Carl Heymanns Verlag AG Köln

[4-30] Statische Elektrizität: Richtlinie Nr. 4 der Berufsgenossenschaft der chemischen Industrie, 1980. Verlag Chemie GmbH, Weinheim

[4-31] Technische Zuverlässigkeit. Herausgeber: Messerschmidt-Bölkow-Blohm. Springer-Verlag, Berlin-Heidelberg-New York 1977

[4-32] Balfanz, H.: Ausfallratensammlung, IRS-W8 (1973). Bericht des Instituts für Reaktorsicherheit, Köln

[4-33] Reactor Safety Study (Wash Report 1400); Appendix III, Failure Data, October 1975

[4-34] Hömke, P. und H. Krause: Der Modellfall IRS-RWE zur Ermittlung von

Zuverlässigkeitskenngrößen im praktischen Betrieb, IRS- W16 (Nov. 1975). Bericht des Instituts für Reaktorsicherheit, Köln

[4-35] Anyakora, S.N., Engel, G.F.M., Lees, F.P.: Some Data on the Reliability of Instruments in the Chemical Plant Environment. The Chemical Engineer, November 1971, P. 396–402

[4-36] Hensley, G.: Safety Assessment—A Method for Determining the Performance of Alarm and Shut Down Systems for Chemical Plants. Measurement and Control, Vol 1, April 1968, P. T72—T79

[4-39] Keil, E., Müller, E.O. und Bettzieche, P.: Zeitabhängigkeit der Festigkeits- und Verformbarkeitswerte von warmfesten Stählen im Temperaturbereich unter 400 °C. Archiv für Eisenhüttenwesen, 43. Jahrgang, Heft 10/72, S. 757–762

Chapter 5

[5-1] Buxhaum, O.: Häufigkeitsverteilung der Förderdruckschwankungen in einer Ölleitung. Bericht des Laboratoriums für Betriebsfestigkeit, Nr. 2165 vom 8. 03. 1968, Darmstadt

[5-2] Buxbaum, O.: Statistische Zählverfahren als Bindeglied zwischen Beanspruchungsmessung und Betriebsfestigkeitsversuch. Bericht des Laboratoriums für Betriebsfestigkeit, Nr. TB-65, Darmstadt, 1966

[5-3] Miner, M.A.: Cumulative Damage in Fatigue. Journal of Applied Mechanics, 1945. SA 159/169

[5-4] Haibach, E. und Olivier, R.: Untersuchung zur Lebensdauerabschätzung von Ölrohren mit Anrissen im Schweißnahtbereich. Bericht des Laboratoriums für Betriebsfestigkeit, Nr. 2481 vom 23. 09. 1970, Darmstadt

[5-5] Haibach, E.: Sinnvolle Festlegung der Kollektiv-Treppung für Betriebsfestigkeitsversuche. Bericht des Laboratoriums für Betriebsfestigkeit, Technische Mitteilung TM-Nr. 47/69, Darmstadt

[5-6] Uebing, D. und Jäger, P.: Lifetime Calculation and Failure Probability— Modern Design Criteria in the Areas of Creep, Fatigue and Embrittlement—. Third International Converence on Pressure Vessel Technology, Tokyo, 1976

[5-7] Drucks, G., Jäger, P. und Kaufmann, H.R.: Zur Abschätzung der Restlebensdauer drucktragender Bauteile im Zeitstandbereich. Zeitschrift 3R-international, 18. Jahrgang, Heft 5, Mai 1979

[5-8] Ullmanns Enzyklopädie der technischen Chemie. Band 3, Verfahrenstechnik II und Reaktionsapparate, S. 489, 4. Auflage 1973, Verlag Chemie, Weinheim

[5-9] Wellinger, K., Sturm, D.: Festigkeitsverhalten von zylindrischen Hohlkörpern mit Fehlstellen. Fortschritt-Berichte der VDI-Zeitschriften Reihe 5, Nr. 13, Mai 1971

[5-10] Kiefner, I.F.: Fracture Initiation, Fourth Symposium on Line Pipe Research. American Gas Association. Catalogue Nr. L 300 75, Nov. 1969

[5-11] Wellinger, K.: Berstdruck von Rohren bei durch Kerben geschwächter Wand. VGB-Mitteilungen 54, S. 309–316

[5-12] Wellinger, K., Sturm, D.: Festigkeitsverhalten von zylindrischen Hohlkörpern mit Fehlstellen. Fortschritt-Berichte der VDI-Zeitschriften, Reihe 5, Nr. 13, Mai 1971

[5-13] Gassner, E. und Olivier, R.: Betriebslastenanalyse und Schwingbruch-sicherheit von Mineralölfernleitungen. Symposium Rohrfernleitungstechnik, 24–25. September 1973, Bad Neuenahr

[5-14] Uebing, D.: Werkstoffverhalten bei ruhender und schwellender Innen-druckbeanspruchung. Verlag Stahleisen, Düsseldorf 1967

[5-15] Schwaigerer, S.: Das Festigkeitsverhalten von Stahlrohrleitungen im Betrieb. Technische Überwachung 13 (1972), Nr. 1

[5-16] Wellinger, K. und Sturm, D.: Nahtlose und geschweißte Rohre bei ruhender und schwelender Innendruckbeanspruchung. Technische Überwachung 10 (1969), Nr. 4

[5-17] Wellinger, K., Geilenkeuser, H. und Zenner, H.: Innendruckschwellverhalten von Großrohren aus St 60.7. Schweißen und Schneiden 23 (1971), Nr. 4

[5-18] Untersuchungsbericht der MPA-Stuttgart Nr. M 29548 vom 1. 3. 1967 (unveröffentlicht)

[5-19] Untersuchungsbericht der MPA-Stuttgart Nr. M 29548/4 vom 4. 10. 1967 (unveröffentlicht)

[5-20] Untersuchungsbericht des Instituts des Instituts für Materialprüfung (IFM), Köln Nr. SB 8/74 vom 3. 5. 1974 (unveröffentlicht)

[5-21] Jäger, P.: Zur Lebensdauerberechnung fehlerbehafteter Rohrfernleitungen unter Innendruck mit Hilfe dimensionsloser Kenngrößen. München 1976

[5-22] Freudenthal, A.: Statistical Approach to Brittle Fracture; Fracture, an Advanced Treatise, Vol. II, Academic Press, New York 1968

[5-23] Ermittlung der Ausfallraten von Dieselaggregaten und Elektronikkarten zur Bestimmung der Ausfallwahrscheinlichkeit von Sicherheitssystemen, P. Sommer, K.-R. Hartung, H. Scholz, S. Wüst, Köln. Qualität und Zuverlässigkeit, Carl Hanser Verlag München, 22. Jahrgang, Mai 1977, Heft 5

[5-24] Technischer Überwachungs-Verein Rheinland. Ausfallraten von Elektronikkarten, Verfasser H. Scholz und P. Sommer, Fachbereich Kerntechnik, Bericht-Nr.: 932-76//08/01

[5-25] Technischer Überwachungs-Verein Rheinland, Technischer Bericht, Mittlere fehlerfreie Betriebszeiten elektronischer Steuerungssysteme, von Dipl.-Ing. Hartmut Scholz. Fachbereich Kerntechnik, September 1974

[5-26] Lörcher: Vollausfall und Driftverhalten von einigen passiven Bauelementen, NTG-Diskussionssitzung 6. und 7. 12. 1966

[5-27] Balfanz, H.P.: Ausfallratensammlung, IRS- W -8 (Dezember 1973). Institut für Reaktorsicherheit der Technischen Überwachungs- Vereine e. V.

[5-28] U.S. Atomic Energy Commission, Reactor Safety Study, Appendix III, Failure Data Wash- 1400, Aug. 1974

[5-29] Kernkraftwerk Biblis, Block A, Ergebnisse der Zuverlässigkeitsuntersuchungen für den Auslegungsstörfall „Bruch einer kalten Hauptkühlmittelleitung", LRA Garching, MRR 168, Dezember 1976

[5-30] TÜV Rheinland, Ausfallstatistik von 100 Diesel-Aggregaten, Bericht-Nr.: 932-78/07/01. Verfasser: Baum, Sommer, Schulz, Wüst

[5-31] Determination of the reliability of emergency Diesel generators, by Peter Sommer, Atomkernenergie, Kerntechnik Bd. 35 (1980) Lfg. 1

[5-32] Unterbrechungsfreie Stromversorgung für EDV-Anlagen, H. Krakowski, etz-b Bd. 27 (1975) H. 2

[5-33] DIN 40042, Vornorm, Zuverlässigkeit elektrischer Geräte, Anlagen und Systeme, Juni 1970, Beuth-Vertrieb GmbH

[5-34] TÜV Rheinland AP1, Handhabung, Input-Beschreibung und Anwendung des analytischen Zuverlässigkeitsprogramms AP1 zur Errechnung von Unverfügbarkeiten und Ausfallwahrscheinlichkeiten von komplexen Systemen, Februar 1975, K.R. Hartung, Fachbereich Kerntechnik

[5-35] Anyakora, Engel, Lees: Some Data on Rehability of Instruments in the Chemical Plant Environment, Chemical Engineer 11/71

[5-36] DIN 2413: Berechnung von Stahlrohren mit kreisrundem Querschnitt, Ausgabe Juni 1971

Chapter 6

[6-1] Baker, P.T., and Weiner, J.S.: The Biology of Human Adaptability, Clarendon Press, Oxford, 1966

[6-2] Benninghof, A., und Goerttler, K.: Lehrbuch der Anatomie des Menschen, Bd. 1, Verlag Urban und Schwarzenberg, Berlin 1961

[6-3] Bücker, J.: Anatomie und Physiologie, Verlag Georg Thieme, Stuttgart 1971

[6-4] Damon, A., Stoudt, H.W., McFarland, R.A.: The Human Body in Equipment Design, Harvard Univ. Press, Cambridge 1966

[6-5] Gauer-Kramer-Jung: Physiologie des Menschen. Verlag Urban und Schwarzenberg, Berlin 1972

[6-6] Grandjean, E.: Physiologische Arbeitsgestaltung, Leitfaden der Ergonomie, Ott Verlag, Thun 1979

[6-7] Hensel, H.: Allgemeine Sinnesphysiologie, Springer-Verlag, Berlin 1966

[6-8] Keidel, W.D.: Physiologie, Verlag Georg Thieme, Stuttgart 1970

[6-9] Lehmann, G.: Praktische Arbeitsphysiologie, Verlag Georg Thieme, Stuttgart 1962

[6-10] Lehmann, G.: Handbuch der gesamten Arbeitsmedizin, Band I, Verlag Urban und Schwarzenberg, Berlin 1961

[6-11] Löhr, R.W.: Ergonomie, Vogel-Verlag, Würzburg 1976

[6-12] Martin, R.: Lehrbuch der Anthropologie, Verlag Fischer, Stuttgart 1957

[6-13] Müller, E.A.: Die Muskelfunktion. Handbuch der gesamten Arbeitsmedizin, Bd. 1, Verlag Urban und Schwarzendberg, Berlin 1961

[6-14] Müller-Limmroth, W.: Elektrophysiologie des Gesichtssinns, Springer-Verlag, Berlin 1959

[6-15] Pauwels, F.: Gesammelte Abhandlungen zur funktionellen Anatomie des Bewegungsapparates. Springer-Verlag, Berlin, Heidelberg, New York 1965

[6-16] Rohmert, W.: Maximalkräfte von Männern im Bewegungsraum der Arme und Beine. Forschungsbericht Nr. 1616 des Landes Nordrhein-Westfalen, Westdeutscher Verlag, Köln und Opladen 1966

[6-17] Rohmert, W., und P. Jenik: Maximalkräfte von Frauen im Bewegungsraum der Arme und Beine, Beuth-Vertrieb GmbH, Berlin, Köln, Frankfurt/M.

[6-18] Schade, J.P.: Die Funktion des Nervensystems, Verlag G. Fischer, Stuttgart 1969

[6-19] Schmidt, R.F., und Thews, G. (Hrsgb.): Physiologie des Menschen, Springer-Verlag, Berlin, Heidelberg, New York 1980

[6-20] Schmidtke, H.: Ergonomie 1, Grundlagen menschlicher Arbeit und Leistung, Carl Hanser Verlag, München 1973

[6-21] Schütz, E.: Physiologie, Verlag Urban und Schwarzenberg, Berlin 1972

[6-22] Spitzer, H., und Hettinger, Th.: Tafeln für den Kalorienumsatz bei körperlicher Arbeit, in: Sonderheft d. REFA-Nachr., 5. Aufl., Beuth-Vertrieb GmbH, Berlin 1969

[6-23] Trendelenburg, W., und Schütz, E.: Lehrbuch der Physiologie. Monje, M., J. Schmidt und E. Schütz: Der Gesichtssinn. Springer-Verlag, Berlin, Göttingen, Heidelberg 1961

[6-24] Trendelenburg, W., und Schütz E.: Lehrbuch der Physiologie, Ranke, O.F., und H. Lullies: Gehör—Stimme—Sprache. Springer-Verlag, Berlin, Göttingen, Heidelberg 1953

[6-25] Trendelenburg, W., und Schütz E.: Lehrbuch der Physiologie, Hensel, H.: Allgemeine Sinnesphysiologie, Hautsinne, Geschmack, Geruch. Springer-Verlag, Berlin Heidelberg, New York 1966

[6-26] Valentin, H.: Arbeitsmedizin. Verlag Georg Thieme, Stutgart 1979

[6-27] Wiener, N.: Mensch und Menschmaschine (Kybernetik und Gesellschaft), Alfred Metzner Verlag, Frankfurt/M. 1966

[6-28] Hauptverband der gewerblichen Berufsgenossenschaften e. V., Bonn, (Hrsg.) Berufsgenossenschaftliche Grundsätze für arbeitsmedizinische Vorsorgeuntersuchungen, Verlag A.W. Gentner, Stuttgart

[6-29] Lewrenz, H. u. Friedel B.: Krankheit und Kraftverkehr, 2. Auflage, Bonn 1979, Heft 67 der Schriftenreihe des Bundesministers für Verkehr

[6-30] Schneider, W. u. Schubert, G.: Die Begutachtung der Fahreignung. In: Handbuch der Psychologie, 11. Forensische Psychologie, Göttingen 1967, 671–739

[6-31] Burkardt, F. u. Hahn, R.: Entwicklung eines Testverfahrens für die psychologische Eignungsuntersuchung bei Fahr-, Steuer- und Überwachungstätigkeiten. Unveröff. Vervielf. Frankfurt 1980 für den Ausschuß Arbeitsmedizin.

[6-32] Metzger, W.: Gesetze des Sehens. (Die Lehre vom Sehen der Formen und Dinge des Raumes und der Bewegung) Verlag von Waldemar Kramer, Frankfurt, 1953

[6-33] Keidel, W.D.: Sinnesphysiologie. Teil 1: Allgemeine Sinnesphysiologie. Visuelles System. Berlin-Heidelberg-New York: Springer-Verlag, 1971

[6-34] Senders, J.W.: Wer ist wirklich schuld am menschlichen Versagen? In: Psychologie heute, 1980, 73–78

[6-35] Rigby, L. V.: The Nature of Human Error, 24th Annual Technical Conference Transactions, American Society for Quality Control, Milwaukee, Wisc, May 1970, 457–465

[6-36] Burkardt, F.: Arbeitssicherheit. In: Handbuch der Psychologie, Bd. 9, 2. Auflage 1970, 385–415

[6-37] Undeutsch, U.: Persönlichkeit und Vorkommenshäufigkeit der „Unfäller" unter den Kraftfahrern. Band 9 der Reihe: „Die Sicherung des Menschen im Straßenverkehr". Bad Godesberg 1962, 1–24

[6-38] Kunkel, E.: Modell Mainz 77. Kurse für Verhaltens- und Einstellungsänderung bei Trunkenheitstätern, Heft 12, 1980 Mensch-Fahrzeug-Umwelt, Verlag TÜV Rheinland GmbH, Köln

[6-39] Mort la Brecque: ,,Denkfehler und Denkverhalten". *In*: Psychologie heute, 1980, 21–27

[6-40] Taylor, D.H.: Driver's galvianic skin response and the risk of accident *In*: Ergonomics, 1964, 7 439–451

[6-41] Wilde, G.J.S.: Verkeersgedrag en veiligheid: uitgangspunten voor een preventieve strategie. Vortrag vor dem Verband ,,Veilig Verkeer Nederland" Amsterdam 1976

[6-42] Ganton, N. u. Wilde, G.J.S.: Verbal ratings of edimated danger by drivers and passengers as a function of driving experience: Im Auftrage des Canadian Ministry of Transport, Ottawa 1971

[6-43] Schneider, W.: Psychische Ursachen und Hintergrundbedingungen bei Unfallverläufen. *In*: Zeitschrift für Verkehrssicherheit, 1977, 140–145

[6-44] Küppers, F.: Untersuchungen über die Beziehung von Verkeersdichten und Unfällen auf Autobahnen unter Berücksichtigung von Ermüdungstheorien des Unfallgeschehens. Diss. Universität zu Köln, Math-Nat. Fak 1970

[6-45] Heller, F.: Entwicklung der Straßenverkehrsunfälle in der Bundesrepublik Deutschland. *In*: Zusammenfassender Bericht der Forschungsgruppe ,,Entwicklung der Straßenverkehrsunfälle in der BRD 1970/71", Köln 1973— S. 20 u. S. 151

[6-46] REFA Verband für Arbeitsstudien—REFA—e. V. (Hrsg.): Methodenlehre des Arbeitsstudiums, Teil 1: Grundlagen. Carl Hanser Verlag, München 1973.

[6-47] Schmidtke, H.: Ergonomie 2: Gestaltung von Arbeitsplatz und Arbeitsumwelt. Carl Hanser Verlag, München, 1974.

[6-48] Neumann, J. und Timpe, K.-P.: Physiologische Arbeitsgestaltung. VEB Deutscher Verlag der Wissenschaft, Berlin, 1976

[6-49] Köck, P. und Odehnal, S.: Ergonomische Prüfliste. Österreichische Arbeitsgemeinschaft für Ergonomie (ÖAfE), Wien, 1977.

[6-50] McCormick, E.J.: Human Factors, Engineering. McGraw-Hill, New York, 1970

[6-51] McCormick, E.J.: Human Factors in Engineering und Design. McGraw-Hill, New York, 1976.

[6-52] Institut für angewandte Arbeitswissenschaft e. V. (Hrsg.): Taschenbuch der Arbeitsgestaltung. Verlag J.P. Bachem, Köln 1977.

[6-53] Richtlinie des BMA über Maßnahmen zum Schutz der Arbeitnehmer gegen Lärm am Arbeitsplatz vom 10. November 1970. Abgedruckt in: Arbeitsschutz (1970) 12, S. 345–352.

[6-54] UVV Lärm, Unfallverhütungsvorschrift der gewerblichen Berufsgenossenschaften VBG 121 vom 1. Dezember 1974 einschließlich Durchführungsregeln und Erläuterungen.

[6-55] Arbeitsstättenverordnung vom 20. März 1975, BGBl. I S. 729.

[6-56] Lange, W.; Kirchner, J.-H.; Lazarus, H. und Schnauber, H.: Kleine ergonomische Datensammlung. (Hrsg.: Bundesanstalt für Arbeitsschutz und Unfallforschung). Verlag TÜV-Rheinland, Köln, 1978.

[6-57] Brigham, F.; Radl, G.W.; Tossing, N. und Wegner, K.: Arbeitsbedingungen bei der Stadtreinigung (Hrsg.: Bundesanstalt für Arbeitsschuz und Unfallforschung, Forschungsbericht Nr. 230). Wirtschaftsverlag NW, Bremerhaven, 1980.

[6-58] Burkardt, F., Schubert, U. und Schubert, G.: (1970) Psychologie der

Arbeitssicherheit. Seminar für Führungskräfte. Koblenz: Bundesinstitut für Arbeitsschutz.

[6-59] Jensch, M. u. Schneider, W.: Fahren lernen mit VW, Wolfsburg 1976.

[6-60] Schneider, W. u. Jensch, M.: Fahren lernen mit VW, 6. Auflage, Wolfsburg 1973

[6-61] Grimm, H.G.: Die Simulation der Nachfahraufgabe mit einem Fahrsimulator mit synthetisch erzeugter Außensicht. Zeitschrift für Verkehrssicherheit, Heft 3/1980

Chapter 7

[7-1]   Lehr- und Handbuch der Abwassertechnik, Band 1, Verlag von Wilhelm Ernst & Sohn (1973), Seite 19

[7-2]   Breuer, G.: Geht uns die Luft aus? Ökologische Perspektiven der Atmosphäre, Deutsche Verlagsanstalt GmbH Stuttgart (1978), Seite 14

[7-3]   Breuer, G.: Geht uns die Luft aus? Ökologische Perspektiven der Atmosphäre, Deutsche Verlagsanstalt GmbH Stuttgart (1978), Seite 107

[7-4]   DIN/ISO 2533: Normatmosphäre, Dezember 1979

[7-5]   Ixfeld, H. und M. Buck; Immissionsüberwachung im Lande Nordrhein-Westfalen Schriftenreihe der Landesanstalt für Immissionsschutz des Landes NW, Essen, Heft 51 (1980)

[7-6]   Ministerium für Arbeit, Gesundheit und Soziales des Landes NW. Luftreinhalteplan Ruhrgebiet West 1978 bis 1982, Duisburg-Oberhausen-Mülheim, Düsseldorf (1977)

[7-7]   Env. Prot. Agency: Air Quality Criteria for Oxides of Nitrogen (Draft). June 1979, p. 5 bis 9

[7-8]   Env. Prot. Agency: Air Quality Criteria for Oxides of Nitrogen (Draft). June 1979, p. 8 bis 16

[7-9]   Machta, L.: Atmospheric Measurement of Carbon Dioxide. Workshop on the Global Effects of Carbon Dioxide from Fossil Fuels, US. Dept. of Energy, CONF 770385 (1979), p. 44

[7-10] Rotty, R.M.: Present and Future Production of CO_2 from Fossil Fuels—a Global Appraisal ibido, p. 36

[7-11] Bundesminister des Innern: Umweltradioaktivität und Strahlenbelastung im Jahre 1978. Drucksache 8/4101, Bonn (Mai 1980)

[7-12] Lehr- und Handbuch der Abwassertechnik, Band 1. Verlag von Wilhelm Ernst & Sohn (1973), Seite 58

[7-13] DVGW- Arbeitsblatt W151 (1957)

[7-14] Deutscher Bundestag—Umweltgutachten 1978. Drucksache 8/1938 vom 19. 09. 1978, Seite 90

[7-15] Brümmer, G.: Funktion des Bodens im Stoffhaushalt der Ökosphäre in G. Olschowy Natur- und Umweltschutz in der Bundesrepublik, Verlag Paul Parey, Hamburg (1978), Seite 111

[7-16] Rororo Pflanzenlexikon 6100 (1979)

[7-17] Meyers Enzyklopädisches Lexikon

[7-18] Dreyhaupt, F.J.: Handbuch für Immissionsschutzbeauftragte, Verlag TÜV Rheinland GmbH, Köln (1978) Seite 173

[7-19] Informations on levels of environmental noise requisite to protect public

health and welfare with an adequate margin of safety. US-EPA, März 1974, 550/9–74–004

[7-20] Ministerium für Arbeit, Gesundheit und Soziales des Landes NW: Schall-ausbreitung in bebauten Gebieten, Düsseldorf (1975)

[7-21] VDI-Richtlinie 2057 (Entwurf, Februar 1975)

[7-22] Vornorm DIN 4150 (September 1975)

[7-23] Deutscher Wetterdienst, Offenbach. Klimatologische Werte für Einzel-monate.

[7-24] Scherhag, R.: Klimatologie (6. Auflage), G. Westermann Verlag, Braun-schweig (1969)

[7-25] Lamb, H.H.: Climate, Present, past and future, Vol. 2, Climatic History. Methuen & Co. Ltd., London (1977), p. 478

[7-26] Klimakunde des deutschen Reiches, Band II, Berlin (1939)

[7-27] Fortak, H.G.: Persönliche Mitteilung

[7-28] Münchener Rückversicherung. Weltkarte der Naturgefahren (1978)

[7-29] Wasser- und Schiffahrtsamt Köln (1980)

[7-30] Dowrick, D.J.: Wie [7-71]

[7-31] Hansen, R.J. ed.:: Seismic design for nuclear powerplants. MIT-press, Cambridge (1970)

[7-32] DIN 4149, Teil 1 (Entwurf, Dezember 1976)

[7-33] Ahorner, L., H. Murawski und G. Schneider: Die Verbreitung von schadenverursachenden Erdbeben auf dem Gebiet der Bundesrepublik. Zs. für Geophysik 36 (1970) 313

[7-34] Ministerium für Arbeit, Gesundheit und Soziales des Landes NW: Luft-reinhalteplan Rheinschiene-Süd 1977 bis 1981 (Köln), Düsseldorf (1976)

[7-35] Minister für Arbeit, Gesundheit und Soziales des Landes NW: Luftverun-reinigungen im Raum Duisburg, Oberhausen, Mülheim, Düsseldorf (1975)

[7-36] Statistisches Jahrbuch der nordrhein-westfälischen Industrie- und Han-delskammern (1979), 26. Jahrgang, Seite 38

[7-37] Ministerium für Arbeit, Gesundheit und Soziales des Landes NW: Luft-reinhalteplan Ruhrgebiet Ost (Dortmund) 1979 bis 1983, Düsseldorf 1978

[7-38] Ministerium für Soziales, Gesundheit und Sport des Landes Rheinland-Pfalz: Emissionskataster Ludwigshafen/Rhein. Verlag TÜV Rheinland GmbH, Köln (1977)

[7-39] Erste Allgemeine Verwaltungsvorschrift zur Durchführung des Bundes-Immissionsschutzgesetzes—Technische Anleitung zur Reinhaltung der Luft (TA Luft) vom 28. 08. 1974 (Gem.M.Bl. Seite 426)

[7-40] Technische Anleitung zur Reinhaltung der Luft (Entwurf TA Luft 1978), Bundesdrucksache 420/78 vom 08. 09. 1978

[7-41] VDI-Richtlinie 2306: Maximale Immissionskonzentrationen (MIK)—Orga-nische Verbindungen, März 1966

[7-42] VDI-Richtlinie 2310: Maximale Immissionswerte, September 1974

[7-43] Lehr- und Handbuch der Abwassertechnik, Band 1, Verlag von Wilhelm Ernst & Sohn (1973), Seite 109

[7-44] Verordnung über Trinkwasser und Brauchwasser für Lebensmittelbe-triebe. Trinkwasserverordnung vom 19. 12. 1974

[7-45] 12. Verordnung zur Durchführung des Bundes-Immissionsschutzgesetzes (Störfall-Verordnung) vom 30. 06. 1980

[7-46] Lefaux: Chemie und Toxikologie der Kunststoffe, Kapitel Industriehygiene. Krausskopf- Verlag GmbH, Mainz 1966

[7-47] Henschler, D.: Gesundheitsschädliche Arbeitsstoffe, Toxikologisch-Arbeitsmedizinische Begründung von MAK-Werten, Verlag Chemie, Loseblattsammlung, Weinheim (1979)

[7-48] Entwurf eines Gesetzes zum Schutz vor gefährlichen Stoffen (Chemikaliengesetz) Bundesrats-Drucksache 8/2319 vom 06. 11. 1979

[7-49] Dreyhaupt, F.J.: Handbuch zur Aufstellung von Luftreinhalteplänen (Abschnitt 3.1.2) TÜV-Handbücher Band 4, Verlag TÜV Rheinland, Köln (1980)

[7-50] Kropp, L.: Zeitliche und räumliche Abhängigkeit der Immissionen in Abgasimmissionsbelastungen durch den Kfz-Verkehr. Herausgegeben BMFT/TÜV Rheinland, Verlag TÜV Rheinland 1978

[7-51] Vorhaben 77 104 03 552 des Umweltbundesamtes: Analyse möglicher Störfälle in industriellen Anlagen im Hinblick auf die Luftreinhaltung (unveröffentlicht)

[7-52] Dornier Systems: Berichte zum prognostischen Modell Neckar, Friedrichshafen

[7-53] Wolf, P.: Die Berücksichtigung neuerer Erkenntnisse bei Sauerstoffberechnungen in Fließgewässern. gwf/wasser-abwasser 112 (1971) 200 und 250

[7-54] Weber, P.: Die Modellierung der menschlichen Einwirkungen auf zwanzig verschiedene Wasserläufe in Luxemburg, Revue Technique Luxembourgeoise 71 (1979) 81

[7-55] Kiefer, D. und H. Sengewein: Die Erstellung eines mathematischen Selbstreinigungsmodells dargestellt am Alzette-Fluß in Luxemburg. Revue Technique Luxemborgeoise 71 (1979) 69

[7-56] Umwelbundesamt: Wasserinhaltsstoffe im Grundwasser, Forschungsbericht 102 03 002 (10/1979) UBA 4/79)

[7-57] VDI-Richtlinie 2714 (E): Schallausbreitung im Freien, Dezember 1976

[7-58] VDI-Richtlinie 2571: Schallabstrahlung von Industriebauten, August 1976

[7-59] TÜV Rheinland (Storch, J. und J. Hennig): Das Evakuierungs-Simulationssystem EVAS und seine Anwendungen, Bericht-Nr.: St. Sch. 684 im Auftrag des BMI (unveröffentlicht), Köln (1979)

[7-60] Structures to resist the effects of accidental explosions, Departments of the Army. Technical Manual TMS-1300, (1969)

[7-61] Ardron, K.H. et al.: On the evaluation of dynamic forces and missile energies arising a water reactor loss of coolant accident. Journal British Nuclear Energy Soc. 16 (1977), 81

[7-62] Canvey-Report: An investigation of potential harzards from operations in the Canvey Island/Thurrock area. Health and Safety Executive, London (1978)

[7-63] Glasstone, S.: Die Wirkungen der Kernwaffen. Heymann-Verlag, Köln 1960)

[7-64] Becker, W.: Brandschutztechnische Anforderungen an Baustoffe. Vortrag auf der Fachtagung Bauen und Brandschutz, Mai 1979 Hannover, veröffentlicht in ,,Brandverhalten von Kunststoffen'', (1980)

[7-65] K.A. van Oeteren: Korrosionsschutz durch Beschichtungsstoffe. Carl Hanser Verlag, München, Wien (1980)

[7-66] Vernon, W.H.J.: Trans. Farad. Soc 31 (1935), 1678

[7-67] Burgmann, G., Grimme, D.: Stahl und Eisen 100 (1980), 641.

[7-68] Braunstein, L., Hochmüller, K.: Vom Wasser 43 (1974), 451

[7-69] Erdbebenauslegung von Betriebsanlagen und Rohrleitungen. Übersetzung aus dem Japanischen, Bericht IRS-A-6 (September 1974)

[7-70] Clough, R.W. Penzien, J.: Dynamics of Structures. McGraw-Hill Mogakuska LTD (1975)

[7-71] Dowrick, D.J.: Earthquake resistant design. A manual for engineers and architects, John Wiley u. Sons (1977)

[7-72] Neuland, W.: Analyse dynamisch belasteter Strukturen. Aus: Festigkeitsberechnung innendruckbelasteter Bauteile. Verlag TÜV Rheinland, Köln (1978)

[7-73] Institut für Unfallforschung des TÜV Rheinland. Studie zur Beurteilung des Kernkraftwerk-Standortes „BASF-Mitte". Abschnitt: Flugzeugabsturzwahrscheinlichkeit IfU 2/70, (1970) (unveröffentlicht)

[7-74] Deutsche Risikostudie Kernkraftwerk. Eine Untersuchung zu dem durch Störfälle in Kernkraftwerken verursachten Risiko. Studie der GRS im Auftrag des BMFT. Verlag TÜV Rheinland, Köln (1979)

[7-75] Riera, J.D.: On the Stress Analysis of Structures Subject to Aircraft Impact Forces. Nuclear Engineering and Design 8 (1968) 415

[7-76] Lorenz, H.: Aircraft Impact Design. Power Engineering, (November 1970) 44

[7-77] Rice, H.S. and L.Y. Bahar: Rection—Time Relationship and Structural Design of Reinforced Concrete Slabs and Shells for Aircraft Impact. 3rd International Conference on SMIRT, Paper J 5/3 London (1975)

[7-78] Drittler, K. und Gruner, P.: Zur Auslegung kerntechnischer Anlagen gegen Einwirkungen von außen. Teilaspekt Flugzeugabsturz. Bericht IRS-W-14, (April 1975)

[7-79] Fuzier, J.P. et al: Crashing of an Aircraft on a Nuclear Power Plant. An Inelastic Analysis. Proceedings York Conference, (1975). CP 8/5, University of York

[7-80] Rebora, B. et al: Dynamic Rupture Analysis of Reinforced Concrete Shells. Paper S 2/3, ELCALAP Berlin (1975)

[7-81] Hahn, Hermann: Umweltschutz im Bereich des Wasserbaus. Erich Schmidt Verlag

Additional References

Richtlinie für den Schutz von Kernkraftwerken gegen Druckwellen aus chemischen Reaktionen durch Auslegung der Kernkraftwerke hinsichtlich ihrer Festigkeit und induzierter Schwingungen sowie durch Sicherheitsabstände. Gemeinsames Ministerialblatt Nr.: 27 vom 03. Sept. 1976

Alexejew, Roitmann, Demidow, Tarasow—Agalkow: Grundlagen des Brandschutzes. Staatsverlag der Deutschen Demokratischen Republik, Berlin 1976

Isterling, Fritz: Vorbeugender Brandschutz—Brandverhütung und Brandbekämpfung. Lexika-Verlag Grafenau 1, Württemberg 1974

ETH-Tagung: Brandschutz und Sicherheit an Hochschulen und Forschungsstätten. Eidgenössische Technische Hochschule Zürich und Brandverhütungsdienst für Industrie und Gewerbe, Zürich 1979

Becker, Wolfgang: Voraussage des Brandgeschehens aufgrund der Ergebnisse von

Brandversuchen. VFDB-Zeitschrift Forschung und Technik im Brandschutz 1/75, Seiten 4 ff. Verlag W. Kohlhammer, Stuttgart

Landesamt für Datenverarbeitung und Statistik, Nordrhein-Westfalen. Statistisches Jahrbuch Nordrhein-Westfalen 1979, 21. Jahrgang

DIN 14011: Begriffe aus dem Feuerwehrwesen. Physikalische und chemische
Vorgänge. Ausgabe September 1977

Timoshenko, S. et al: Vibration Problems in Engineering. John Wiley and Sons.
New York, London 1974

Harris C.M. and Ch. E. Crede: Shock and Vibration Handbook. McGraw-Hill
Book Company 1976

USNRC Regulatory Guide, 1.102: Flood Protection for Nuclear Power Plants.

American National Standard: Industrial Security for Nuclear Power Plants. ANSI
N18.17–1973

Tychonoff, A.N.; Samarski, A.A.: Differentialgleichungen der mathematischen
Physik. VEB, Berlin 1959

Chapter 8

[8-1]   GVBl NW, S. 241
[8-2]   Gesetzsammlung für die königlich-preußischen Staaten, S. 243
[8-3]   Gesetzsammlung für die königlich-preußischen Staaten, S. 244
[8-4]   BVerfGE 20, S. 155, 157
[8-5]   Wolff, Verwaltungsrecht I, 8. Aufl., München 1971, § 48 IIa) 1)
[8-6]   NJW 1979, S. 364
[8-7]   NJW 1977, S. 1645
[8-8]   NJW 1977, S. 1649
[8-9]   BGBl 1980 I, S. 772 ff.
[8-10] NJW 1980, S. 759
[8-11] NJW 1979, S. 362
[8-12] NJW 1976, S. 2360
[8-13] NJW 1978, S. 1450
[8-14] NJW 1978, S. 439
[8-15] NJW 1979, S. 359

Additional References

Backherms, J.: JuS 1980, S. 9 ff.: Zur Einführung: Recht und Technik

Bender, B.: NJW 1979, S. 1425 ff: Gefahrenabwehr und Risikovorsorge als
Gegenstand nukleartechnischen Sicherheitsrechts

Bothe, M., Gündling, L.: Tendenzen des Umweltrechts im internationalen Ver-
gleich, Umweltbundesamt: Berichte 7/78, Berlin 1978

Böttger, J.: Ursachen und Wirkungen des Vertrags zwischen der Bundesrepublik
Deutschland und dem DIN, Deutsches Institut für Normung e. V., in:
Technische Normung und Recht, DIN-Normungskunde, Heft 14, Hrsg.
DIN, Deutsches Institut für Normung e. V., Berlin, Köln 1979, S. 31 ff.

Breuer, R.: AöR 101 (1976), S. 46 ff.: Direkte und indirekte Rezeption technischer
Regeln durch die Rechtsordnung
DVBl 1978, S. 829 ff.: Gefahrenabwehr und Risikovorsorge im Atomrecht

(zugleich ein Beitrag zum Streit um die Berstsicherung für Druckwasser-reaktoren)

Bungarten, H.H.: Umweltpolitik in West-Europa, EG, Internationale Organisa-tionen und nationale Umweltpolitiken, Schriften des Forschungsinstituts der Deutschen Gesellschaft für auswärtige Politik e. V., Bonn 1978

Eberstein, H.H.: BB 1977, S. 1723 ff.: Technik und Recht

Friauf, K.: Das Standortplanfeststellungsverfahren als Rechtsproblem in: Rechts-fragen des Genehmigungsverfahrens von Kraftwerken, Veröffentlichungen des Instituts für Energierecht an der Universität zu Köln, Heft 41/42, herg. v. B. Börner, Düsseldorf 1978, S. 63 ff.

Hanning, A., Schmieder, K.: DB Beilage Nr. 14/77 zu Heft Nr. 46 vom 18. November 1977: Gefahrenabwehr und Risikovorsorge im Atom-und Im-missionsschutzrecht

Herschel, W.: Technische Regelwerke—ein Beitrag zur Staatsentlastung, in: Technische Regelwerke—ein Beitrag zur Staatsentlastung, Gemeinschafts-ausschuß der Technik GdT Schriften Nr. 4, Düsseldorf 1972, S. 6 ff. Rechtsfragen der technischen Überwachung, Schriftenreihe Recht und Technik, Band 1, Heidelberg 1972

Jacchia, E.: Atom, Sicherheit und Rechtsordnung, Freudenstadt 1965

Korbion, H.: Rechtsvorteile für den Anwender von Normen, in: Technische Nor-mung und Recht, DIN-Normungskunde, Heft 14, Hrsg. DIN, Deutsches Institut für Normung e. V., Berlin, Köln 1979, S. 169 ff.

Krüger, H.: NJW 1966, S. 617 ff.: Rechtsetzung und technische Entwicklung

Lukes, R.: Reform der Produkthaftung: Bestrebungen, insbesondere auf europäi-scher Ebene, und ihre Verwirklichung im deutschen Recht, Köln u. a. 1979 Überbetriebliche technische Normung in den Rechtsordnungen ausgewähl-ter EWG und EFTA-Staaten; Frankreich, Großbritannien, Italien, Öster-reich, Schweden; Schriften zum Wirtschafts-, Handels-, Industrie-Recht, Band 24; Köln u. a., 1979 WB 1978, S. 317 ff: Die Verwendung von Risikoanalysen in der Rechtsordnung unter besonderer Berücksichtigung des Kernenergierechts
NJW 1978, S. 241 ff.: Das Atomrecht im Spannungsfeld zwischen Technik und Recht

Lukes, R., Vollmer, L.: Ersetzung der Genehmigungsverfahren für kerntechnische Anlagen durch ein atomrechtliches Planfeststellungsverfahren, in: Lukes, Vollmer, Mahlmann, Grundprobleme zum atomrechtlichen Verwaltungs-verfahren, Schriftenreihe Recht und Technik, Band 3, Heidelberg 1974, S. 13 ff.

Mahlmann, W.: Grundzüge, neue Entwicklungen und ausgewählte Fragen des Genehmigungsverfahrens für Kernenergieanlagen in den Vereinigten Staaten von Amerika, verglichen mit dem deutschen Recht, in: Lukes, Vollmer, Mahlmann, Grundprobleme zum atomrechtlichen Verwaltungs-verfahren, Schriftenreihe Recht und Technik, Band 3, Heidelberg 1974, S. 163 ff.

Nickusch, A.: Die Normativfunktion technischer Ausschüsse und Verbände als Problem der staatlichen Rechtsquellenlehre, Diss. 1964
NJW 1967, S. 811 ff.: § 330 StGB als Beispiel für eine unzulässige Verweisung auf die Regeln der Technik

Ossenbühl, F.: Die gerichtliche Überprüfung der Beurteilung technischer und wirtschaftlicher Fragen in Genehmigungen des Baus von Kraftwerken, in:

Rechtsfragen des Genehmigungsverfahrens von Kraftwerken, Veröffentlichungen des Instituts für Energierecht an der Universität zu Köln, Heft
41/42, hrsg. v. B. Börner, Düsseldorf 1978, S. 39 ff.

Pietzner, R.: JA 1973, S. 691 ff.: Das Verbot mit Erlaubnisvorbehalt

Plischka, H.P.: Technisches Sicherheitsrecht, die Probleme des technischen
Sicherheitsrechts, dargestellt am Recht der überwachungsbedürftigen
Anlagen (§24 Gewerbeordnung), Schriften zum öffentlichen Recht, Band
109, Berlin 1969

Redeker, K.: Die anerkannten Regeln der Technik als Rechtsbegriff im öffentlichen Recht, in: Technische Normung und Recht, DIN-Normungskunde,
Heft 14, Hrsg. DIN, Deutsches Institut für Normung e. V., Berlin, Köln
1979, S. 19 ff.

Rengeling, H.W.: NJW 1978, S. 2217 ff.: Vorbehalt und Bestimmtheit des
Atomgesetzes

Schenke, W.R.: NJW 1980, S. 743 ff.: Die verfassungsrechtliche Problematik
dynamischer Verweisungen

Scholz, R.: Das Verhältnis von technischer Norm und Rechtsnorm unter besonderer Berücksichtigung des Baurechts, in: Technische Normung und Recht,
DIN-Normungskunde, Heft 14, Hrsg. DIN, Deutsches Institut für Normung e. V., Berlin, Köln 1979, S. 85 ff.

Soell, H.: ZRP 1980, S. 105 ff.: Aktuelle Probleme und Tendenzen im Immissions-
schutzrecht

Sonnenberg, G.S.: 100 Jahre Sicherheit, Beiträge zur technischen und administra-
tiven Entwicklung des Dampfkesselwesens in Deutschland 1810–1910,
Technik-Geschichte in Einzeldarstellungen, Nr. 6, Düsseldorf 1968

Strecker, A.: Rechtsfragen bei der Verknüpfung von Rechtsnormen mit technischen Normen, in: Technische Normung und Recht, DIN-Normungskunde,
Heft 14, Hrsg. DIN, Deutsches Institut für Normung e. V., Berlin, Köln
1979, S. 43 ff.

Vogel, H.-J.: Eröffnungsansprache, in: Technische Normung und Recht, DIN-
Normungskunde, Heft 14, Hrsg. DIN, Deutsches Institut für Normung e.
V., Berlin, Köln 1979, S. 9 ff

v. Mohrenfels, B.W.: ZRP 1980, S. 86 ff.: Errichtung und Betrieb von Kernkraftwerken

Wagner, H.: NJW 1980, S. 665 ff.: Die Risiken von Wissenschaft und Technik als
Rechtsproblem

Winckler, R.: Normung und Harmonisierungsbestrebungen der Europäischen
Gemeinschaften, in: Technische Normung und Recht, DIN-
Normungskunde, Heft 14, Hrsg. DIN, Deutsches Institut für Normung e.
V., Berlin, Köln 1979, S. 59 ff.

Chapter 9

[9-1] U.S. Congress, House Document 360 „Technological Trends and National
Policy". Report of the Subcommittee on Technology of the National
Resources Commission, 75th Cong., 1st sess., June 1937.

[9-2] U.S. House of Representatives, Statement of Emilio Q. Daddario, Chair-
man of the Subcommittee on Science, Research, and Development of the
Committee on Science and Astronautics. 90th Cong., 1st sess., Ser.I
(Revised, August 1968).

[9-3]   Coates V.T., Chemical Engineering Progress, Vol. 70, No. 11 (1974) S. 41-45.

[9-4]   Mayo L.H.: Social Impact Evaluation—Some Implications of the Specific Decisional Context Approach for Anticipatory Project Assessment with special reference to Available Alternatives and to Techniques of Evaluating the Social Impacts of the Anticipated Effects of such Alternatives, Program of Policy Studies in Science and Technology, The George Washington University, Occasional Paper No. 14, Washington, D.C. November 1972.

[9-5]   MITRE Corporation, A Technology Assessment Methodology, Project Summary and 6 Volumes/Some Basic Propositions, Automotive Emissions, Computers-Communications Networks, Enzymes (Industrial), Mariculture (Sea Farming), Water Pollution: Domestic Wastes), Washington, D.C., 1971.

[9-6]   Jochem, E.: Die Motorisierung und ihre Auswirkungen—Untersuchungen zur Frage der Realisierbarkeit der Technikfolgen-Abschätzung (technology assessment) anhand von ex post-Projektionen. Heft 108 der Kommission für wirtschaftlichen und sozialen Wandel. Verlag Otto Schwartz & Co., Göttingen, 1976.

[9-7]   Paschen H., Gresser K., Conrad F.,: Technology Assessment: Technologiefolgen-abschätzung: Ziele, method. u. organisatorische Probleme, Anwendungen. Herausgegeben von der Stiftung Gesellschaft und Unternehmen. 1 Aufl. Campus Verlag, Frankfurt/Main, New York, 1978.

[9-8]   MITRE Corporation, A Comparative State-of-the-Art Review of Selected U.S. Technology Assessment Studies, prepared for NSF, Washington, D.C. 1973

[9-9]   Altenpohl D.: TP—Die Zukunftsformel, Möglichkeiten und Grenzen der Technologie-Planung, Umschau Verlag, Frankfurt am Main, 1975

[9-10] U.S. House of Representatives: Public Law 92–484, 92nd Cong., H.R. 10243: An Act to establish an Office of Technology Assessment for the Congress . . . , October 13, 1972

[9-11] Kash, Don E. et al: Energy Under the Oceans—A Technology Assessment of Outer Continental Shelf Oil and Gas Operations, 1st ed. 1973; 2nd print. 1974; 3nd print. 1975. University of Oklahoma Press.

[9-12] University of Oklahoma, Energy Under the Oceans—A Summary Report of A Technology Assessment of OCS Oil and Gas Operations—The Technology Assessment Group Science and Public Policy Program. First printing August, second printing September 1973. University of Oklahoma Press.

[9-13] Bentham, J. (1789): ,,An Introduction to The Principles of Morals and Legislation'' in: The Hafner Library of Classics, Hafner Publishing Company, Inc., New York, 1961

[9-14] Quételet, L.A.J. (1835): »Sur l'homme et le développement des ses facultés«, 2 Bde. Bachelier, Paris

[9-15] Quételet, L.A.J. (1870): »L'anthropométrie ou mesure des différentes facultés de l'homme«, C. Muquardt, Bruxelles, 1870; 2 Auflage 1871

[9-16] Buckle, H. Th. (1841): ,,On Liberty'' in: ,,Essays'' Bd. 1, D. Appleton und Co., New York, 1877

[9-17] Wagner, A. (1864): ,,Gesetzmäßigkeiten in scheinbar willkürlichen Handlungen'', Hamburg

[9-18] Starr, C. (1969): ,,Social Benefit versus Technological Risk", Science *165*, 19. September 1969, S. 1232

[9-19] Fischhoff, B. et al. (1977): ,,How Safe is Safe Enough? A Psychometric Study of Attitudes Towards Technological Risks and Benefits" School of Engineering and Applied Science, University of California, Los Angeles, Report UCLA-ENG-7717, January 1977

[9-20] Maslow, A.H. (1954): ,,Motivation and Personality", Harper & Row, New York

[9-21] Jaeger, Th.A. (1978): ,,Beurteilung technischer Risiken" Vortrag auf der Tagung ,,Technisches Sachverständigenwesen—Entscheidungshilfe für Rechtsprechung, öffentliche Hand und Wirtschaft" des Verbandes Deutscher Elektrotrechniker (VDE), Braunschweig, 13–14. April 1978

[9-22] Destojewskij, F.M. (1864): ,,Zapiski iz podpol'ja" in: Epoha, russ. in: Poln. sobr. soč. F.M. Dostoevskogo 1882, deusch: ,,Aufzeichnungen aus dem Untergrund" in: Fjodor Dostojewski. Sämtl. Werke in 10 Bänden, ed. E.K. Rahsin, Bd. 9 ,,Der Spieler. Späte Romane"; Wissenschaftliche Buchgesellschaft, Darmstadt, 3. Aufl. 1974

[9-23] Lauth, R. :1950): ,,Die Philosophie Dostojewskis in systematischer Darstellung", R. Piper und Co., München

[9-24] Steinbuch, K. (1979): ,,Zur Akzeptanz von Risiken in der modernen Zivilisation", Vortrag auf dem I. Sommersymposon der Gesellschaft für Sicherheitswissenschaft (GfS), Wuppertal, 10.–13. Juni 1979 in: ,,I. GfS-Sommer-Symposon '79, 10.–13. Juni 1979, Gesamthochschule Wuppertal: Risiken komplizierter Systeme—ihre komplexe Beurteilung und Behandlung". Gesellschaft für Sicherheitswissenschaft (Hrsg.), Wuppertal, 1980

[9-25] Stoll, W. (1978): ,,Technischer Fortschritt oder heile Umwelt—eine Frage an unser Gewissen?" Manuskript vom 18. 07. 1978, Hanua, ALKEM GmbH
Stoll, W. (1980): ,,Warum wird gerade an der Kernenergie die Technophobie exemplifiziert?" Manuskript vom 21. 04. 1980, Hanau, ALKEM GmbH

[9-26] Lauth, R. (1969): ,,Ethik—in ihrer Grundlage aus Prinzipien entfaltet", W. Kohlhammer Verlag, Stuttgart

[9-27] Lindackers, K.H. (1980): ,,Risiken aus der Energiebedarfsdeckung" Vortrag auf der Jahrestagung des Deutschen Atomforums, Mainz, 22. 01. 1980

[9-28] USNRC-Regulatory Guide 1.110: Cost-Benefit-Analyses for Radwaste Systems for Light Water Cooled Nuclear Power Plants. Ausgabe 3/76

[9-29] Helms, E. (1971): ,,Ökonomische Grundlage zur Erfassung der Unfallkosten im Straßenverkehr" Dissertation, Bonn

[9-30] Jäger, W. (1977): ,,Verkehrssicherheitsplanung mit Hilfe von Kosten-Nutzen-Analysen" Buchreihe des Institut für Verkehrswissenschaft an der Universität zu Köln. ed. Rainer Willeke, Verkehrs-Verkehrs-Verlag J. Fischer, Düsseldorf 1977

[9-31] Péguy, Ch. (1970): »Notre jeunesse« Oevres complètes, Bd. IV Edition de la NRF, Paris, 1917 ff.

[9-32] Schmidt, H. (1979): Ansprache vor der Deutschen Physikalischen Gesellschaft am 28. September 1979 in Ulm, in: Phys. Bl. *35*, 544 (1979)

Index